Handbook of

HEAT TRANSFER MEDIA

PAUL L. GEIRINGER

ROBERT E. KRIEGER PUBLISHING COMPANY
HUNTINGTON, NEW YORK
1977

Original Edition 1962
Reprint 1977

Printed and Published by
ROBERT E. KRIEGER PUBLISHING CO., INC.
645 NEW YORK AVENUE
HUNTINGTON, NEW YORK 11743

Printed in the United States of America

Library of Congress Cataloging in Publication Data

Geiringer, Paul L.
Handbook of heat transfer media.

Reprint of the edition published by Reinhold, New York.
1. Heat-transfer media. 2. Fluids.
I. Title.
[TP159.H4G44 1977] 660.2'8427 76-57170
ISBN 0-88275-498-X

FOREWORD

For the transport of heat from one place to another, combustion gases and steam have been used as heat carriers for thousands of years. However, the desire for improved processes and increased temperatures has brought about a quest for more efficient fluids. Thus, the twentieth century has seen the development of new media, especially liquids, for transmitting heat to machinery and equipment.

The development of high temperature water heating has been due to the recognition of its many advantages as a heat carrier. However, the excessively high pressures at elevated temperatures have created a demand for more suitable heat transfer fluids. These fluids should operate at temperatures from 400° up to 1000°F and higher, but still maintain low pressures. The search for new liquids for the lubrication of high temperature gas turbines and the cooling and moderating of nuclear reactors has forced both European and American chemical industries to investigate and compound new fluids that are able to serve dual purposes of heat transfer and lubrication. The American chemical industry, in the forefront, developed most of these liquids, the various giant companies utilizing different approaches to this problem.

For years, vapors and gases in general, were considered suitable fluids for this service. However, they are much more cumbersome to use than fluids in the liquid form.

For a time, great hopes were held for fluorine-containing compounds. Certain ethers also appeared to be highly promising at first, as liquids which could withstand temperatures of 1000°F for long periods and still melt at moderate temperatures; and various hydrocarbons, derivatives of benzene combined with halogens, or modified in other ways, seemed to have potential. Currently, however, silicon derivatives seem to be in the forefront.

The primary known liquids produced in Europe and in the United States are considered in this volume, the first treatise of its type. This data could be compiled, and these many liquids evaluated and classified according to merit for heat transfer applications, only by some one who has had considerable experience in the field. The author, who has devoted more than thirty years of his life to the design of thousands of high temperature plants all over the world, has personal knowledge of most of these heat transfer fluids, and is eminently qualified to describe their properties.

This volume, a comprehensive accumulation of information concerning more than sixty heat transfer fluids, makes these data available to engineers interested in the design, utilization, and application of heat transport under various conditions. The number of liquids increases continually, and additional fluids and data will have to be added later. However, the basic thought and arrangement of the material will remain, to assist the interested reader in his continuing search for the medium best suited for the particular application.

CHARLES F. BONILLA
Chairman of Dept. of Chemical Engineering
Columbia University

PREFACE

This Handbook is probably the first collection of data on this new and rapidly developing technology. The data compiled concerning the most important properties of the different fluids have been supported by numerous curves showing the relation of the different properties, in dependence of temperature, or of other variables. In order to make this book more useful to the reader, the curves have been prepared on graph paper. Numerical values have been shown on most curves to support the accuracy of a graphic interpretation. These graphs should provide a valuable tool for determining accurately the corresponding physical properties at any desired temperature level. These data are essential for the computation of heat transfer and flow resistance values.

The trend in industry to augment and speed up production has led to an increase in operating temperatures. This has necessitated a search for fluids other than steam and water, which could be utilized for accelerating and improving many chemical, physical, and nuclear processes. Most of these new fluids were investigated, developed, and introduced to industry only during the last thirty years.

The desire to develop a pressureless nuclear reactor design gave new impetus to finding and formulating better fluids, extending the temperature limit of organic heat carriers, and studying liquid metals for similar applications. In addition to satisfying conventional heat transfer requirements, reactor coolants must exhibit high radiolytic stability in order to withstand intense ionizing radiation. Many fluids have been investigated for such service: liquid metals, inert gases, inorganic salts, and various organic substances (aromatic hydrocarbons, polyphenyls, and silanes).

High temperature lubrication of bearings and transmission of power in jet aircraft engines is another problem which modern engineers must solve. Elevated temperature lubricants and hydraulic fluids must resist stringent oxidizing and pyrolytic environments, possess a favorable viscosity temperature relationship, display a high flammability resistance, and have a pour point well below room temperature. Standard petroleum-based lubricants have failed to meet Air Force requirements. A concerted effort was made to solve this problem; the entire field of organic chemistry has been explored. Among the materials investigated are: triester, diester, and disiloxane base stocks; silicate esters, alkyl, aryl and aryloxy silanes; alkyl substituted and unsubstitued aromatic ethers; triaryl esters of phosphoric acid; organo derivatives of titanium and of such substances as phosphorus-boron and tin-silicon compounds, fluoroalkoxysilanes, and perarylated silanes. All of these fluids failed to satisfy one or more requirements, thus numerous compromise structures have been proposed. The long-term effects of these studies on future heat transfer applications cannot be overemphasized.

The area of substances presently under discussion is limited to a temperature range of 200° to 2000°F, which covers most heating and high temperature cooling applications. Fluids employed primarily for cooling have not been included in this Handbook.

Also, fluids transferring heat only by combustion have been excluded. However, flue gases—the products of combustion—which also transfer heat by radiation and convection, are included.

The Handbook does not cover the technology of applying the media, since adequate coverage of this topic would require a book twice the size.

PAUL L. GEIRINGER
June 1962

ACKNOWLEDGMENTS

In 1959, the Reinhold Book Division of Reinhold Publishing Corporation, approached Mr. Malcolm McEwen, Project Engineer in the Development Department of Monsanto Chemical Company, Saint Louis, Missouri, and asked him to suggest somebody qualified to prepare a composite book on Heat Transfer Materials.

Mr. McEwen was aware of the extended work and wide experience in this field of my company, American Hydrotherm Corporation, and of our previous publications on this subject. He felt that we would have a large collection of data on heat transfer materials which would prove valuable in the preparation of such a volume.

I agreed to undertake the preparation of a Handbook of Heat Transfer Media, provided Mr. McEwen, an outstanding engineer well versed in the development of certain heat transfer compounds, would be willing to contribute an important section describing the heat carriers found in nature, and to discuss how the chemical engineer of today can approach the task of developing new and better heat carriers. I wish to thank Mr. McEwen sincerely for his valuable contribution, Section II, and for his constructive criticism of the entire text.

A large part of the great collection of data on heat transfer media of American Hydrotherm Corporation, containing material on substances produced both in the United States and abroad, has been assembled by Mr. Bernard S. Breitman, Vice President of our company. Without these data, and without his great personal assistance, the book would not have been complete. My gratitude is expressed for his valuable help.

Mr. Hugh Payne, Technical Director, and Mr. Eric Beanland, Chief Chemist, of Hygrotherm Engineering Limited, London, England, reviewed the text and studied the graphs. Their comments and suggestions were most valuable and I am grateful for their assistance.

Mr. G. M. Rapp, Associate Director, and Mr. Herbert Kaufman, Senior Research Chemist, of the John B. Pierce Foundation of New Haven, Connecticut, also reviewed the manuscript and made valuable recommendations for which I am much indebted.

My appreciation is extended to Dr. Charles F. Bonilla, Professor and Chairman of the Department of Chemical Engineering at Columbia University, for preparing the Foreword and for reviewing and revising the chapter on nuclear engineering.

With my warm personal thanks, I would like to mention my assistants, Mr. Perry Bergman, a graduate of The Cooper Union School of Engineering, presently engaged in graduate work at Purdue University, and Mr. Donald Messing, now studying at The College of the City of New York. Without their diligent research and labor, completion of this Handbook would have been impossible.

Finally, thanks to the publishers for their fine job of putting the work together into a most presentable volume.

PAUL L. GEIRINGER

CONTENTS

INTRODUCTION

Heat transfer fluids, or heat carriers, are defined as fluids which absorb thermal energy delivered from a source and then, by cooling or change of phase, surrender heat to a place of utilization called a sink. The reverse flow of heat from the sink to the source is equally possible; it occurs if the sink is at a higher temperature and thereby becomes the source. The fluid is then said to function as a coolant. During this process heat can be transmitted by conduction, convection, radiation, boiling, and condensation. To complete the thermal circuit, the cooled fluid returns to its initial position, where it obtains a fresh supply of heat. It is important that at the conclusion of this cycle, the fluid revert to its original physical or chemical structure without suffering any significant changes. This book discusses only those means of heat transmission which are reversible. Any chemical process which absorbs or ejects heat (endothermic or exothermic) is excluded, even if it is a reversible one. Heat produced by combustion, solution, dissociation, etc., is not within the scope of this text.

In so defining the function of a heat carrier, it is presumed that all such fluids must maintain their original chemical properties and composition. No process is mentioned whereby heat is created or absorbed solely by chemical reactions. However, the reversible heating and cooling process may indirectly cause chemical changes, such as solution and partial evaporation, or splitting off of lower-boiling constituents. The small amount of heat so produced or absorbed is not considered in this discussion. Only a very few heat carriers, such as water, air, and some of the basic elements, approach the criterion of maintaining an unchanged chemical structure. For industrial applications, the ability of the heat carrier to resist physical or chemical changes for an extended period is of paramount importance. In case of superficial deterioration, the heat transfer fluid can usually be treated and eventually returned to the system. If the change in properties is more serious, the fluid must be replaced after certain periods of use.

During the heat transfer process, most heat transfer materials may undergo a change of physical state. For instance, under certain equilibrium conditions of temperature and pressure, vapor, liquid and solid phases prevail alone or even coexist. Each phase is characterized by a unique set of properties. Therefore, it is necessary to examine carefully the properties of each fluid relative to its physical state.

The purpose of this Handbook is to present an over-all picture of the most important heat transfer fluids suitable for use at temperatures above 200°F. It is intended only to cover commercially available fluids, not all possible or conceivable media. A detailed coverage of low temperature coolants such as refrigerants has been omitted, since they are treated in several comprehensive handbooks on refrigerants.

Section I of this book is devoted to a thorough discussion of the key physical properties of heat transfer fluids. These properties are divided into three groups:

1. Design Properties (Heat Transfer and Transport Properties)
2. Construction Properties (Limiting Properties)
3. Operation Properties (Handling Properties)

A basic understanding of all features of these three groups of properties will assure successful design, construction, and operation of a heat transfer system.

Section II explains how the fundamental criteria established in the previous section are met by materials found free in nature and by man-made synthetic fluids.

Section III offers a survey of commercially available gaseous, vapor, and liquid heat transfer media.

Section IV, the concluding section, discusses the possible application and range of utilization of the different heat transfer fluids for process, space heating, and nuclear reactors.

NOMENCLATURE

The following system of nomenclature has been adopted. The symbols and abbreviations are those which have gained fairly general recognition in engineering practice. The English system of units has been adhered to; however a number of properties such as surface tension, viscosity, and vapor pressure, are quite often expressed in terms of metric units. As a concession to convention and for ease of data presentation, a number of graphs were plotted with metric ordinates. Should the reader desire to have a particular property in a different set of units, a number of the most useful conversion factors are given in Table I.

SUBSCRIPTS AND SUPERSCRIPTS

c critical state
p constant pressure
r reduced state
t constant temperature
$-$ mean value
h constant enthalpy

Symbol	Quantity	Units
A	area of surface	ft^2
c	specific heat of given substance	Btu/(lb)(°F)
c_p	specific heat at constant pressure	Btu/(lb)(°F)
c_v	specific heat at constant volume	Btu/(lb)(°F)
D	diameter	ft
D_h	hydraulic diameter	ft
e	base of natural logarithm	Dimensionless
f	friction factor	Dimensionless
g	acceleration of gravity	32.174 ft/sec^2
g_c	dimensional constant	32.174 ft/sec^2
G	mass velocity	$lb/(ft)^2(sec)$
h	film coefficient of heat transfer	$Btu/(hr)(ft)^2(°F)$
h_{fg}	latent heat of vaporization	Btu/lb
H	head loss, pressure drop	ft of fluid
H	enthalpy	Btu/lb
k	thermal conductivity	$Btu/(ft)^2(hr)\left(\frac{°F}{ft}\right)$
L	length	ft
ln	natural logarithms (base e)	Dimensionless
log	natural logarithms (base 10)	Dimensionless
n	mole—number of pound molecular weights	Dimensionless
Nu	Nusselt's number	Dimensionless

Symbol	Quantity	Units
p	pressure	$lb/(ft)^2$
P	power	(ft)(lb)/hr
q	rate of heat transfer	Btu/hr
Q	heat quantity	Btu
r	radius	ft
r_h	hydraulic radius	ft
R	gas constant	(ft)(lb)/(lb)(°R)
Re	Reynolds number	Dimensionless
S	cross-sectional area	ft^2
t	temperature	°F
t	time	sec or hr
T	absolute temperature	°R
v	specific volume	ft^3/lb
V	total volume	ft^3
V	average velocity	ft/sec
w	mass flow	lb/hr or lb/sec
W	weight	lb
x	distance in flow direction	ft
y	depth	ft
z	compressibility factor	Dimensionless
α alpha	angle of contact	degrees
β beta	coefficient of volumetric expansion	$°F^{-1}$
Δ delta	difference	Dimensionless
ε epsilon	height of roughness projection	ft
ε epsilon	dielectric constant	Dimensionless
K kappa	specific heat ratio (c_p/c_v)	Dimensionless
λ lambda	enthalpy per unit volume	Btu/ft^3
λ lambda	enthalpy per unit mass	Btu/lb
μ mu	absolute viscosity	lb/(ft)(sec)
ν nu	kinematic viscosity	ft^2/sec
$1/\nu$ nu	fluidity	sec/ft^2
ϕ phi	function of	Dimensionless
π pi	a constant	Dimensionless, 3.1416...
ψ psi	function of	Dimensionless
ρ rho	density	lb/ft^3
ρ rho	resistivity	ohm-cm
σ sigma	surface tension	lb/ft
τ tau	shearing stress	lb/ft^2

TABLE I. CONVERSION TABLES

MULTIPLY	BY	TO OBTAIN
atm	760	mm Hg
atm	14.696	psi
Btu	252	cal (gram-calorie)
Btu/(lb)(°F)	1	cal/(gm)(°C)
Btu/(hr)(ft)(°F)	4.13×10^{-3}	cal/(sec)(cm)(°C)
cal	3.968×10^{-3}	Btu
cal/(gm)(°C)	1	Btu(lb)(°F)
cal/(sec)(cm)(°C)	241.9	Btu/(hr)(ft)(°F)
centipoise	2.419	lb/(ft)(hr)
gm/cm^3	62.428	lb/ft^3
gm/cm^3	8.345	lb/gal
lb/ft^3	0.13368	lb/gal
lb/gal	7.48	lb/ft^3
lb/gal	0.1198	gm/cm^3
lb/(ft)(hr)	0.41338	centipoise
mm Hg	1.3158×10^{-3}	atm
mm Hg	0.019337	psi
psi	0.068	atm
psi	51.715	mm Hg

TEMPERATURE CONVERSIONS

$T_c = (5/9)(T_F - 32)$

$T_F = (9/5)\,T_C + 32$

T_c = Temperature centigrade

T_F = Temperature Fahrenheit

$T_K = T_C + 273$

$T_R = T_F + 460$

T_R = Temperature Rankine

T_K = Temperature Kelvin

Section I

DEFINITION OF PROPERTIES

A. GENERAL

In order to apply a heat transfer fluid intelligently, the engineer must first examine the physical, chemical, and physiological properties which limit and determine its engineering application. To facilitate future analysis, these properties, which might also be called "criteria," have been organized into the following three subdivisions. Undoubtedly, there are a number of equally worthy methods of classifying the sundry characteristics of a heat transfer fluid. The arrangement given is certainly a flexible one.

Group One—Design Properties (Heat Transfer and Transport Properties). In approaching a heat transfer problem, the engineer first performs a series of thermal computations and then develops the design of the system. To execute this phase of his work, he must be familiar with a group of properties designated as thermal and transport properties.

1. Density, Specific Gravity, Expansion Coefficient
2. Heat Capacity (Specific Heat)
3. Heat Content (Enthalpy)
4. Heat Conductivity
5. Viscosity

Group Two—Construction Properties (Limiting Properties). The next step is to translate the thermal design from the drawing board into reality. Now the engineer must deal with another group of properties which unalterably fix the range of application of the heat carrier. This group is known as the limiting properties.

1. Physical Stability
2. Chemical Stability
3. Thermal Stability
4. Radiolytic Stability
5. Cost

Group Three—Operation Properties (Handling Properties). Finally, in the actual day-to-day operation of the system, the engineer must contend with a third set of properties involving the compatibility of the fluid and its environment. This set of properties is called handling properties.

1. Surface Tension
2. Corrosion
3. Flammability
4. Physiological Effects
5. Less Important Properties:
   - Electrical Properties
   - Refractive Index
   - Appearance

B. DESIGN PROPERTIES (TRANSFER AND TRANSPORT PROPERTIES)

In the theoretical computation and design of a heating or cooling system, the resistance of the circuit to flow and the ability of the heat carrier to store and deliver thermal energy to its destination are of primary importance.

Pumping. To determine fluid friction or head loss, H, sustained in isothermal turbulent flow through straight pipes, the empirical relationship most frequently employed is the Fanning equation:

$$H = \frac{\Delta P}{\rho} = \frac{4fLV^2}{2g_c D_h} = \frac{4fLG^2}{2g_c D_h \rho^2} \qquad (1\text{-}1)$$

f, the dimensionless friction factor, depends upon the Reynolds number, $\frac{DV\rho}{\mu}$, and to some extent on the diameter and the roughness of the given pipe, ε, or in analytical notation:

$$f = \phi\left[(Re), \left(\frac{\varepsilon}{D}\right)\right] \tag{1-2}$$

Much experimental and theoretical evidence has been combined in an effort to represent the relationship between f and the two independent variables.

The pumping power, P, needed to transport the heating medium through the thermal circuit is obtained by multiplying both sides of equation (1-1) by the mass flow rate, w:

$$P = Hw = \frac{4fLV^2(\rho VA)}{2g_c D_h} = \frac{4fL\rho AV^3}{2g_c D_h} \tag{1-3}$$

Heat Transfer. The primary equation (1-4) for heat transfer from a wall at temperature t_s to a fluid flowing by a temperature t_b has been formulated by Newton. According to his Law, the rate of heat transfer through a surface is directly proportional to the wall area, A, and the difference in temperature potential $(t_s - t_b)$.

The heat transfer coefficient, h, is defined as the proportionality factor between the above quantities:

$$q = hA\,(t_s - t_b) \tag{1-4}$$

h is an extremely complex function of the physical properties and state of the fluid, the geometry of the apparatus, and the fluid velocity. The four basic equations for the heat transfer coefficient, expressed in terms of dimensionless groups, as cited by McAdams, are:

Forced Convection:

$$\frac{hD}{k} = \phi\left[\left(\frac{DV\rho}{\mu}\right), \left(\frac{c_p\mu}{k}\right)\right] \tag{1-5}$$

Natural Convection:

$$\frac{hD}{k} = \phi\left[\left(\frac{D^3\rho^2 g\beta\,\Delta t}{\mu^2}\right)\left(\frac{c_p\mu}{k}\right)\right] \tag{1-6}$$

Condensing Vapor:

$$\frac{hD}{k} = \phi\left[\left(\frac{D^3\rho^2 g}{\mu^2}\right), \left(\frac{\mu\lambda}{k\,\Delta t}\right)\right] \tag{1-7}$$

Boiling Liquids:

$$\frac{hD}{k} = \phi\left[\left(\frac{D^2\rho\Delta tkJg_c}{\mu^3}\right), \left(\frac{c_p\mu}{k}\right), \left(\frac{\mu^3}{\rho D\sigma g_c}\right)\right] \tag{1-8}$$

Heat Transport. Compared to the phenomenon of heat transfer, heat transport can be represented in much simpler mathematical terms. If $\Delta t = (t_1 - t_2)$ is equal to the change in temperature, the basic relationship for thermal transport is the oft-encountered equation:

$$q = w\,c_p\,\Delta t = V\,S\,\rho\,c_p\,\Delta t \tag{1-9}$$

where w is the fluid weight in lb/hr or lb/sec, V the fluid velocity, S the cross-section of the conduit, and ρ and c_p the properties of the fluid, density, and specific heat.

As may be observed, the volumetric heat capacity, ρc_p, is of value as an index of the ability of a fluid to convey heat directly, by virtue of its own motion, to a given destination.

Summary. From a brief inspection of these key equations, it is evident that for the successful building of a thermal system, the following properties must be considered: density (ρ); specific heat (c_p); thermal conductivity (k); latent heat (λ); absolute viscosity (μ).

B-1. Density, Specific Gravity, Expansion Coefficient

Density and specific gravity are often used interchangeably in textbooks.

The *density* of a homogeneous material is its weight per unit volume at a given temperature and pressure. The reciprocal of density is called *specific volume*, represented by the letter, v. Density is represented by the Greek letter, ρ, and is expressed in lb/ft^3.

Specific gravity is defined as the ratio of the density of a given substance to the density of a reference substance. Air, either at the same temperature and pressure or at 60°F and 14.7 psia, is used as a reference substance for gases. For liquids, water at 60° or 32°F and 14.7 psi is used as a reference substance. In the English system, 60°F is used primarily as reference temperature.

Specific gravity is a dimensionless number universal for all systems of units, while the value of a density varies with the system of units chosen.

Gases. Gas density is generally computed from the ideal gas law:

$$pv = RT \tag{1-10}$$

$$\frac{1}{v} = \rho = \frac{p}{RT} \qquad (1\text{-}11)$$

In deriving the *perfect gas equation,* two simplifying assumptions are necessary:

The gas molecules are presumed to be so widely spaced that (a) they have zero volume, and (b) their intermolecular binding forces may be ignored. However, in many instances, real gases do not follow this thesis and substantial deviations from ideal gas behavior result.

Many investigators have attempted to formulate equations of state which accurately describe the P-V-T relationships of real gases. These studies have led to a series of intricate and cumbersome equations too time-consuming for engineering purposes.

A convenient approach to the problem is to modify the ideal gas law by introducing a correction, or compressibility factor, called "z":

$$pv = zRT \qquad (1\text{-}12)$$

Another useful tool for analyzing this discrepancy is the principle of corresponding states. This hypothesis proposes the use of so-called reduced states, indicated by the index "r," whereby the gas properties are expressed as fractions of their critical property values:

$$P_r = p/p_c, \; T_r = T/T_c, \; V_r = V/V_c \qquad (1\text{-}13)$$

Through this artifice, the distinction between the individual gases vanishes, since all gases possessing the same reduced state are in identical physical conditions. Finally, uniting these two concepts of compressibility factor and the reduced state, the compressibility factor z may be obtained empirically and then plotted versus the reduced pressure for different reduced temperatures. Such a chart establishes a nearly universal P-V-T relation for all gases. This method is accurate within 5 to 10 per cent for all gases with the exception of H_2 and He, which are nearly ideal gases. The reduced-state concept is a general one and can be easily related to the prediction of other physical properties. For example, reduced viscosity and reduced thermal conductivity correlations produce fairly accurate results for any gas not in the vicinity of its critical point.

Liquids. All real liquids are compressible and expansible. Complete incompressibility and inexpansibility do not exist in nature. Variations in liquid density caused by temperature changes may be quite substantial. In most cases, the density of liquids is a linear function of temperature and is only slightly affected by pressure.

The coefficient of thermal expansion, β_T, specifies the relative change in volume per degree temperature difference, and is expressed in reciprocal degrees Fahrenheit or reciprocal degrees centigrade. The expanded volume, V_T, is given by the equation:

$$V_T = V_o\,(1 + \beta_T\,\Delta T) \qquad (1\text{-}14)$$

which expresses the change of volume due to temperature. From 0° to 100°C, these coefficients range from 2×10^{-4} to 16×10^{-4}.

Similarly, the compressibility of a material is the fractional change in volume per unit increase of pressure, and its units are those of reciprocal pressure. The compressibility coefficient β_p is used to calculate the new volume, V_p, as follows:

$$V_p = V_o\,(1 + \beta_p\,\Delta p) \qquad (1\text{-}15)$$

The compressibility of a liquid due to pressure is small. For example, 300,000 cu ft of water, when compressed by 1 psi, will reduce its volume by 1 cu ft. Consequently, it will require only 299,999 cu ft of space after adding 1 psi of pressure. The cubical compressibility of water is:

$$\frac{1}{300{,}000}\,\frac{\text{(cu ft)}}{\text{(cu ft)(psi)}} \text{ or } 0.000{,}003{,}3 \text{ psi}^{-1}$$

B-2. Heat Capacity, Specific Heat

The specific heat, c_x, of a material is defined as the quantity of thermal energy required to produce a unit temperature rise in a unit mass of a certain material at a given temperature. The subscript x denotes the fact that the specific heat of the process under investigation may have any value from zero to infinity, depending on the path of the process pursued. Specific heat is represented mathematically as:

$$c_x = \frac{1}{W}\left(\frac{dQ}{dT}\right) \qquad (1\text{-}16)$$

In accordance with the nomenclature previously adopted, the units of specific heat are Btu/(lb)(°F). W is the weight in pounds. However, units are usually omitted because specific heat has identical values in the English and metric systems. The specific heat usually varies with the temperature level of the material. It is therefore important, in giving specific heat values for the different materials, to denote the temperature limits of their application.

Most industrial processes are performed at either constant pressure or constant volume. Therefore, two kinds of specific heat, the specific heat at constant pressure, c_p, and the specific heat at constant volume, c_v, are generally tabulated. The specific heat of all fluids, both gases and liquids, is thermodynamically defined by:

$$c_p = \left(\frac{\partial h}{\partial T}\right)_p \tag{1-17}$$

$$c_v = \left(\frac{\partial U}{\partial T}\right)_V \tag{1-18}$$

For ideal gases, a series of useful relationships between c_p and c_v have been derived. For one mole or gram-molecular weight of an ideal gas, $(c_p - c_v) = R$. According to the kinetic theory of gases, c_p and c_v, for an ideal monatomic gas enjoying only translational freedom are equal to 5/2 R and 3/2 R, respectively. The kinetic gas theory is such a powerful tool that with it one can predict the heat capacities of complex polyatomic gases possessing many degrees of rotational and vibrational freedom. The ratio of c_p to c_v is a dimensionless quantity called κ (kappa), and equals R plus 1, for an ideal monatomic gas. Again, the kinetic theory foretells correctly the value of κ for different gaseous molecular arrangements.

Contrary to theoretical considerations, c_p and c_v of a real gas vary with pressure, primarily at low reduced temperatures. As for the influence of temperature, gaseous specific heat increases with temperature at constant pressure. Empirical formulas of the type:

$$c_p = a' + b'T + \frac{c'}{T^2} \tag{1-19}$$

$$c_p = a + bT + cT^2 + dT^3 \tag{1-20}$$

are widely used, where a', b', c', a, b, c and d are experimental constants. In many practical situations a mean specific heat,

$$\bar{c} = \int_{t_1}^{t_2} \frac{c_x\,dt}{t_2 - t_1} \tag{1-21}$$

is employed at all temperatures within the given range.

The specific heat of liquids is generally greater than that of gases and vapors. Some exceptions to this rule are hydrogen, helium, ammonia, and nitrogen peroxide. For liquids, the heat computations are greatly simplified because the specific heat of liquids is unaffected by pressure variations; c_p is approximately equal to c_v for all liquids since the volume changes which accompany heating are rather small. The specific heat of liquids, with the rare exceptions of Hg and alkali metals, increases with temperature. However, despite accurate correlations of specific heat values for many solids and vapors, a theoretical analysis of liquid specific heat is lacking. Empirical evidence only is available for many engineering materials such as oils and petroleum fractions. Recent research seems to indicate that the specific heat of organic liquids can be computed from a knowledge of molar refraction and the structural parachor, or from a simplified model of the liquid state and use of data on the velocity of sound, infrared and Raman spectra.

B-3. Heat Content, Enthalpy

The vast majority of industrial heating applications are conducted under constant pressure conditions; therefore, the need arises for a property which will specify the heat absorbed or discarded by a system at constant pressure. Heat, which is the energy in transit due to a temperature difference, is not a property of matter, but a path function dependent on the process examined. Hence, a new derived property called 'enthalpy,' or heat content, H, also designated by h, which equals the heat absorbed or rejected in an isopiestic process, has been proposed. Enthalpy has the dimensions of energy and is usually expressed in Btu/lb.

Heat transfer media undergo enthalpy variations as a result of changes of phase or temperature variations. The enthalpy change accompanying a phase transition at constant temperature and pressure in a one-phase region is numerically identical with the latent heat accepted or rejected by the surroundings. As for its dependence on temperature, the enthalpy rises with increasing temperature as prescribed by the thermodynamic relation:

$$\Delta h = \int_{t_1}^{t_2} c_p \, dt \qquad (1\text{-}22)$$

where t_1 is the reference temperature and t_2 the temperature of the medium at which the enthalpy is expressed.

If c_p is not a function of temperature changes, then equation (1-22) becomes

$$\Delta h = c_p \Delta T \qquad (1\text{-}23)$$

Enthalpy cannot be assigned a definite absolute value; however, this limitation is not a handicap, since the engineer is concerned only with enthalpy changes incurred, not with the absolute values. The usual procedure is to calculate enthalpy variations from a selected zero reference datum. Keenan and Keyes selected liquid water at 32°F as datum level for their steam tables. For air and other permanent gases, the datum level chosen is in the vicinity of absolute zero. Ammonia tables measure enthalpies from a reference point of −40°F. Thus, when dealing with a particular material, one should be aware of the conditions at which zero enthalpy is taken for that material.

B-4. Heat Conductivity

One of the primary methods of transmitting thermal energy is by molecular conduction. The study of heat conduction commences with Fourier's Law, which states: "*The rate of heat transfer between two regions is directly proportional to the thickness and the size of the area perpendicular to the direction of the flow and to the temperature gradient or driving potential.*" The ensuing energy transfer is expressed by the following equation:

$$\frac{dQ}{dt} = -kA\frac{dT}{dx} \qquad (1\text{-}24)$$

The negative sign indicates that heat flows in the direction of decreasing temperature. Thermal conductivity, k, is defined as the proportionality constant employed in Fourier's Law. From a physical standpoint, k is the thermal energy transferred in a unit time interval through a specimen of unit area and unit thickness across which a unit temperature drop occurs. The units of thermal conductivity are Btu/(ft)^2(°F/ft)(hr).

The property of thermal conductivity is usually associated with solids which act as insulators between a hot region and a cold region. Unfortunately, thermal conductivity data for liquids and gases are quite sparse, since only a few workers have performed successful measurements. This failure is a consequence of the difficulty of segregating heat losses by conduction from radiation and convection losses in the same system. In general, liquid thermal conductivity is approximately ten to twenty times greater than the thermal conductivity of gases because there are more molecules per unit volume in the liquid state.

The thermal conductivity of liquids is a function of temperature and pressure. Liquid thermal conductivity diminishes with rising temperature at constant pressure, with the rare exceptions of water, glycerol and mercury. As for variation with pressure, liquid thermal conductivity shows a negligible increase with pressure at constant temperature.

In contrast to liquids, gases have greater thermal conductivity at high temperature than under ambient conditions. When gases are subjected to increased pressures, their thermal conductivity appears to increase.

B-5. Viscosity

An inherent property of all fluids is their viscosity. As moving layers of fluids slide one over another, shearing or viscous forces are exerted at the surface of contact. From a descriptive viewpoint, viscosity is a measure of intermolecular fluid friction.

A precise definition of viscosity is extracted from the following physical system. A layer of fluid is assumed to be placed between two plates, one movable and the other fixed, at a distance, dy, from each other. A shearing force, F, now imparts a change in velocity, dV, to the movable plate. (See Fig. 1)

The fluids deform continuously under the shear force. Empirical studies have concluded that the shearing stress, τ, is directly proportional to the velocity gradient, dV/dy. The viscosity coefficient, or the absolute viscosity, μ, is the

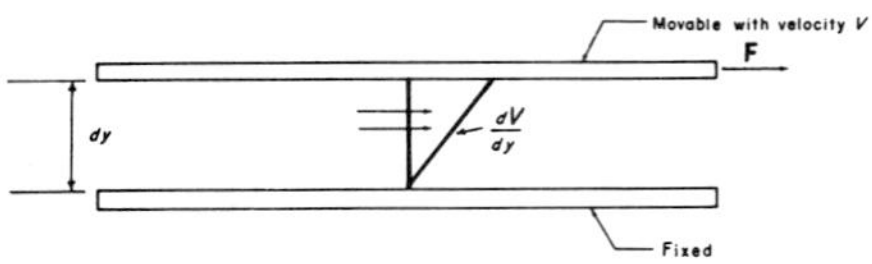

Figure 1

proportionality factor equating both quantities. The fundamental relationship is:

$$\tau = \mu \frac{dV}{dy} \text{ and } \frac{dV}{dy} = \frac{V}{y};\ \tau = \mu \frac{V}{y} \quad (1\text{-}25)$$

and when V (velocity), y (distance) and τ (stress) obtain unit values, μ expresses the absolute viscosity:

$$\mu = \tau \frac{y}{V} \quad (1\text{-}26)$$

Absolute viscosity may be physically interpreted as the tangential force per unit area required to move a fluid layer a unit distance away, with a unit velocity.

The units of absolute viscosity are lb/(ft)(hr) or lb/(ft)(sec), or centipoises.

Kinematic viscosity is defined as the ratio of absolute viscosity to density at the same temperature and shall be used in the dimensions of ft^2/sec. The kinematic viscosity, ν, is a derived fluid property first recognized by Reynolds in 1880 and included in his famous dimensionless relation, the Reynolds number, $\frac{DV}{\nu}$.

Viscosity is a function of temperature, pressure, and the velocity gradient. Any fluid in which μ is a constant at a given temperature and pressure is classified as a Newtonian fluid. Analytically, τ versus dV/dy is a straight line, with μ as the slope (Figure 2). On the other hand, for a non-Newtonian fluid, the viscosity either increases (dilatant) or decreases (pseudoplastic) with the rate of shear.

The viscosity of liquids usually rises following compression at constant temperature. Water is an exception to this rule; its viscosity decreases with pressure at constant temperature. As the temperature of a liquid increases under isobaric conditions, its viscosity drops.

The kinetic theory of gases has estimated the effect of temperature and pressure on gas viscosity. According to the hypothesis, viscosity is directly proportional to average velocity, mean free path, and the density of the gas molecules. The mean free path varies inversely with pressure, while gas density varies directly with pressure. The two opposing effects cancel each other out and viscosity becomes independent of pressure. However, the viscosity of real gases increases with pressure, and the above prediction is only fulfilled under ideal conditions (high reduced temperatures and low-reduced pressures).

The kinetic theory asserts that as the gas temperature rises, the net effects of a mean free path and velocity increase outweigh a gas density decrease and result in a viscosity increase. For this situation, experimental data seem to verify the theoretical kinetic approach.

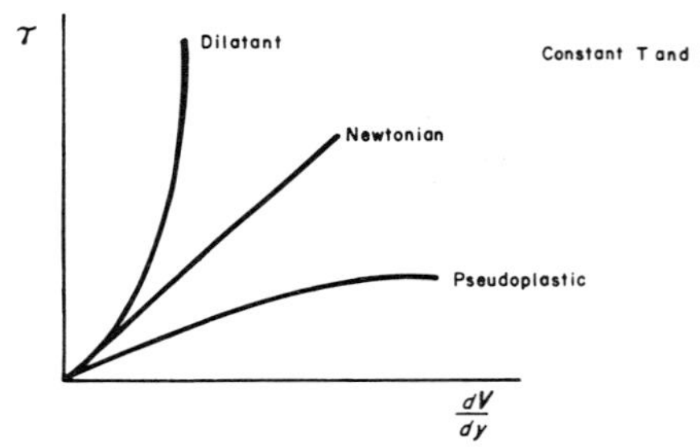

Figure 2

C. CONSTRUCTION PROPERTIES (LIMITING PROPERTIES)

A heating medium often undergoes substantial physical and chemical changes while circulating in a thermal network. A critical step in selecting a heat carrier is to assess a set of limiting properties which delineate the boundaries of safe application. These properties or criteria are: physical stability, thermal stability, chemical stability, radiolytic stability, and cost.

C-1. Physical Stability

All substances exist in one of three states—solid, liquid, or vapor. Each phase is physically distinct and enjoys a unique set of chemical and physical properties. The discontinuous change of state from solid to liquid is termed 'melting', while the reverse process is called 'freezing'. Evaporation, or boiling, is the transition from liquid to vapor, while condensation is the conversion of vapor to liquid. To motivate a phase change at an equilibrium temperature and pressure, a fixed quantity of latent heat per unit mass, h_{fg}, must be absorbed from, or rejected to the surroundings. For example, when liquid water vaporizes under atmospheric pressure, 970 Btu are required for each pound of water transformed into steam. The addition of latent heat energy does not raise the water temperature. Instead, it furnishes the kinetic energy required for the molecules of the liquid to overcome and escape the restraint of their neighbors and to gain the increased mobility of the gaseous state.

To clarify the entire phenomenon, assume that water is a typical fluid. Figure 3 is a Pressure-Temperature (P-T) plot for water. In a pure water system, three kinds of two-phase equilibria (liq-gas, sol-liq, sol-gas) and one kind of three-phase equilibria (liq-sol-gas) are possible. For two phases in equilibrium, the Gibbs' Phase Rule, $F = C - P + 2$, where F is the number of degrees of freedom, C is the number of components and P is the number of phases in equilibrium, predicts a single degree of freedom.

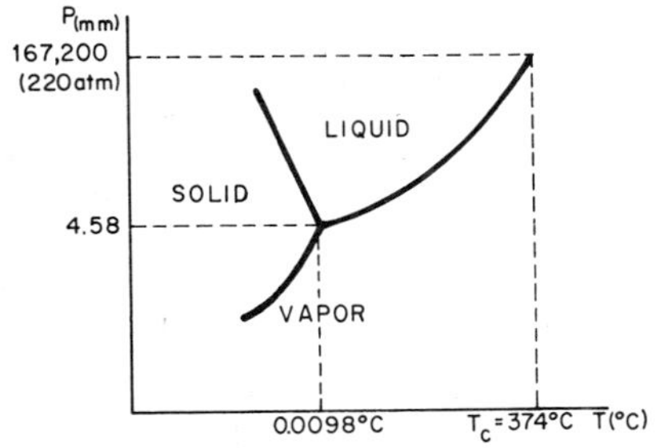

Figure 3 (not to scale)

Therefore, at any given temperature the pressure is single valued. This fact is indicated graphically by the formation of a number of P-T curves, along which two-phase equilibria prevail. Correspondingly, only under one single temperature and pressure condition can three-phase equilibria exist. As one traverses the vapor pressure curve in an upward direction, liquid and vapor properties gradually approach one another. In the limiting case, or at the critical point, the liquid and vapor phases coalesce into a single phase. Above the critical temperature only gas remains and liquid-vapor equilibrium is impossible.

Several analytical and graphical methods of estimating the vapor pressure curve as a function of temperature have been formulated. However, all are modifications of the fundamental Clausius-Clapeyron equation, which applies for solids, liquids, and gases. The equation, written in differential form, is:

$$\frac{dp}{dT} = \frac{h_{fg}}{T\Delta V} \tag{1-27}$$

where h_{fg} is the latent heat/mole, and ΔV is the volume difference between two states per mole of the gas. The slopes of the P-T curves are equal to $\frac{\Delta H}{T\Delta V}$ and may be directly computed from the Clausius-Clapeyron equation. For the special case of liquid-vapor equilibrium, the equation may be vastly simplified by making a few straightforward assumptions. If the volume of the liquid phase is regarded as negligible in comparison to the volume of the vapor phase, and if the perfect gas equation can be applied to the vapor under consideration, the equation can be converted to a more useful form:

$$\frac{d(\ln p)}{dT} = \frac{h_{fg}}{RT^2} \tag{1-28}$$

If the latent heat of vaporization is independent of temperature, equation (1-28) can be integrated to:

$$\ln p = \frac{-h_{fg}}{RT} + C \tag{1-29}$$

where C is determined from the experimental data. If this pressure, P, is plotted versus $(1/T)$ on semi-log paper, a straight line of slope, $\frac{-h_{fg}}{R}$, is obtained. Thus, the latent heat of vaporization can be computed from the Clausius-Clapeyron equation.

In order to warn the engineer and designer that he is entering a danger zone and to set a restriction on low temperature applications, a property called the 'pour point' has been arbitrarily formulated. The pour point is defined as the temperature at which an oil or other liquid ceases to flow when cooled at a standard rate in a standard container as designated by the ASTM.

It is important to determine the distinction between freezing point and pour point. The pour point relates to a considerable increase of viscosity with temperature, while the freezing point refers to the change of state from liquid to solid. At the freezing point, a sudden discontinuity in fluid properties appears, while no such effect accompanies attainment of the pour point. The pour point is always located above, but sometimes coincides with (water), the freezing point. Even before reaching the pour point, the viscosity generally increases to such an extent that the liquid loses its 'pumpability'. It would therefore be practical to create a 'pumpability' point. This, however, is difficult, since it would vary with the type of pump employed.

C-2. Thermal Stability

A number of laboratory methods have been developed for determining thermal stability. The data given in Section III are those published by the different manufacturers. However, the method of determining thermal stability has been indicated in only a very few instances.

There is no standard test for stability. Special considerations, such as test temperature, duration of test, susceptibility of the fluid to traces of oxygen and water, choice of proper characteristics for evaluating degradation, and the reproducibility of the test, are all factors making thermal stability difficult to express in absolute terms.

In order to assure some understanding of the various laboratory methods, a brief description of them follows.

The Isoteniscope. The use of this instrument is based on the assumption that the most general change in property during the thermal decomposition of organic molecules is an increase in the vapor pressure of the system, brought about by cleavage of the molecules into smaller, more volatile fragments. This physical change has the advantage over others, such as weight loss, viscosity change, and heat of reaction, since it is related to the true rate of thermal decomposition by a simple although unknown proportionality constant. On the other hand, it has the disadvantage that it is not necessarily related to a practical rate of decomposition. The rate of increase in vapor pressure can be measured comparatively easily with a modification of the static isoteniscope of Smith and Menzies. According to this method, the basis of comparison is the isothermal rate of pressure rise. Experimentally, the variation of pressure with time is obtained, and log $\frac{dp}{dt}$ is plotted against $\frac{1}{T}$. A decomposition temperature, T_D, is arbitrarily defined so that $dp/dt = 0.014$ mm Hg/sec. This rate would require ten hours to develop a pressure of 500 mm Hg.

T_D affords a convenient single number for comparing stability of different compounds but must not be taken as a direct indication of a practical use temperature. Corrections can be made for the number of fragments, and differences of molar volume and thermal expansion. T_D is always in excess of the maximum use temperature for a given substance as a heat transfer fluid, possibly by as much as 150°F.

The method is excellent and is the first real attempt to compare thermal stabilities by the measurement of a parameter related to bond strengths. It gives useful preliminary information, but it does not give a measure of the performance that would be obtained under given practical conditions. In particular, it has these disadvantages:

1. The temperature indicated is higher than would normally be tolerated and therefore does not give a maximum use temperature.

2. T_D may be so much higher than the normal operating temperature that bonds are broken which are not normally disrupted in commercial application. Therefore, the method can be misleading, as in the case of Hydrotherm 750–200, which is a phenyltriphenoxysilane. The phenyl silicon link might be the one indicated as breaking by the isoteniscope method whereas in use, normal deterioration may occur as a result of reactions of the phenoxy groups.

3. The method gives an indication of "decomposition point" of a vapor-phase reaction, whereas in liquid heat transfer systems, reactions in solutions are involved.

4. The method does not indicate whether a reaction is likely to be stabilized or accelerated by an accumulation of decomposition products.

5. In practice, the method may not give an assessment of the effect of additives due to differences in vapor pressure.

6. The isoteniscope method does not indicate thermal stability in the presence of materials of construction encountered in commercial heat transfer systems.

Thus, the isoteniscope gives an indication of thermal stability of equilibrium mixtures of liquids and vapors. For practical assessment, procedures are required to compare materials under conditions similar to those that would be economically feasible in commercial installations.

Test Bombs Incorporating a Pressure Gauge. These are used for vapor phase and liquid-phase heat transfer media. They have the advantage of being able to follow decomposition by increase in pressure and by a periodic sampling of the measure parameter indicating decomposition such as viscosity, density, polymer content, low boiling products, or chemical analysis. The most pertinent property in any case is likely to vary with the chemical type of the material under test.

Glass Test Apparatus. These are useful for comparing liquid phase heat transfer media. They may be sealed, closed to the atmosphere through a mercury trap, or continuously purged with an inert gas. The test may be of long duration, one to two years, simulating temperature conditions in plants, or accelerated by heat to higher temperatures to obtain results in two to three months. Change in viscosity is the parameter normally determined, sometimes together with an increase in polymer content. The apparatus used for this type of test will not give an assessment of absolute stability, but can be used to give a comparative stability over limited temperature ranges.

A number of variations of this method have been suggested, including heating the fluid in open-tared vessels under a stream of nitrogen. It is clear from a description of these methods that they can give relative evaluation of thermal stability but, in practical operation, other forces acting on the liquid may suggest a lower thermal decomposition point.

References

1. Beanland, E., "Report on Thermal Stability," Hygrotherm Engineering Ltd., London, England.
2. Willard, Myra, "Determination of the Thermal Stability of Disiloxane-Base Hydraulic Fluids," ASTM Bulletin, October 1960.
3. "Lubrication under Extreme Conditions," Symposium sponsored by the Division of Petroleum Chemistry, *American Chemical Society,* **5,** 2-B, (April 5-14, 1960).

C-3. Chemical Stability

Numerous heat transfer fluids are mixtures of two or more pure substances. By selectively blending different chemical compounds, the favorable properties of each substance can be united to form a so-called "optimum" mixture. This behavior may be easily visualized by two rather fundamental examples:

One of the foremost problems in utilizing water as a thermal liquid is its relatively high freezing point. This problem can be solved by introducing certain soluble chemical additives, such as ethylene glycol (CH_2OHCH_2OH) or methanol (CH_3OH) into the water. These additives depress the freezing point without exerting harmful effect on other fluid properties.

The technology of liquid metals provides another interesting example of the importance of solubility. Despite their outstanding heat transfer properties, liquid metals have a number of major drawbacks. Perhaps the most notable one is their exceptionally high melting point. In general, homogeneous mixtures of two or more metals always have lower melting points than the pure constituents alone. For metals having mutual liquid solubility and mutual solid insolubility, there always exists a particular composition called the eutectic composition, which has the lowest attainable melting point for the given metal pair. Cadmium and bismuth form mixtures of the eutectic type (Figure 4).

As may be readily seen from the phase diagram, the eutectic composition of 60 per cent Bi-40 per cent Cd is undoubtedly the best heat

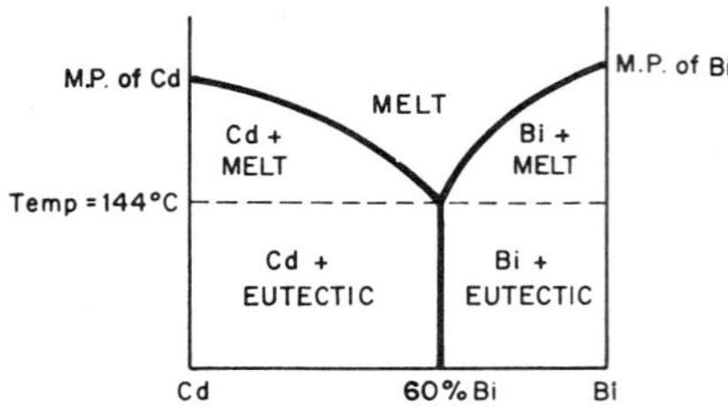

Figure 4

transfer alloy. The freezing point curve for the Cd-Bi mixture may be considered a solubility curve. Similar eutectic mixtures can be produced by mixing different nitrate and nitrite salts of sodium and potassium, forming the important heat carrier called "heat transfer salt." Another example is the eutectic mixture of diphenyl and diphenyl oxide, known as Dowtherm A in the United States.

It is now quite evident that solubility effects are highly significant in the selection of heat transfer fluids.

C-4. Radiolytic Stability

In a nuclear reactor, high energy neutrons and gamma rays continually strike materials which are not in contact with the fuel. The neutrons do the most damage to solid reactor components, but gamma radiation seems to have the greatest effect on organic materials in particular. The two primary effects on fluids are polymerization and ionization.

Polymerization caused by radiation generally causes an increase in fluid viscosity; sometimes a solid sludge is formed. Other possible effects are gassing, coking, and foaming. All these factors can lead toward increased corrosion, which is the cardinal concern of the engineer. In several reactors, the liquid is circulated from an external reservoir in order to dilute the effects of irradiation. However, this presents the additional problem of shielding the radioactive material when it is outside the reactor.

Corrosive or otherwise undesirable products can emanate from ionization reactions within the reactor fluids. To illustrate, ionizing radiation decomposes water into H and OH radicals. These radicals can react with dissolved solutes, if they are present, or they can react with themselves. If they do combine with one another, then hydrogen and hydrogen peroxide are formed.

Tests indicate that biphenyl is more stable to radiation than any other organic liquid. Other polyphenyls and condensed ring structures are also very radiation stable. It is interesting to note that there has never been a change in the physical character of liquid metals by radiation.

C-5. Cost—Price Class (1961)

Undeniably, the cost of a fluid is an important factor in determining the scope of its application. However, since this Handbook is concerned more with scientific aspects, detailed prices of the various commercial media should not be divulged. As a compromise, therefore, the heat carriers considered have been categorized into several broad price classes:

Class A: Under $1.00 per gallon at 60 °F and 14.696 psia
Class B: Under $5.00 per gallon at 60 °F and 14.696 psia
Class C: Under $20.00 per gallon at 60 °F and 14.696 psia
Class D: Over $20.00 per gallon at 60 °F and 14.696 psia

The price classes for gases are given either per cubic foot or per pound; for materials sold as solids, the price class is per pound. This breakdown gives a general idea of the expense involved in purchasing the natural and synthetic fluids presented herein.

D. OPERATION PROPERTIES (HANDLING PROPERTIES)

A third group of properties is intimately associated with handling a heat transfer agent in an engineering system. Ease of handling is vitally important in the actual day-to-day operation of a thermal system. This group includes surface tension, corrosion, physiological effects, flammability, and special properties.

Surface Tension. Surface tension, along with viscosity, is a measure of the 'wettability' of a liquid. Materials with low surface tension can leak through porous iron or other metals, or through any defects in glands, joints, or welds.

A scrupulous choice of valves and fittings is necessary. Wettability is also an important heat transfer consideration, determining whether film or drop-wise condensation takes place. For gases, the diffusion coefficient is analogous to surface tension in that it, too, determines the ability of a fluid to seep through a container. Diffusion of light gases, such as hydrogen and helium, through the walls of a system can cause considerable expense.

Corrosion. The tendency of a fluid to attack the structural components in which it circulates can have serious consequences. System failures or blockage of equipment passages may result. For certain media, such as liquid metals, the provision of corrosion-resistant materials has become a limiting factor. The corrosive properties of heat carriers are often highly dependent on the presence of dissolved impurities.

Flammability. The ability of fluids, particularly petroleum oils and organic liquids, to ignite and burst into flame constitutes a hazard. However, a closed thermal circuit, with inert gas blankets over expansion and sump tanks provides a measure of protection. Fire-resistant organic fluids are commercially available.

Physiological Effects. As human beings are directly responsible for plant maintenance and performance, any poisonous or harmful effects of a fluid must be known. Adequate safety precautions can successfully counteract this danger.

Special Properties. Certain special properties may exert a subtle influence on the ultimate selection of a heat carrier. Electrical properties, such as dielectric constant and electrical resistivity determine the feasibility of employing a medium in an electrical environment. Refractive index is a property which helps to describe the molecular structure and the chemical composition of a substance. Physical appearance, color, and odor are properties which affect consumer reaction to a given product.

D-1. Surface Tension

Unusual events take place at the interface between two immiscible fluids. The molecules at the surface of the fluid are drawn inward by the neighboring molecules because of the absence of any opposing force. Liquids tend to fall in spherical drops; dense objects float on a fluid surface; certain liquids may rise to tremendous heights in capillaries. The state of stress responsible for these bizarre effects is called *surface tension.* Surface tension can be viewed as the desire of a fluid to reduce its surface potential energy to an absolute minimum. The most stable equilibrium configuration for a fluid film is a spherical one because a sphere has the least possible area per unit volume. The units of surface tension (σ) are lb/ft or energy/area.

A quantitative definition of surface tension is derived from the system, as shown in Figure 5.

If a slide wire is placed on a U-shaped piece of wire and the system is dipped into a soap solution, two surface films are formed with liquid between them. The slide wire will move to the left because of the unbalanced surface force. The force required to hold the slider stationary is $2\sigma l$. The surface tension can be defined as the ratio of the force exerted by the liquid to the length of the side perpendicular to the action line of the force. It can also be thought of, as the work input necessary to increase the fluid surface area by a unit amount.

Another quantity related to surface tension is the angle of contact, α, between a fluid interface and a solid wall (Figure 6). Fluids such as CH_3I, with contact angles less than 90°, are said to wet glass and are called adhesive, while fluids like Hg, which do not wet glass, are cohesive. The contact angle usually depends on the fluid involved, the wall material, and the past history of the system. Wetting or detergent materials are employed to diminish the angle of contact.

Surface tension is a function of fluid composition and temperature, and is totally unaltered by pressure changes. Impurities are often re-

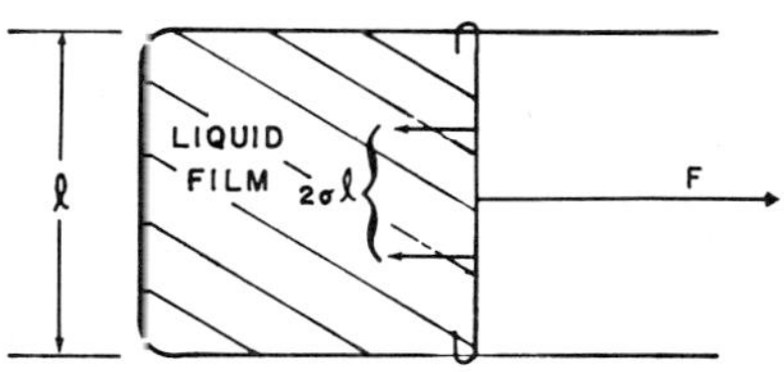

Figure 5

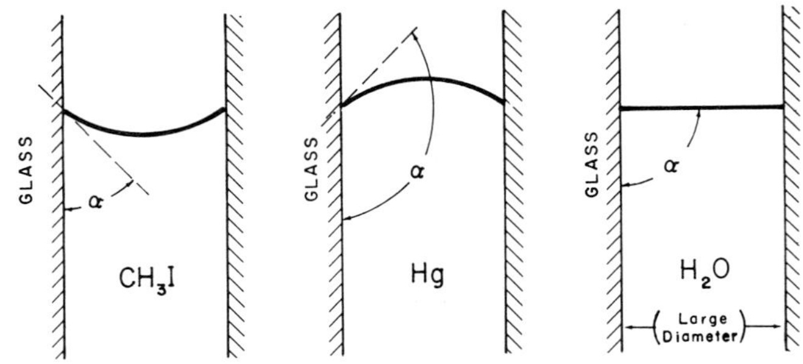

Figure 6

sponsible for surface tension deviations. The surface tension of all liquids decreases with rising temperature and finally becomes zero at the critical point, because the interface vanishes and a single phase is formed. The surface tension of all gases is approximately alike and surface tension is usually regarded strictly as a property of liquids.

D-2. Corrosion

A serious engineering problem arises when a heat carrier attacks and decays common structural materials such as metals and alloys. Corrosion can be defined as the destruction of a metal by chemical or electrochemical reaction with its environment. Under ordinary circumstances, the usual products are oxides, carbonates, hydrated oxides, hydroxides and sulfides, with oxides predominating.

The fundamental mechanism of corrosion is best explained by a typical example, that of iron in contact with water or an acid solution. For iron to corrode, an inherent tendency or an electromotive driving force must exist for the iron to attain the plus 2 or plus 3 state, adding two or three positive ions by dissolving in water. However, once positive ions enter the solution, electroneutrality is maintained by removing an equivalent number of negative ions from the solution. In the iron $-H_2O$ system under investigation, the above condition is met by the formation of a hydrogen ion as a catalyst. The initial reaction is:

$$2H^+ + Fe_{(s)} \rightleftharpoons Fe^{++} + H_{2(g)} \qquad (1\text{-}30)$$

The hydrogen film then retards further corroin two ways:

1. It insulates the metal from attack by the solution.
2. The increase in H_2 concentration forces a reversal of the corrosion reaction.

If the destruction is to continue, the H_2 film must be eliminated either by escaping as gas bubbles or through the formation of H_2O as a secondary reaction. Ultimately, the divalent iron accumulating in solution is oxidized and precipitated as rust by the reactions:

$$4Fe_{(s)} + 3O_{2(g)} \xrightarrow{H^+} 2Fe_2O_3 \qquad (1\text{-}31)$$

$$2Fe_{(s)} + O_{2(g)} \xrightarrow{H^+} 2FeO \qquad (1\text{-}32)$$

However, many complex variables, too numerous to be discussed here, influence the outcome of a corrosion reaction.

Protection of Steel by the Use of High pH Values. From the above, it can be seen that the hydrogen ion is required to catalyze the corrosion reaction and dissolve the resulting oxide into solution. With a pH of 9.5–10.5, the hydrogen ion concentration is reduced sufficiently so that the reaction is not catalyzed and therefore will not proceed. In this way the iron will be protected and any oxide formed will not dissolve, thus safeguarding the iron.

Industry is continually seeking new ways to prevent rust formation. Most rust inhibitors act by erecting a physical barrier between the metal and the solution. The surface is coated with a protective film such as paint or metal, or a very thin corrosion-resistant adherent film such as an oxide, a chromate, or a phosphate.

Another mode of protection involves a chemical alteration of the metal surface itself to make it corrosion resistant. Studies with rust inhibitors such as technetium, a synthetic element formed in uranium fission, supports a new inhibition theory. For example, the pertechnate ion, TcO_4^-, consists of a technetium atom surrounded by four oxygen atoms in the form of a tetrahedron. This ion has a strongly positive central atom with a valence of plus 7. Electrons are readily drawn from the metal surface, electron movement is immobilized, and corrosion thwarted.

Another type of corrosion, which is most prominent in liquid metal systems is mass transfer. This phenomenon is the dissolving of materials from the confining walls into the liquid metal and the subsequent deposition of the solvated material in a zone of lower solubility. The continuous removal of substance from one part of the system may accelerate corrosion in that area. Of course, the precipitated material will eventually clog the flow passages. Inhibitors, such as titanium and zirconium, have been used with some success in certain systems.

D-3. Flammability

Many disastrous fires and accidents result from the improper handling and storage of flammable materials. All possible fire risks must be ascertained prior to employing a heat transfer fluid. To evaluate quantitatively the fire and explosion hazards associated with the use of a given fluid, a set of physical quantities must be defined. These properties are empirical constants, invented for engineering convenience. They are highly dependent upon the system examined and the apparatus employed.

The *flash point* of a material is the temperature at which the liquid or the combustible vapor above the liquid ignites and burns; however, the ensuing fire is transient and self-extinquishing. The liquid is heated in a reference vessel at a standard rate, as formulated by the ASTM. The three most common flash point tests are: the Pensky-Martens, the Tagliabue Closed Cup, and the Cleveland Open Cup. All three methods yield different flash points for the same substance, with experimental disagreements occurring primarily in the procedures of determination.

The *fire point* of a liquid or solid is the lowest temperature at which its combustible vapors flash and burn continuously. The fire point is usually a few degrees above the flash point, and similar experimental techniques are used to determine both properties. The concept of a fire and flash point also applies to flammable solids such as camphor and naphthalene, which burn at room temperature.

The *ignition point,* also called 'self ignition point', is the temperature at which the *self-sustained* ignition and combustion in ordinary air take place, independent of the heating source. As may be anticipated, the size, shape, and composition of the standard container exert a significant effect. Ignition points are usually considerably higher than flash or fire points. For example, acetone has a flash point of 0°F and an ignition point of 1000°F, whereas aniline has a flash point of 168°F and an ignition point of 1000°F.

The *explosion range* is the highest and lowest limiting composition of mixtures of flammable vapors and air which can burn. At the lower limit the molecules of the flammable vapors within the air-vapor mixture are widely dispersed so that those molecules already aflame cannot spread the fire to their distant neighbors. When the molecules of the flammable vapor within the air-vapor mixture are packed tightly together so that the required oxygen is excluded, no burning is possible. Within the explosion range various types of slow and rapid combustion can occur.

All these properties describing flammability are of practical importance because they serve as indicators of the volatility of the material and may also aid in detecting impurities. The lower the flash point of a material, the greater the probability of a fire. Of course, the most desirable fluids are those like water, which do not have any flash, fire, or ignition point. Nevertheless, materials with high flash points are often quite acceptable. For example, most chlorinated biphenyls used for heat transfer have a flash point of 348° to 384°F and yet do not have fire or ignition point, nor do they support combustion. Several analytical methods of determining flash points are available, but actual experimental data are the most reliable approach.

D-4. Physiological Effects

In the course of daily plant operation, human beings must handle and manipulate a selected heat transfer fluid. By definition, toxic materials have harmful effects, hence toxicology, or the study of industrial poisoning, plays a vital role in the choice of a heat transfer medium. Many unique and challenging problems arise in this branch of science. At times, it is extremely difficult to ascertain exactly the amount of fluid which constitutes a maximum allowable concentration or a fatal dosage. Determination of allowable concentrations requires a comprehensive series of chemical analyses, physiological investigations, and calculations.

Toxic materials can harm the body in different ways. Most poisonings occur as a consequence of inhaling the toxic material. Upon being inhaled, toxic gases or vapors such as CCl_4, CO, and Hg, have deleterious effects on the respiratory organs. However, liquids of low volatility, such as phenol or cresol, present an equal or greater hazard if absorbed through the skin. Oh course, direct physical contact with the toxic agent, or with a fluid at an elevated temperature, may also cause skin injury. In certain isolated cases, toxic materials enter the body by ingestion, along with food. Numerous cases of industrial poisoning result from the insidious accumulation of small concentrations of the toxic agent over long periods of time.

In any process or heating plant, the engineer must be cognizant of possible physiological problems and skilled in their prevention.

D-5. Less Important Properties

Refractive index. The refractive index is the ratio of the velocity of light in a vacuum to the velocity of light of a particular wave length in a given substance. The refractive index of a substance is a pure number larger than one. Air offers a good example of the variation of refractive index with light wave length. The refractive index of air when exposed to violet light of 4359 A is 1.0002957; while for red light of 6563 A, air has an index of 1.0002914. The refractive index is a valuable property in confirming molecular structure, since it is a function of the number and the arrangement of the atoms of a compound.

Dielectric Strength. A substance in which relatively few particles are free to travel under the influence of an electric field is called an insulator or dielectric. For any dielectric there is an electric intensity at which the dielectric property breaks down and some conduction takes place. The maximum electric intensity which a dielectric can withstand prior to breakdown is termed the *dielectric strength.* Hence, dielectric strength is a quantitative measure of the insulating ability of a material. The units of dielectric strength are newton/coulomb. The dielectric strength of a heat transfer fluid is an essential property if the fluid comes into contact with electrical or electronic equipment, in which case fluids with a high dielectric strength are advantageous. Many heat transfer systems are designed to improve electric transformer cooling; here, this property is of paramount importance.

Another property often encountered in the literature is the *dielectric constant,* ε, which is the ratio of the capacity of a condenser with the compound to be tested between its plates, to the capacity of the same condenser in a vacuum under identical conditions of temperature and electric intensity. Compounds with large dielectric coefficients such as water, ammonia, and glycerol, have relatively high melting points, are usually ionic, and have high chemical reactivities; while compounds with low dielectric coefficients, such as CCl_4, C_6H_6, and C_6H_{14}, are nonpolar materials with low boiling points, and are relatively chemically inert.

Electric resistivity, ρ, is the ratio of potential gradient, E, in a conductor to the current density, J, thereby produced, $\rho = \frac{E}{J}$. It is a measure of the resistance of a unit quantity of a given material. A unit such as the ohm-centimeter is usually preferred. High resistivity is desirable for a dielectric coolant. However, for all other heat carriers, a high electrical conductivity is desirable, since this is associated with high thermal conductivity.

Appearance. Color and odor are properties which the human senses can perceive directly and distinguish qualitatively. Manufacturers are often prejudiced in favor of a product with a pleasant color and fragrant aroma since the response of the customer is more likely to be positive. However, from a purely scientific standpoint, these properties are minor factors in evaluating a heat transfer fluid. Nevertheless, research has been undertaken on the subject of color and odor improvement. Additives which do not harmfully modify the key fluid properties, have been used. The odor of some liquids is so disagreeable that it is virtually impossible to employ them without an additive.

Section II

NATURAL AND SYNTHETIC SUBSTANCES AS HEAT TRANSFER MEDIA*

A. HEAT TRANSFER SUBSTANCES IN NATURE

Life on the earth, as we know it, is made possible and certainly more pleasant by the great abundance and widespread distribution of the two naturally occurring heat transfer substances, water and air. Air, having a low thermal conductivity, insulates the surface of the earth from the tremendous quantities of radiant heat emanating from the sun, and effectively disseminates by convection to cooler regions, that portion of its heat which is absorbed by the land masses and oceans, and vice versa. The evaporation of water in hot areas, the distribution of humid air by air currents set up by temperature differentials, and the condensation of water as rain or snow in cooler areas assist greatly in moderating temperatures. Because of its high latent heat of vaporization and high specific heat, water is ideally suited for the transfer of heat by this process.

The sun heats land faster than water because the specific heat of land is lower; but water holds more heat and cools more slowly. For this reason, large bodies of water have a great influence on the climate of adjacent land areas. In areas with marine climates, the seasons are slowed down. The highest temperature is observed in autumn instead of summer, and the lowest in late winter or early spring. Also, the difference between summer and winter temperature along sea coasts is less than inland.

There are currents in the ocean, set up by the temperature differences between equatorial and polar waters, which tend to equalize these temperatures and affect the adjacent coastal areas. These moderating influences are demonstrated by the fact that the Gulf Stream, surging north out of the Gulf of Mexico at the rate of a trillion gallons per minute, maintains a mild temperature of 45 °F in Norway, which lies within the Arctic Circle. Seaports not affected by the warmth of this stream are icebound in winter, although some are as much as a thousand miles south of the Arctic Circle.

Evidence of the temperature conditions that would prevail on the earth without the generous bestowal of air and water is offered by the moon, which receives essentially the same quantity of solar energy per unit area as the earth; yet without the heat transfer benefits provided by water and air, the temperature on its sunlit side exceeds that of the boiling point of water (212 °F) and on its dark side it falls as low as - 240 °F; a temperature cold enough to liquefy air, if any were present.

Although the animal and vegetable oils are not critical requisites for human life, they do provide a needed heat transfer service to man. Animal fats and oils, which consist normally of the glyceryl esters of fatty acids, having low thermal conductivities, insulate the body from radical heat losses or gains, and because they have a heat of combustion higher than that of

*By Malcolm McEwen, Project Manager, Functional Fluids Section, Development Department, Monsanto Chemical Company, St. Louis, Mo.

the proteins and carbohydrates, they provide the body with a means of storing heat energy in concentrated form. These oils also provide a heat transfer function in cooking, protecting the food from localized overheating, and allowing faster cooking at higher temperatures than is possible with water.

Mineral or petroleum oils were formed by bacterial action on plant and animal remains of prehistoric times, by heat, pressure, and perhaps other natural forces. They have been used effectively by man to transfer heat under temperature conditions above and below the range of most natural products. Only within recent times have products based on petroleum enjoyed widespread usage, although crude petroleum has been known and used since ancient times (King Nebuchadnezzar used asphalt to pave the streets of Babylon in 600 B.C., and the boatmen on the Euphrates River made vessels of woven reeds and asphalt). The birth of the oil industry in the early 1850's, and its growth since 1900 is probably the most important factor in our modern technical society. Improved refining techniques have produced a wide variety of petroleum-based fluids essential to the efficient conversion of chemical energy to useful mechanical and electrical energy. The availability of such products for combustion, hydraulic, and lubricative applications has made possible the automotive, aircraft and utilities industries. Of equal importance is their ability to transfer and dissipate heat generated by the friction and shear of mechanical parts under conditions of high loading and close tolerances.

Limitations. Water, air, and oils, indispensable as they are to modern civilization and indeed to the mere existence of man, do have limitations. There is only a 180°F temperature difference between the melting point and boiling point of water. Although water, as steam, is commonly used up to its critical temperature of 705.4°F, and in a few supercritical power-generating plants above this temperature, its vapor pressure increases rapidly with temperature, reaching 3200 psia at its critical point. The use of dissolved salts in water (brines) depresses its melting point to about −40°F, but these solutions can be extremely corrosive, thus requiring the use of corrosion inhibitors such as sodium dichromate and buffers to maintain the alkalinity needed to protect the materials which contain them.

Air has a useful temperature range of over 2000°F but can be a strong oxidizing agent at high temperatures. Because of its low thermal conductivity and heat capacity, it is a relatively poor heat transfer medium at the pressure levels encountered on earth.

Animal and vegetable oils are susceptible to rancidity due to hydrolysis by bacterial decomposition and to oxidation, yielding free fatty acids and glycerol. Glycerol, when strongly heated, dehydrates into acrolein which has a highly penetrating odor. The engineering application of these materials, therefore, has been limited almost exclusively to short-term uses in cooking and to a maximum temperature of 400° to 450°F.

The variety of hydrocarbon compounds contained in petroleum that can be used for heat transfer and other applications is almost limitless. The most commonly used petroleum products are high-boiling paraffinic and aromatic distillation fractions having an operating range of 40°–550°F. However, some highly refined, dewaxed, degummed, hydrogenated oils have been produced that can be used at higher temperatures. The main disadvantage of these products, which has severely restricted their use, is their flammability.

Because of the many serious limitations imposed by the chemical characteristics of natural products, synthetic heat transfer fluids have been developed over the last few decades which supplement natural products, allowing more efficient and safe heat transfer at both lower and higher operating temperatures.

B. SYNTHETIC HEAT TRANSFER FLUIDS

The important characteristics to be considered in designing new chemical compounds for heat transfer purposes are: chemical properties, physical properties, and sometimes special properties such as nuclear, electrical, and toxicological. Although understanding of the relationship between chemical structure and these properties is far from perfect, what is known has been and can be used profitably in the development of new synthetic fluids. The following is a discussion of the important theoretical and empirically derived relationships between structure and properties.

Chemical Properties Important for the Production of New Compounds

These properties are important, as they relate to the ability of fluids to withstand chemical change which, in turn, can affect their usefulness as heat transfer fluids. Chemical change can manifest itself in various ways, depending upon the basic structure involved. Some chemical structures, when their thermal or oxidative stabilities are exceeded, tend to polymerize, with consequent drastic increases in viscosity. Others decompose into products of lower molecular weight, causing increased vapor pressure and lower viscosities. Decomposition can also produce acidic products which, in the presence of even trace amounts of water, can be extremely corrosive.

Thermal Stability. Thermal stability is obviously one of the important specifications to be met in choosing or designing a fluid for a particular heat transfer application. The thermal stability of a chemical compound or compounds making up a heat transfer fluid is its resistance to alteration or decomposition by the absorption of thermal energy. The rate of thermal decomposition of most chemical structures varies with temperature approximately according to the Arrhenius rate law:

$$K = Ae^{\frac{-E}{RT}}$$

where K is the rate constant; A the frequency factor; E the activation energy; R the gas constant, and T the absolute temperature. The activation energy, E, is the amount of molecular energy required for the decomposition reaction. The interchange of energy from one molecule to another occurs as a result of molecular collisions. The frequency factor, A, which represents the total frequency of encounters between reactant molecules, does not vary greatly for most decomposition reactions and normally is assumed to be constant. The exponential factor, $\frac{-E}{RT}$, accounts for the marked influence of the activation energy and temperature on decomposition reactions.

Molecules in a system do not all have the same energies; a small percentage have energies above and below the average. It is only those molecules that have an energy in excess of the activation energy of reaction that are capable of taking part in the reaction. As the temperature of a system is raised, that portion of the molecules having an energy in excess of the activation energy increases greatly.

In diatomic molecules the energy of activation for decomposition is equivalent to the bond dissociation energy. In more complex molecules the decomposition activation energy is determined by the weakest intermolecular bond, unless there are low-energy decomposition paths available, which unfortunately is usually the case. The ability of certain cyclic molecules to resonate, thereby disseminating the energy received by one ring atom to all other connected ring atoms, greatly enhances the bond strength between them. The high thermal stability of aromatic compounds, such as benzene, is due to this resonance, which transforms the structure from one of alternating single-double bonds to one of six equivalent bonds, each having a bond strength of 1½ times that of the normal carbon-carbon bond. Most organic compounds have thermal stabilities less than would be predicted from bond energy considerations, decomposition occurring at lower activation energies than the weakest bond in the molecule. There are a number of explanations for this phenomenon, one of which is that a chemical bond, being electrical in character, can be influenced by electrical fields surrounding it. The amount of this influence that is effective in straining, distorting, or even breaking chemical bonds depends on the field strength, polarity, and proximity. By such processes, molecules can form activated complexes and transition states which shorten decomposition routes.

The effect of these processes on the thermal stability of a fluid is extremely complicated and difficult to predict when considering the molecular structure of most fluids. Environmental factors such as the effect of catalytic impurities, metal surfaces, etc., further complicate these processes to the point where reliable predictions on thermal stability are virtually impossible. For this reason, fluid thermal stability studies must necessarily be empirical in nature.

One of the most practical ways of determining the relative thermal stabilities of organic compounds is by using an instrument called the isoteniscope. This method is based on the principle that, when organic compounds decompose, products of lower molecular weights are formed; these increase the vapor pressure of the system, the measurement of which can be made easily and accurately. The procedure involves measuring, under isothermal conditions, the pressure of a closed system containing the fluid at increasing temperature steps. At a certain temperature (the decomposition temperature) the pressure begins to depart from the normal vapor pressure of the fluid because of the build-up of decomposition fragments. The effect of metals on fluid stability can be readily ascertained by having metal coupons present during the above operation. A recent systematic study[1] of the thermal stability of various classes of organic compounds using this technique shows the following:

Decomposition Point Range (°F)	Chemical Class
Over 800	Unsubstituted Polyaromatic compounds
	Unsubstituted Polyphenyl ethers
	Unsubstituted Aromatic amines
	Unsubstituted Aryl silicates
	Unsubstituted Aryl silanes
	Unsubstituted Aryl borates
700–800	Lower alkylaromatic ethers
	Aromatic phosphates
	Aromatic sulfones
	Aromatic ketones
	Silicones
	Halo-substituted aromatic ethers
	Halo-substituted polyphenyls
600–700	Alkyl borates
	Alkyl silicates
	Higher alkylaromatic ethers
	Highly refined mineral oils
	Fluorinated esters
	Neopentylpolyol esters
500–600	Sebacate esters
	Methyl aryl esters
	Aliphatic tertiary amides

Other techniques are available for determining the relative thermal stability on a laboratory or bench scale of operations, such as the use of bomb calorimeters and rocking autoclaves. Also, criteria other than vapor pressure are commonly used for evaluating thermal stability. These include change in neutralization number and viscosity, and direct measurement of the amount of unchanged fluid remaining after exposure to various temperature levels in the test equipment described above or in the operating heat transfer apparatus itself.

Oxidation Stability. The oxidative stability of materials that are to be used as heat transfer media depends on many factors, some of which are associated with their molecular structures, and others which are related to environmental conditions such as temperature, pressure, and the presence of catalysts. Inorganic compounds vary widely in their susceptibility to oxidation. These range from fused salts ($NaNO_3$, KNO_3) which are completely stable, being already in the oxidized form to alkali metals (sodium, potassium, etc.) which readily oxidize (to form extremely corrosive oxides) at low temperatures when in contact with moist air.

Several general rules[2] apply to the oxidative stability of hydrocarbons:

1. Oxidation is a free radical reaction with chain propagation.

2. Oxidation reaction rates are accelerated by increases in pressure and by increases in chain length, and are greater for straight-chain compounds than for those with branched chains.

3. Oxygen attacks first the carbon atom of aliphatic compounds to which the fewest hydrogen atoms are attached.

4. In alkyl-aromatic compounds, the carbon atom attacked first by oxygen is the one immediately adjacent to the aromatic ring.

A wide variety of reaction products (alcohols, aldehydes, ketones, acids, peroxides, polymers, etc.) result from the oxidation of organic compounds. These products can have a deleterious effect on the physical properties of the virgin fluid by increasing viscosity, vapor pressure, and corrosivity.

Dozens of laboratory or bench tests have been used to evaluate the oxidative stability of fluids. Most of these are accelerated tests designed to evaluate military lubricants and hydraulic fluids, and to develop improved versions of these. In one such test,[3] the fluid to be evaluated is contained in a tall cylinder which is kept at test temperature by means of a heat bath. Air, or other gaseous media whose composition can be closely controlled, is bubbled through the test fluid at a specified rate (usually 5 liters per hour) and for a specified time (usually 48 hours). Coupons of metals to which the fluid would be exposed in practice are lashed together and suspended in the test fluid. The weight and dimensional changes of the coupons and the changes in such characteristics of the fluid as viscosity, neutralization number, etc., indicate the combined effect of this controlled environment on the materials present.

Compatibility. The compatibility of heat transfer fluids with the materials making up the environment in which they are used has to do with corrosion and solubility. The question of suitability of these materials of construction for various fluids is classically determined by exposing one to the other.

Theoretical approaches to the problem of predicting the resistance of polymeric materials usually present in gaskets, packing, etc., in heat transfer systems to the solvent action of fluids, have been of interest to investigators for many years. The usual "like dissolves like" philosophy has been used, but many instances of failure of this approach have led to more refined approaches. One very interesting technique[4] is the so-called "solubility parameter," sometimes referred to as the "cohesive energy density method." By this technique, solvents and polymer systems can be rated on a numerical scale, those with the same level of rating having a high degree of solubility. Cohesive energy densities can be calculated by using the basic physical and chemical properties (molecular weight, boiling point, and density), and the Clausius-Clapeyron equation.

Another effect of the solubility of materials of construction in heat transfer fluids is the mass transfer problem. Mass transfer occurs as a result of containment materials dissolving in the fluid and depositing out in lower temperature regions, in boiling sections, or where flow is restricted. This problem appears to be more pronounced with liquid metal heat transfer fluids. Even in situations where the degree of solubility is very low, mass transfer can present serious difficulties.

The selection of the materials of construction of heat transfer systems obviously must be made with an eye to providing the necessary resistance to the corrosive effects of the fluid under the conditions of use. In many cases, this selection must be made not only for new fluid, but also for the fluid contaminated with thermal and oxidation decomposition products, and other impurities impossible to eliminate from the system.

Flammability. Although considerable effort has been expended in developing theories to explain and predict the flammability characteristics or fire resistance of heat transfer fluids, a generally applicable theory has yet to be discovered. The most reliable techniques are empirical observations which have brought to light the fact that a number of compounds are fire-resistant. These include organophosphates and halogenated hydrocarbons. Fire resistance in a fluid does not mean that the fluid will not burn (organic compounds and many inorganic compounds will burn under the right conditions), but that the fluid is difficult to ignite and is self-extinguishing once the source of ignition is removed.

Physical Properties

The need for detailed knowledge of the physical properties of heat transfer fluids is obvious, not only to the engineers designing a heat transfer system, but also to the chemists attempting to develop new fluids. Generally speaking, complete experimentally determined data covering all the conditions of temperature, pressure, and composition to which the fluid would be exposed are not available. For this reason, theoretically and empirically derived relationships between atomic and molecular structure and physical (and thermodynamic) properties are now widely used. The literature contains many contributions made to this field, starting at the beginning of this century with the birth of the petroleum industry. The fast growth of the synthetic organic chemicals industry since the early 1930's greatly accelerated this work, so that now very reliable correlations are available for some properties, and moderately accurate correlations are available for others.

The physical properties of a fluid may be divided roughly into three categories: *additive*, *constitutive*, and *colligative*.

An *additive* property is one which is the sum of the corresponding properties of the constituents making up the compound or mixture. The only strictly additive property is mass, the mass of a mixture being exactly equal to the sum of masses of constituent parts.

A *constitutive* property is one that depends primarily on the arrangement of the atoms of a molecule and to a lesser extent on their nature and number. Most properties are partly additive and partly constitutive.

A *colligative* property is one that depends primarily on the number of molecules concerned and not on their nature, the volume of a gas at constant temperature and pressure being an example.

It should be recognized that none of the theoretical, semi-theoretical, and empirical approaches used to relate structure with property, although useful in answering many questions, are complete in themselves. In most cases, it is very difficult to select a single method of estimating properties that is most suitable for all situations; therefore, the following discussion of structure versus property relationship must be very general in character. Also, their availability does not mean that experimentally determined physical property data are no longer needed.

Thermal Conductivity. One of the most difficult aspects in the development and evaluation of thermal conductivity correlation is the fact that this property can be very difficult to measure experimentally, and many such values reported in the literature are doubtless in error by as much as 20 per cent. Errors in measuring thermal conductivity are due primarily to natural convection currents that are almost impossible to eliminate in the cells commonly used.

The most widely used equation for estimating thermal conductivity is:

$$k = c_p \rho L U_s \tag{2.2}$$

where thermal conductivity (k) is related to specific heat (c_p), density (ρ), intermolecular distance (L), and the acoustical velocity in the fluid (U_s). This method is relatively complex because of the L factor, the calculation of which is based on X-ray diffraction measurements, and the U_s factor, which is calculated from an equation involving the density, molecular weight and a constant which is an additive, temperature-independent function. At least eleven other estimating methods have been proposed for predicting thermal conductivity.

Specific Heat. The possible mathematical relationships in which this property is a part are extremely varied, involving not only thermodynamics, but also molecular kinetics and empirical experiments, either separately or in combination. There are a number of thermodynamic relationships that include heat capacity at constant pressure (c_p) and at constant volume (c_v). To use these relationships one must know the equation of state and either c_p at one pressure and temperature, or c_v at one volume. The accuracy of calculations using this method depends on the accuracy and applicability of the equation of state, and the accuracy of the initial heat capacity values.

Another method[6] developed to calculate c_p and c_v of organic liquids as a function of temperature requires generalized spectra data and the velocity of sound in the liquid. This approach involves the classification of the contribution of translational, external, rotational, internal rotational, and vibrational energy to the total energy of the molecular system. Probably the best general method for estimating heat capacities is the use of empirically derived correlation equations, of which there are literally dozens. However, care must be taken in selecting the proper correlation as they are usually highly specific for certain kinds of molecular structures and conditions.

Heat of Vaporization. A widely used equation[7] for calculating heat of vaporization for a wide variety of nonpolar liquids relates heat of vaporization to the normal boiling temperature. It can be used only to estimate heat of vaporization at this temperature. Over moderate temperature ranges (about 100 °C), the latent heat of vaporization is practically a linear function of temperature (latent heat = $A - BT$). For larger temperature ranges, several correlation methods are available that relate latent heat to critical temperature. One such is:

$$\text{Latent Heat} = K(T_c - T)^n \qquad (2.3)$$

where K and n are constants, and T_c is the absolute critical temperature.

Surface Tension. The most useful methods for estimating surface tension are those involving the parachor. The parachor[8] is the constant in the following equation relating surface tension with the density of the liquid and vapor, at the same temperature:

$$\frac{(\text{Surface Tension})^{1/4}}{\rho_L - \rho_V} = \text{Constant (Parachor)} \qquad (2.4)$$

A review of the data for isomeric substances and of members of homologous series showed that the parachor is primarily an additive, but also a constitutive property. The parachor equivalents of a number of atomic and structural factors (double bonds, triple bonds, etc.) have been determined which, when added properly, yield the total parachor for a particular molecular species. This parachor then can be used in the above equation along with density data to calculate surface tension.

Viscosity. Without experimental data, it is usually impossible to calculate viscosity without large errors. Viscosity measurements make use of fluid flow equations such as that of Poiseuille, which describes fluid flow through a narrow tube, or Stokes Law, which describes the fall of a small sphere through a liquid.

The effect of temperature on viscosity can be best calculated using empirical relationships such as that of Batschinski[9]:

$$\text{Viscosity} = \frac{C}{v - \omega} \qquad (2.5)$$

where C and ω are constants and v is the specific volume of the liquid measured at the same temperature as the viscosity. From the van der Waals' equation, it can be shown that the quantity $(v - \omega)$ is equivalent to the free space in a liquid. The viscosity of a fluid is therefore approximately inversely proportional to the number of "holes" it contains.

Density. There are a large number of theoretical and empirical relationships that contain the term density and thus can be used to calculate this property if values for the other terms are known. One such equation relates refractive index and density as follows:

$$\frac{N^2 - 1}{N^2 - 2} + \frac{1}{\rho} = \text{Constant} \qquad (2.6)$$

where N is the refractive index and ρ is the density.

Vapor Pressure and Boiling Point. The following equation, like the original Clapeyron-Clausius equation, can be employed to calculate vapor pressure from latent heat of vaporization:

$$\log \frac{p_2}{p_1} = \left(\frac{h}{4.576}\right)\left(\frac{T_2 - T_1}{T_1 T_2}\right) \qquad (2.7)$$

where p_1 and p_2 are the vapor pressure at T_1 and T_2, respectively, and h is the latent heat. The influence of pressure on the boiling point can also be determined by this relationship. Over a narrow temperature range, where h can be considered constant, a simplified form can be used:

$$\log p = C - \frac{A}{T} \qquad (2.8)$$

where A and C are constants for a given substance.

The van der Waals' vapor-pressure equation can be useful in calculating vapor pressure and normal boiling point from critical data, as follows:

$$\log \frac{p}{p_c} = K\left(1 - \frac{T_c}{T}\right) \qquad (2.9)$$

where p is the vapor pressure at temperature T, T_c the critical temperature, p_c the critical pressure, and K is a constant.

Melting Point. This property of a compound or mixture is undoubtedly the most difficult to predict by any method. Although many attempts have been made to explain the phenomenon of melting (and freezing) and to account quantitatively for the entropy and other changes, none has been fruitful. Fortunately, this property is relatively easy to determine experimentally. Many liquids, however, fail to freeze at any definite temperature but gradually stiffen and become rigid as the temperature is reduced.

Critical Constants. The critical properties, T_c, p_c, and v_c, appear in many generalized correlations of other properties. Because these correlations are useful in calculating these other properties, knowledge of critical constants of a fluid is important.

A number of methods have been developed to estimate critical temperatures such as that of Lydersen.[10] By this method, the critical temperature (T_c) is calculated from the following equation:

$$\frac{T_b}{T_c} = 0.567 + \Sigma\Delta_T - (\Sigma\Delta_T)^2 \qquad (2.10)$$

where T_b is the normal boiling temperature. The term $\Sigma\Delta_T$ is evaluated by the summation of the atomic, bond, and group contribution values. In a like manner,[10] the critical pressures and critical volumes can be calculated from:

$$p_c = \frac{M}{(0.34 + \Sigma\Delta_p)^2} \quad \text{and} \quad V_c = 40 + \Sigma\Delta_v \qquad (2.11)$$

where M is the molecular weight and $\Sigma\Delta_p$ and $\Sigma\Delta_v$ are the sum of the atomic and structural contributions.

As indicated above, there are literally hundreds of pertinent correlations relating property to structure. Many have limited applicability and vary widely in accuracy and utility. They usually contain combinations of properties so that the same correlations can be used to calculate more than one property, the others being known. An excellent comprehensive compilation[11] of articles in this field has recently been published and has proven to be very useful in predicting and estimating engineering properties of heat transfer fluids.

References

1. Blake, E. S., Hammann, W. C., Edwards, J. W., Reichard, T. E., and Ort, M. R., "Thermal Stability as a Function of Chemical Structure," presented before the Division of Petroleum Chemistry, *American Chemical Society*, Cleveland Meeting, (April 5–14, 1960).
2. Gilman, H., "Organic Chemistry—An Advanced Treatise," John Wiley & Sons, Inc., 1947.
3. "Federal Test Method Standard 791," Method 5308.4, (December 15, 1955).
4. Small, J., "Applied Chemistry," Vol. 3, p. 71, 1953.
5. Sakiadis, B. C., and Coates, J., *Am. Inst. Chem. Engrs. J.* 1, 275 (1955).
6. Sakiadis, B. C., and Coates, J., *Am. Inst. Chem. Engrs. J.* **2**, 88 (1956).
7. Kistiakowski, Z., *Physical Chemistry*, Vol. 107, p. 65, 1923.
8. Glasstone, S., "Textbook of Physical Chemistry," D. Van Nostrand, 2d Edition, pp. 526–527, 1946.
9. Batschinski, Z., *Phys. Chem.*, **84**, 643 (1913).
10. Lydersen, A. L., University of Wisconsin," Engineering Expt. Sta. Report #3, Madison, Wisconsin (April 1955).
11. Gambill, W. R., "Estimate Engineering Properties," Chemical Engineering, New York, McGraw-Hill Book Co., 1957–1958.

Section III

HEAT TRANSFER MEDIA

The evaluation and eventual selection of an appropriate heat transfer medium is far from a precise quantitative process; the final choice is usually a combination of theory and empiricism. A number of unrelated and frequently nebulous parameters must be studied and weighed against one another. There is no substance, either found in nature or compounded by man, which can fulfill all the heat transfer requirements imposed on it by an exploding technology. The ultimate answer is usually a compromise between the contending factors of optimum heat transfer properties, successful handling, ability to withstand severe operating conditions, low cost, and ready availability. Ironically enough, despite the arduous efforts of chemists to synthesize new and esoteric materials, water and liquid metals still possess an unsurpassed set of thermal and transport properties.

Clearer understanding is certainly enhanced by a knowledge of the merits and disadvantages of various fluids. Whereas earlier sections dealt with general concepts, the aim of this section is to present in concise form, the primary physical and chemical properties of industrial heat transfer media, based on information furnished by the supplier. Additional pertinent material has been secured from the literature. In cases where the manufacturer's claims, particularly those of thermal stability, are at variance with the experience of the author and his collaborators, they have been toned down accordingly. Pages 2 and 3 of Section IV clarify this point more fully. More than sixty fluids are discussed in accordance with the criteria formulated earlier.

Heat transfer fluids may be separated into a number of convenient groups. By definition, a fluid may exist in three distinct states: vapor, liquid, or gas. Since each physical state has a unique set of properties, three different branches of technology center around gas-, liquid-, and vapor-phase heat transfer; hence classification by phase is quite logical. Because it is universally familiar to scientists and engineers, chemical composition has been another means of differentiating between heat carriers. Where a fluid falls into two chemical groups, it has been inserted in the group which is most compatible with its characteristic properties, and then cross-referenced.

Within each chemical group, the individual fluids are carefully described. There is a general discussion, followed by a summary of properties for each fluid. Several fundamental properties such as density, specific heat, viscosity, and vapor pressure are usually presented in graphic form. Not all properties are available for all fluids. When a property does not appear, it is either because the manufacturer has not determined the necessary data, or because the property is not important for the particular material.

Classification of the fluids according to gases, vapors, and liquids is somewhat arbitrary since fluids classified as vapors can also be used as liquids, and some of the gases, such as aluminum chloride, could be listed among either vapors or liquids. The chemical classifications are also somewhat arbitrary, therefore cross references have been employed. For instance, a silicate ester such as Coolanol 45 could be listed either in the esters group or the silicate group.

G-GASES

Permanent gases are utilized as media for heat transport at elevated temperatures, and are the only substances capable of spanning the entire range of industrial application. At very high temperatures, (above 2500°F) at which heat energy is plentiful, diatomic and polyatomic gas molecules become increasingly excited; eventually, chemical bonds are broken and ions and free radicals appear. All dissociation reactions are highly endothermic. Because of this voracious absorption of thermal energy on dissociation, the possibility of using gases such as nitrogen and hydrogen for high temperature heat transmission presents some interesting potentials. For gaseous mixtures such as air and flue gases, many unique molecular species prevail in equilibrium at elevated temperatures, thus making the picture an extremely chaotic one.

Excellent thermal stability, inertness toward environment below the dissociation limit, and relative ease of handling, are the major advantages of gaseous coolants. Unfortunately, from both a hydrodynamic and thermal viewpoint, gases have rather poor properties. In fact, the major features of gas-cooled systems, compared to systems cooled by liquids and vapors, are the lower heat transfer coefficients, the larger coolant volumes and conduits needed, and the larger expenditure for pumping power required to circulate gaseous coolants. Low densities, an inherent characteristic of the random incoherent gaseous state, are primarily responsible for this behavior. High densities favor low frictional power losses (E is proportional to ρ^{-2}) and high film coefficients (h is proportional to $\rho^{0.8}$). The deleterious effects of low density can be partially cancelled by compressing a gas to moderate pressures up to 100 atm. With a greater number of molecules per unit volume, both thermal and power characteristics show remarkable improvements. Attempts have been made at a number of nuclear installations to minimize pressure drops in gas coolant circuits and to improve internal heat transfer by use of various finned tube configurations.

The choice of a particular gaseous medium is usually a compromise between one or two alternatives: free, inexpensive gases, such as air and flue gases, with poor thermal properties; commercially prepared gases, such as carbon dioxide (high molar heat capacity), helium and hydrogen (high thermal conductivity and low viscosity), with auspicious thermal properties. As reactor coolants, nitrogen, carbon monoxide, air, and especially argon, require too large a quantity of pumping power, and the field has currently narrowed to carbon dioxide and helium. Application of hydrogen still remains in the speculative stage; the gas is still regarded as hazardous, and little is known about its influence on the strength of metals. Gaseous aluminum chloride, which dimerizes on cooling, possesses an extraordinarily high molar heat capacity, and is presently under investigation. Thermodynamic considerations alone do not constitute sufficient grounds for selecting a gaseous coolant, and such problems as the compatibility of carbon dioxide with graphite moderators, and leakage of helium through the walls of the system, must also be examined. Chemical reactivity and ready availability are often controlling factors.

Accurate predictions of the thermal properties of gases at high temperatures and high pressures, particularly viscosity, and to an even greater extent, thermal conductivity, are noticeably lack-

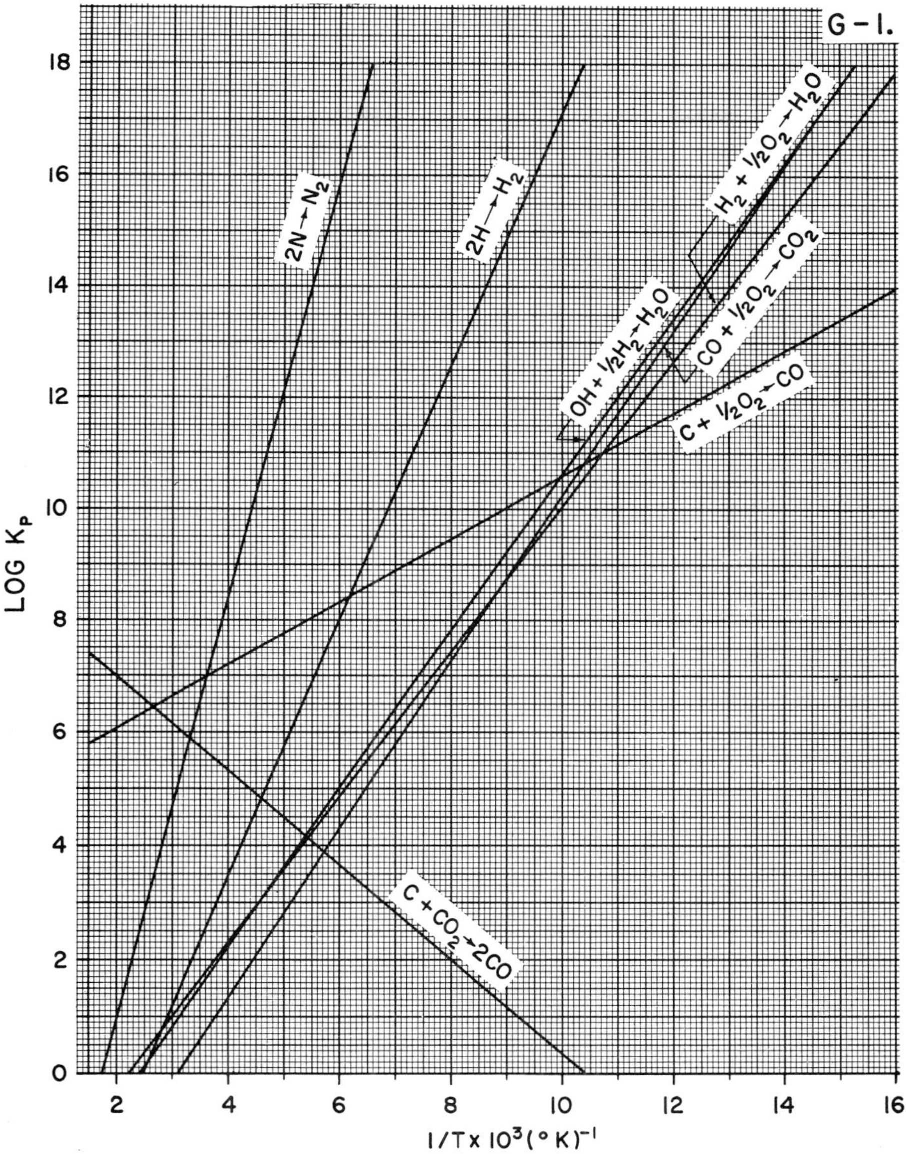

Equilibrium Constants of Dissociation Reaction for Gases

ing. The establishment of a Thermophysical Properties Research Center at Purdue University, devoted to the collection and correlation of all available information on transport properties of gases, liquids and solids, is a step forward. To estimate properties at 1000° to 20,000°K, far beyond the realm of laboratory work, both thermodynamic equations and experimental results from molecular-beam scattering experiments have been used. Nevertheless, the properties of certain gases such as air have already been investigated and tabulated at temperatures up to 24,000°K. Thermally induced effects, such as excitation, dissociation, and ionization, necessitate an exceedingly intricate theoretical approach at these high temperature levels.

G-1. AIR

Air is perhaps the most plentiful and inexpensive heat transfer material in existence. As is typical of practically all gases, air is distinguished by low thermal conductivity and density. Compression betters the situation but complicates plant operation. Air can be applied where low heat transfer coefficients are acceptable, where the duty is heating rather than heat transmission, and at elevated temperatures where other media undergo pyrolysis. The molecular composition of dry air is 78.084 per cent N_2, 20.946 per cent O_2, 0.934 per cent A, 0.033 per cent CO_2, and 0.003 per cent Ne. The small amount of helium present is included in the neon total.[1]

Properties of Importance

Density: See Figure G-1.1 and G-1.2.

Specific Heat: See Figure G-1.2.

Thermal Conductivity: See Fig. G-1.3.

Viscosity: See Figure G-1.4. Air has a higher viscosity than other permanent gases.

Physical Stability: The P-T relationships of the individual gases determine the behavior of the multicomponent mixture. In fact, air is an important commercial source of a number of the cryogenic fluids.

Thermal Stability: The thermal decomposition of air at high temperatures consists of a plethora of different chemical reactions: oxidation, vaporization, dissociation, electron capture, ionization of molecules, and ionization of atoms. An exceedingly complicated system of equations must be solved to determine the equilibrium compositions at various temperature levels.

Chemical Stability: Despite the reactivity of oxygen, air exhibits excellent resistance to chemical change. The solubility of air in water by volume is: 32°F—0.032, 68°F—0.020, and 212°F—0.012.

Corrosion: None.

Flammability: None.

Physiological Effects: None.

Appearance: Colorless and odorless.

Price Class: B per 100 cu ft at 2100 psi.

References

1. Gilmore, F. R., "Equilibrium Composition and Thermodynamic Properties of Air to 24,000°K," Rand Corp., RM-1543, August 24, 1955.
2. Marks, L. S., Ed., "Mechanical Engineers' Handbook," 547, New York, McGraw-Hill Book Co., 1941.

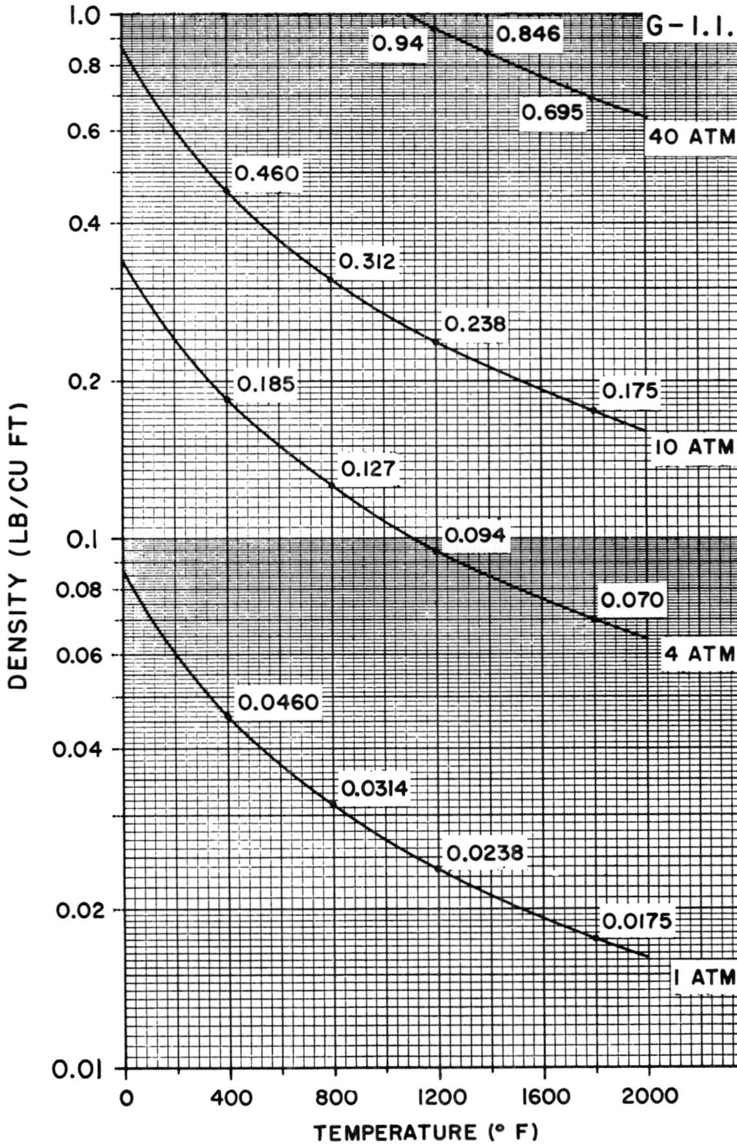

Density vs Temperature

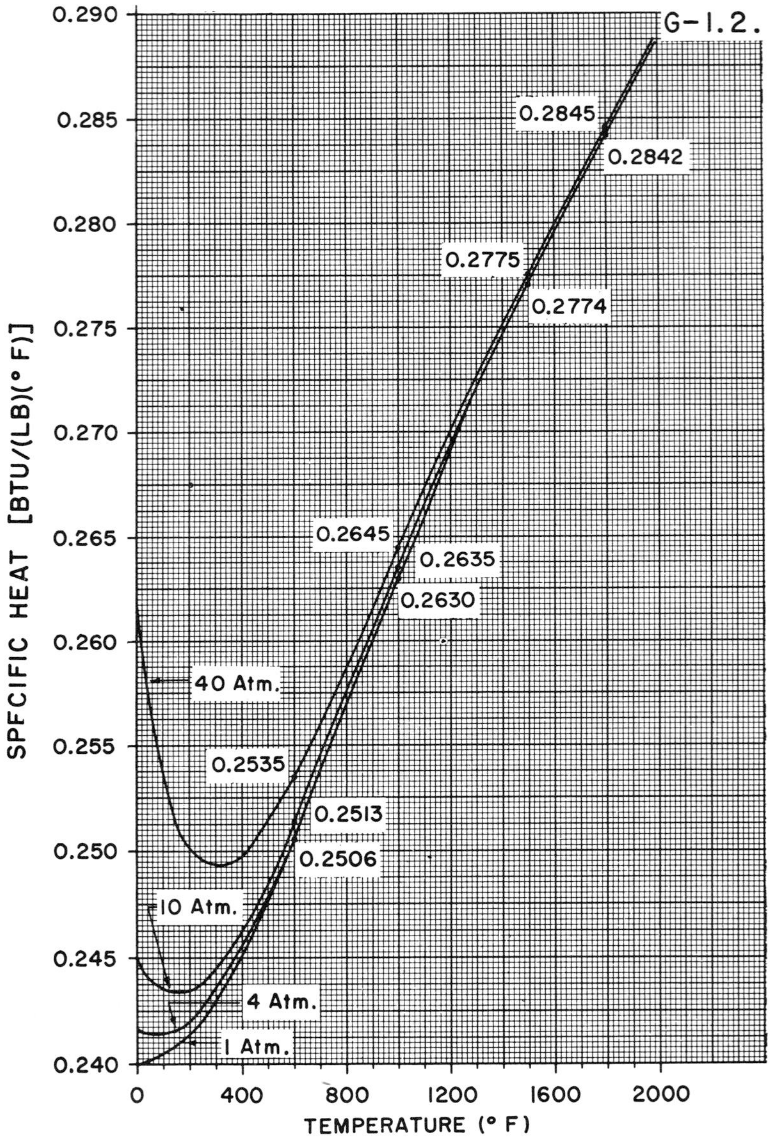

Specific Heat vs Temperature

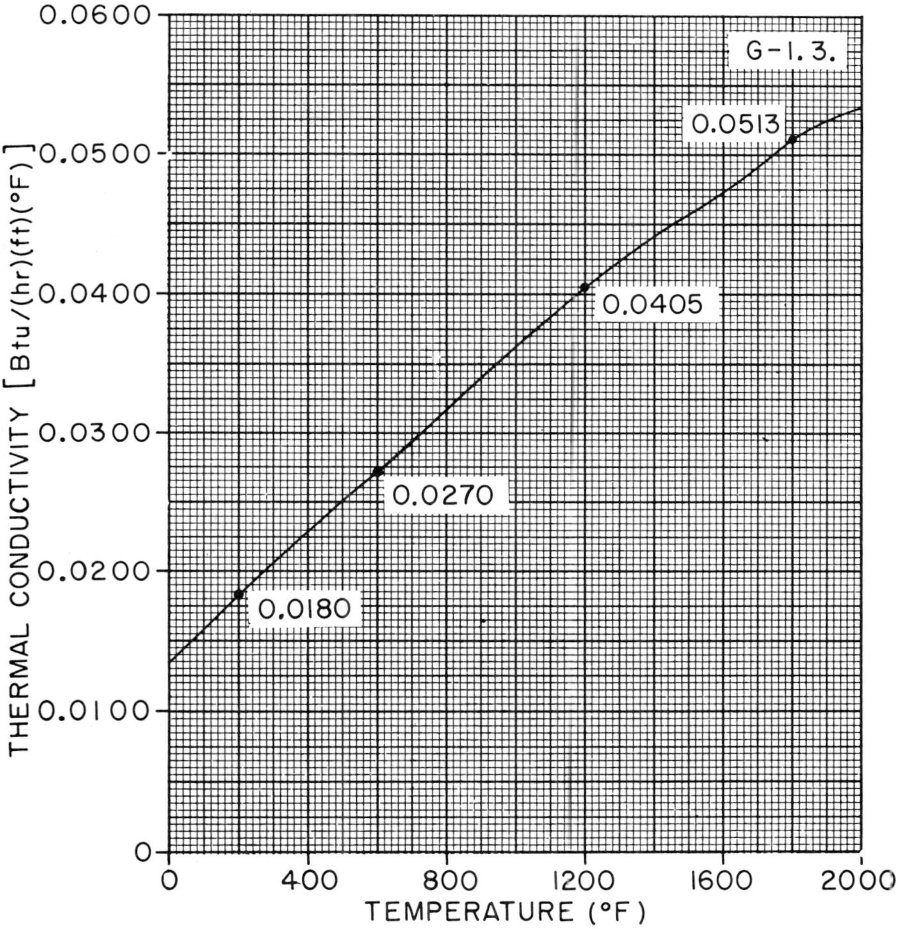

Thermal Conductivity vs Temperature

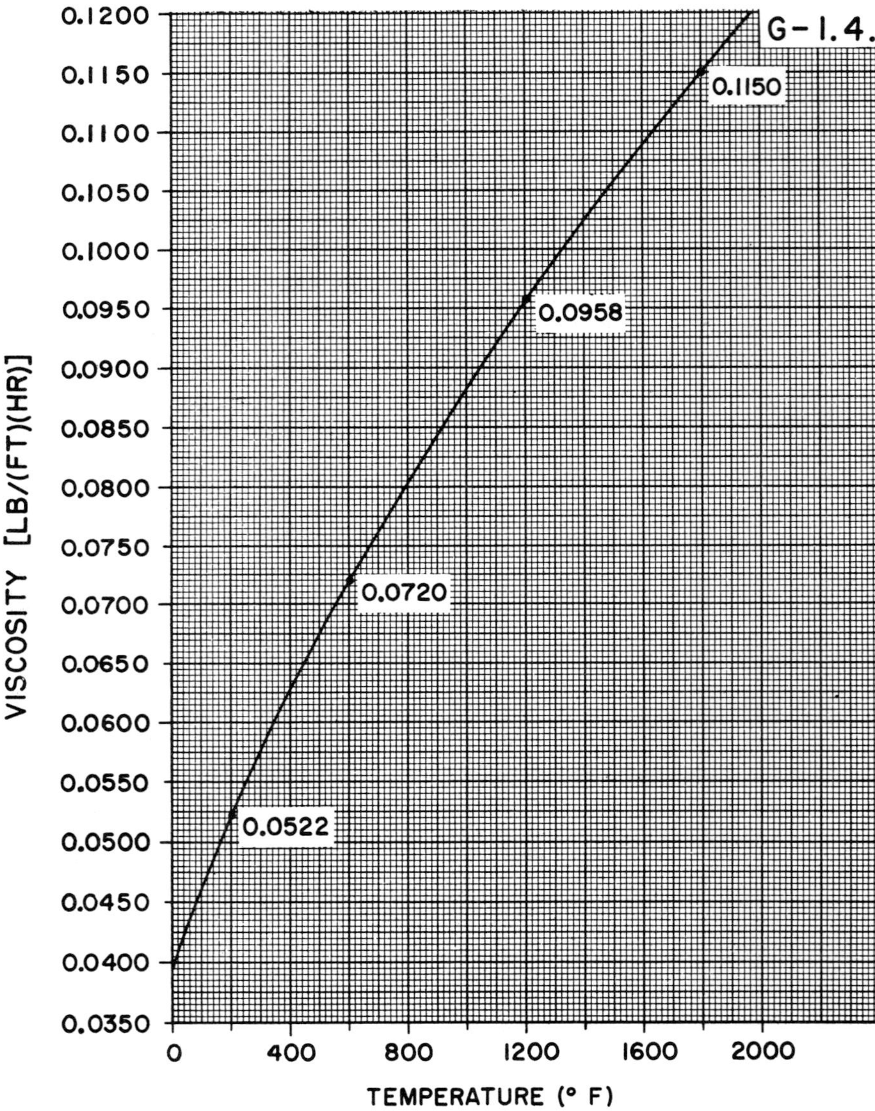

Viscosity vs Temperature (1 atm)

G-2. ALUMINUM CHLORIDE

Gaseous $AlCl_3$ is currently attracting intense interest as a heat transfer medium and a thermodynamic working fluid for a graphite moderated, molten salt reactor at Oak Ridge National Laboratory.

Aluminum chloride is particularly appealing because it exists as a monomer, $AlCl_3$, at high temperatures, and as a dimer, Al_2Cl_6, at low temperatures. At temperatures where appreciable fractions of monomer and dimer coexist in equilibrium, the effective specific heat and thermal conductivity of the mixture is greatly enhanced. The explanation is quite simple. Cooling the gas yields not only the sensible heat of the gas ordinarily expelled when gas composition remains "frozen" but also, an additional contribution of chemical heat caused by the union of simpler monomer molecules into more complex dimer molecules, as the lowered temperature shifts the equilibrium direction. Plots of effective specific heat and effective thermal conductivity against temperature (Figure G-2.1.) show the effect of high temperature dissociation on heat transfer properties.

Viscosity data for monomer and dimer are given in Figure G-2.2. A normal boiling point of 450° F fixes the lower end of the application scale. As for corrosion, high nickel alloys are satisfactory containing materials for $AlCl_3$. Investigators are still examining the compatibility of $AlCl_3$ with various structural metals, as well as possible mechanisms for corrosion of metals by $AlCl_3$. The material must be kept dry, otherwise strong corrosive action may occur.

It has been concluded that the temperature range of molten salt heat transfer is too low to employ $AlCl_3$ effectively. Nevertheless, special applications, such as gas turbine cycles and binary fluid cycles, will surely be devised in which the unique properties of $AlCl_3$ can be exploited.

References

1. Blander, M., and Newton, R. I., *M.S.R. Quar. Prog. Rep.* ORNL-2626 (October 31, 1955).
2. Blander, M., Epel, L. G., Frass, A. P., and Newton, R. F., $AlCl_3$ as a Thermodynamic Working Fluid and Heat Transfer Medium," ORNL-2677 (October 5, 1959).

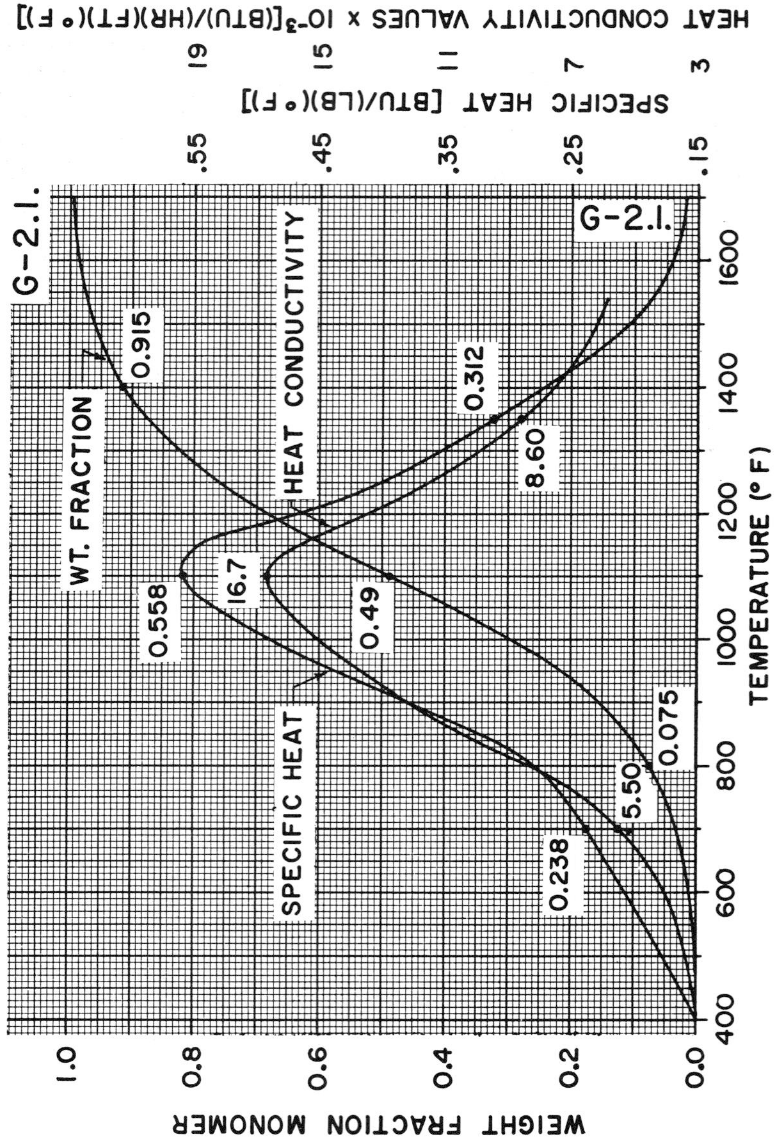

Weight Fraction of Monomer
Specific Heat, Heat Conductivity vs Temperature

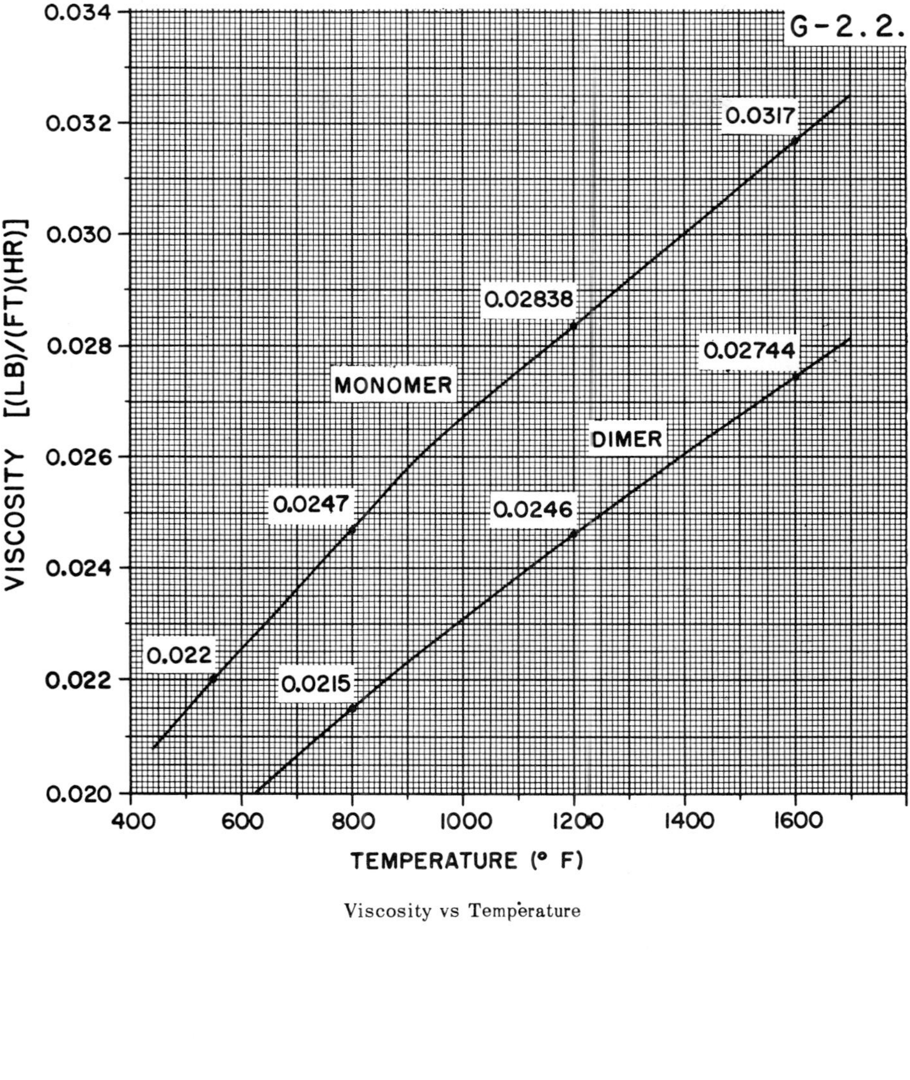

Viscosity vs Temperature

G-3. CARBON DIOXIDE

Carbon dioxide has superior transport properties but poor transfer properties. It is one of the major products of fuel combustion and is obtained industrially by flue gas recovery. Carbon dioxide is shipped in pressurized cylinders in either the solid or liquid state.

Properties of Importance

Density: For a gas, carbon dioxide has a high density. See Figure G-3.1.

Specific Heat: Carbon dioxide has a highly variable molar specific heat of 8.5 to 12.5. Btu/(lb)(mole)(°F). See Figure G-3.2.

Thermal Conductivity: See Figure G-3.3.

Viscosity: See Figure G-3.4.

Physical Stability:
Critical Constants: T_c = 88°F
P_c = 72.8 atm
Triple Point: −69.8°F, 5.11 atm
Sublimation Point: −109.3°F

Thermal Stability: At about 3200°F, temperatures are sufficiently elevated to reduce carbon dioxide to its component parts, CO + O_2. See Figure G-1 for the equilibrium constants of the various dissociation reactions. A few representative values for the per cent dissociation of carbon dioxide are given below.

TABLE 1

$CO_2 \longrightarrow CO + \frac{1}{2}O_2$

Per Cent Dissociation

Temp. (°F)	Pressure (atm)			
	0.1	1.0	10	100
2730	0.104	0.048	0.0224	0.01
3630	4.35	2.05	0.96	0.445
4530	33.5	17.6	8.63	4.09
5430	77.1	54.8	32.2	16.9

Chemical Stability: The compatibility of carbon dioxide and graphite is doubtful at temperatures above 930°F. Therefore helium is preferred as a coolant in a graphite moderated nuclear reactor. Graphite oxidation limits the use of carbon dioxide to temperatures below 1000°F. Several complex schemes have been proposed to describe the gas phase decomposition.

Radiolytic Stability: Intense radiation induces a chemical reaction between carbon dioxide and graphite, which would not otherwise occur except at higher temperatures. Ionized and excited gas molecules are formed, that is, active species of short-lived intermediates of carbon dioxide, and the prevailing equilibrium is quite complicated. Between 1050° and 1110°F, thermal reaction of carbon dioxide with graphite becomes the controlling one and overshadows the radiation-induced reaction.

Surface Tension: No leak-tight precautions are necessary for a carbon dioxide system.

Corrosion: A carbon dioxide atmosphere can be oxidizing or reducing, depending on carbon dioxide to carbon monoxide ratio. Hence the behavior of structural materials is quite difficult to predict. Carbon dioxide is highly corrosive in presence of water vapor. Austenitic stainless steels are particularly noted for their oxidation resistance. Stainless steel in a carbon dioxide cooled reactor may be heavily carburized if the carbon monoxide concentration exceeds 1 per cent. Stainless steel has a longer life in pure carbon dioxide than in air. Magnox C (1 per cent Al, 0.05 per cent Be, remainder Mg) is quite promising for use as a cladding material. Beryllium has excellent oxidation resistance below 1300°F.

Flammability: Carbon dioxide does not support combustion and is used as a fire extinguishing agent.

Physiological Effects: Normally non-toxic. Carbon dioxide is heavier than air, and there is danger of suffocation if escaping gas is permitted to accumulate in low or closed areas.

Appearance: Odorless and colorless.

Price Class: C per 50 lb at 2100 psi.

References

1. Chen, L. H., "Properties of Gaseous CO_2," Thermodynamic and Transport Properties of Gases, Liquids and Solids, p. 358, ASME, 1959.
2. General Dynamics Corp., "Evaluation of Coolants and Moderations for the Maritime Gas-Cooled Reactor GA-570," December 10, 1958.
3. Lewis, W. K., Radasch, A. H., Lewis, H. C., "Industrial Stoichiometry," p. 86, New York, McGraw-Hill Book Co., 1954.
4. Linde Engineering Corp., Marks, L. S., Editor, "Physical Property Equivalents of Cryogenic Fluids," in "Mechanical Engineers' Handbook," p. 368, New York, McGraw-Hill Book Co., 1941.
5. Oak Ridge National Laboratories, "Gas-Cooled Coolant Choice," ORNL-2699, June 15, 1959.

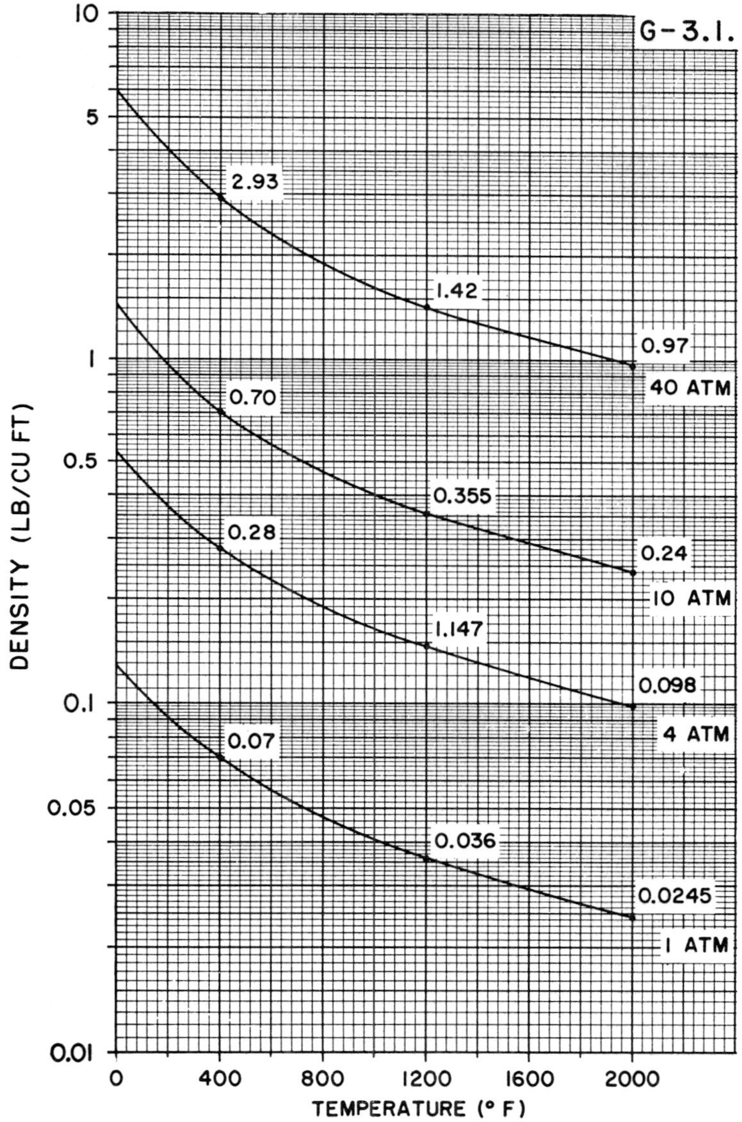

Density vs Temperature

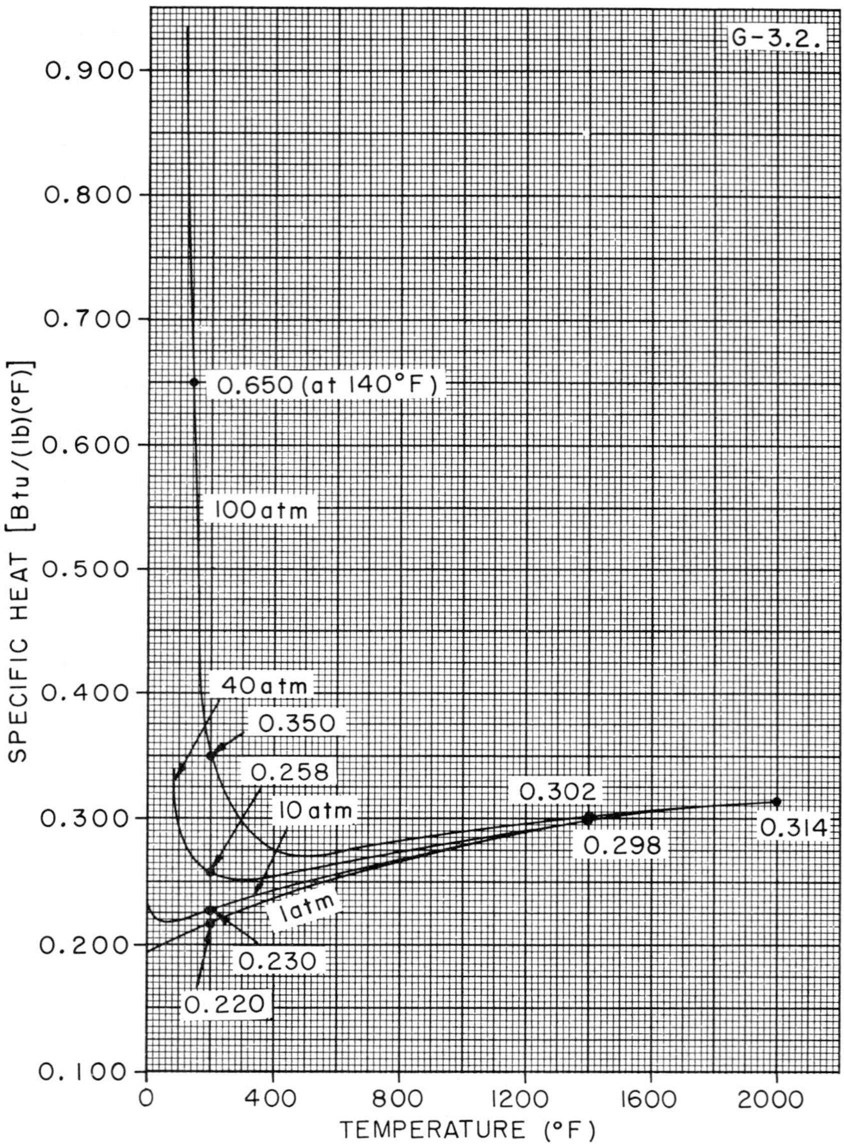

Specific Heat vs Temperature

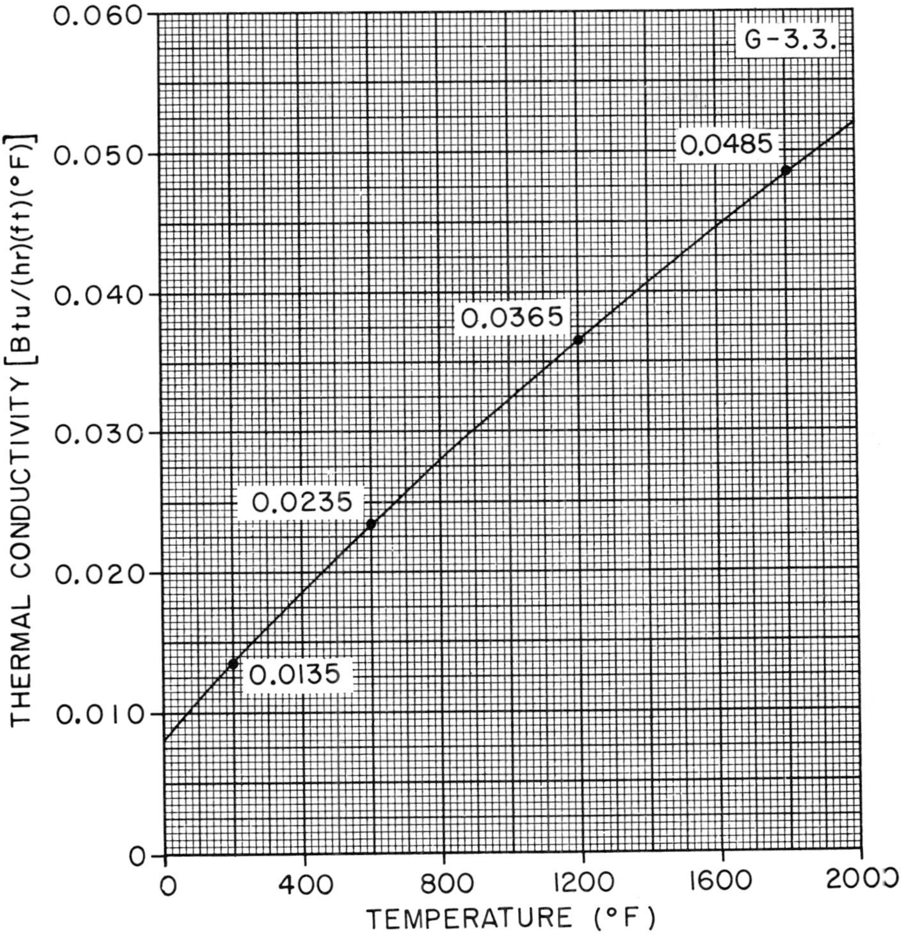

Thermal Conductivity vs Temperature (1 atm)

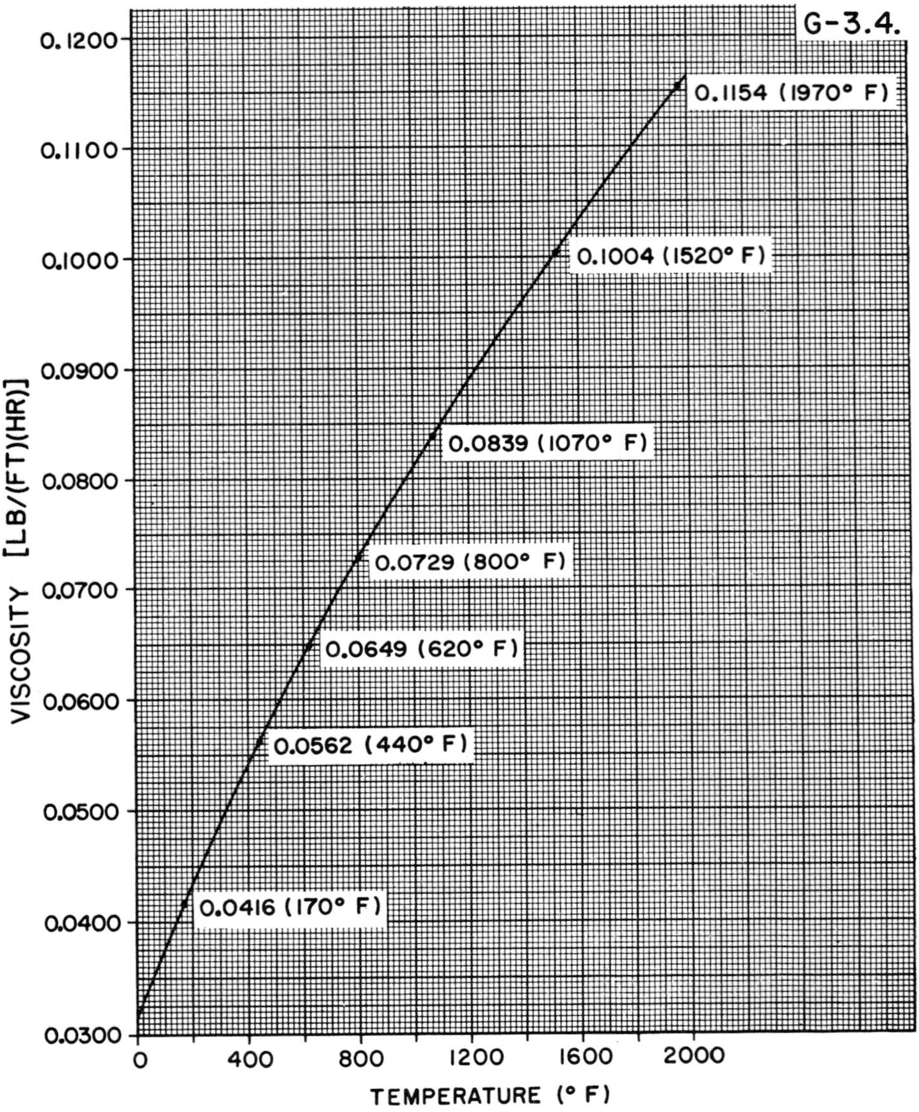

Viscosity vs Temperature (1 atm)

G-4. FLUE GASES

Flue gases are those gases which leave a furnace after combustion has taken place. They include the gaseous products of combustion, water vapor, and the excess air. There is a significant amount of heat in these gases, and to waste it would be a sacrifice of thermal efficiency. In a steam power plant the flue gases are usually employed either to preheat the combustion air in an air preheater, or to preheat the feedwater in an economizer.

Stack gases can be classified according to the state of the fuel which produces them, i.e., solid, liquid, or gaseous. They have an infinite number of possible compositions, depending on the conditions of combustion. Fortunately, all the gases produced by one kind of fuel have very similar properties. The properties of flue gases are additive and can be predicted by using equations of the following type:

$$c_p = (x_{CO_2})(c_{p\,CO_2}) + (x_{N_2})(c_{p\,N_2}) + (x_{O_2})(c_{p\,O_2}) + (x_{H_2O})(c_{p\,H_2O}) + \cdots \quad (3\text{-}1)$$

where c_p is the specific heat and x is the weight fraction of the individual component. Density, specific volume, and enthalpy can also be calculated in the same way. Obviously, in order to determine the properties of a particular flue gas, the exact composition of the gas must be known. There is considerable literature concerning methods of analyzing flue gases and it is not the purpose of this book to describe them. It might be noted that there is a paucity of collected information on the properties and heat transfer applications of flue gases.

Properties of Importance

Density: Figure G-4.1. gives the density of a flue gas.

Specific Heat: See Figures G-4.2. and G-4.3. Values given in these curves closely parallel values for industrial flue gases.

Enthalpy: See Figure G-4.4.

Thermal Conductivity: See Figure G-4.5.

Viscosity: See Figure G-4.6.

Thermal Stability: At combustion gas temperatures above 3200 °F, dissociation takes place. Hydrogen, oxygen, and carbon monoxide are the principal dissociation products formed. This effect is not important from the standpoint of heat transfer.

Chemical Stability: The reactivity of a flue gas is determined by the properties and concentrations of the component gases. At the high temperatures encountered, chemical reaction and dissociation may take place within the flue gas. For example, many oxides of nitrogen can be formed.

Corrosion: Some of the by-products of combustion, especially sulfur compounds, and some of the gases formed by reaction within the flue are corrosive to many metals. This is important in equipment design utilizing waste heat. However, since the flue gases are most often used as soon as they leave the furnace, there is no problem of transporting the hot gases. The stack or flue through which the gases travel is usually made of brick, concrete or some other type of stoneware; hence there is no corrosion problem.

Flammability: Flue gases are products of combustion and are not flammable themselves, provided complete combustion takes place.

Physiological Effects: Some flue gases may contain carbon monoxide, sulfur dioxide or other toxic gases. However, the main danger is the intense heat contained in the gases.

Appearance: These characteristics are dependent on the composition of the gas. The majority of flue gases are odorless; only the sulfur compounds have distinct odors. Lack of odor makes it quite difficult to detect toxic components.

Price Class: A per 1000 cu ft at 60 °F and atmospheric pressure.

References

1. General Electric Co., "Properties of Combustion Gases," New York, McGraw-Hill Book Co., 1955.
2. The Babcock and Wilcox Co., "Steam-Its Generation and Use," 8-10, 7-5, -- 1955.
3. North American Mfg. Co., "Combustion Handbook," 55, 1957.
4. Ledinegg, M., "Dampferzeugung," 137, Wien, Springer-Verlag, 1952.
5. Keenan, J. H., and Keyes, F. G, "Gas Tables," New York, John Wiley and Sons, Inc., 1945.

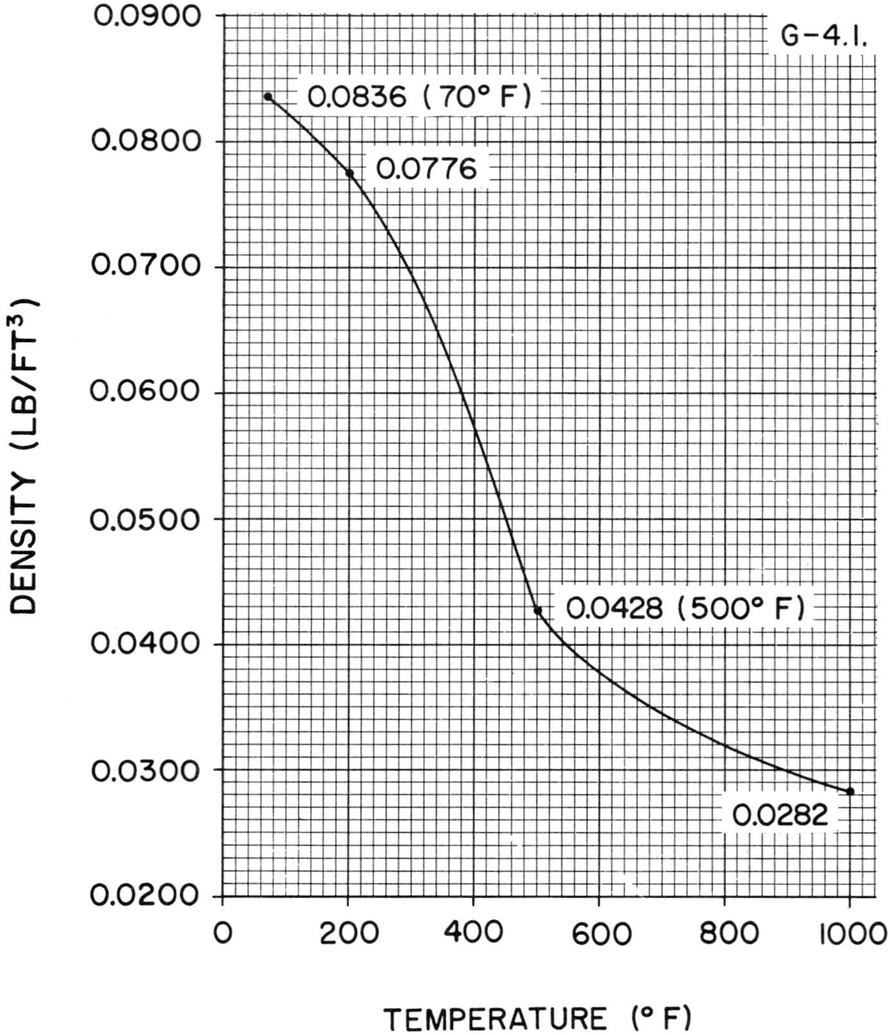

Density vs Temperature

Data are for a gas produced by the combustion of coal with 20% excess air. The molecular weight of the flue gas is taken to be 30.

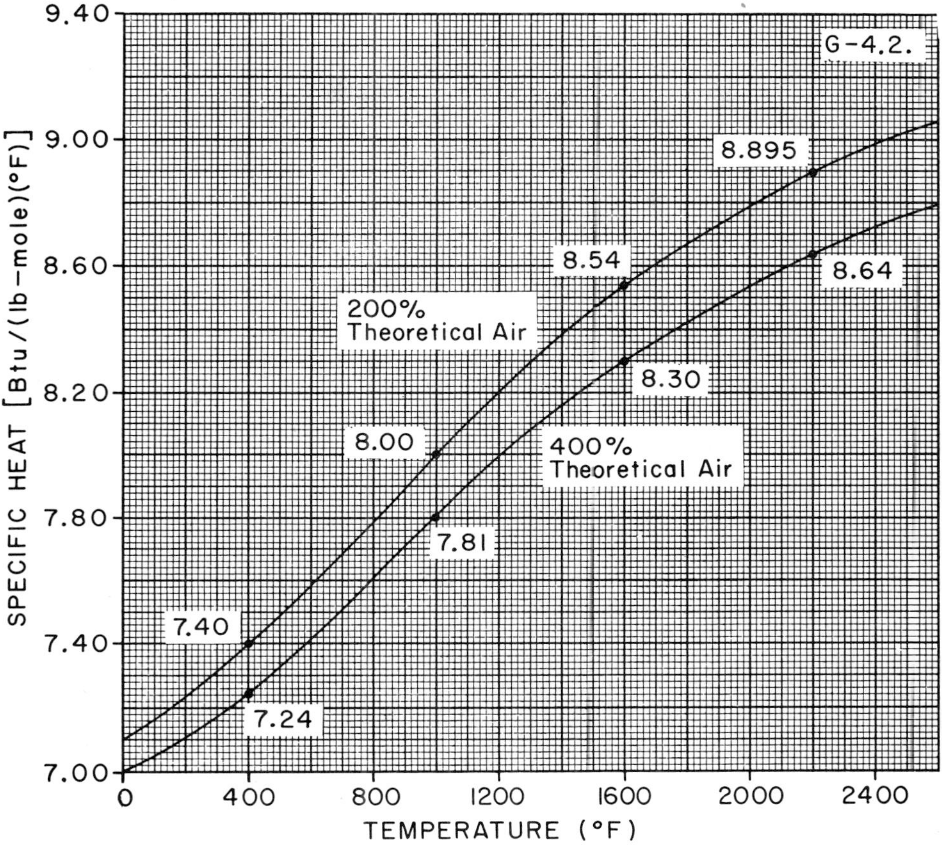

Specific Heat vs Temperature for Combustion Products of a Typical Hydrocarbon Fuel

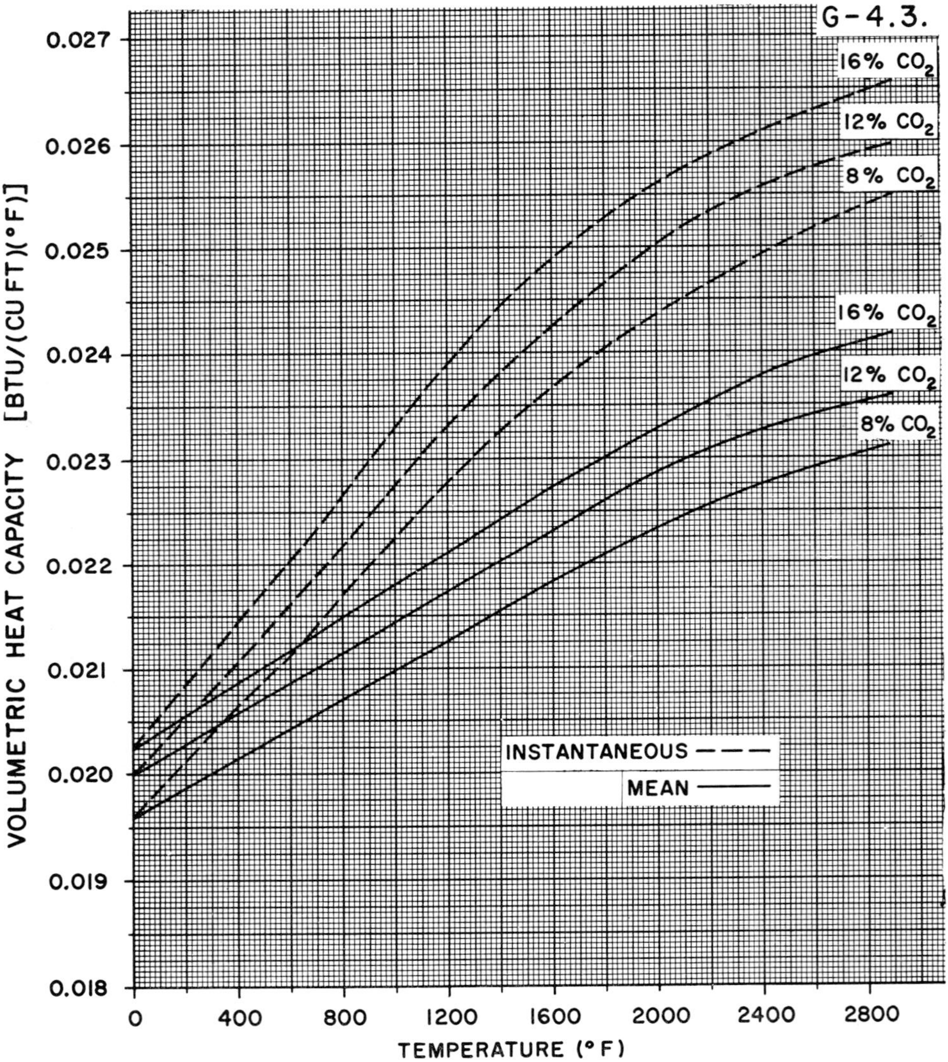

Flue Gas From Bituminous Coal
Instantaneous Specific Heat and Mean Specific Heat vs Temperature
(Mean values are calculated between 0° and selected temperature)

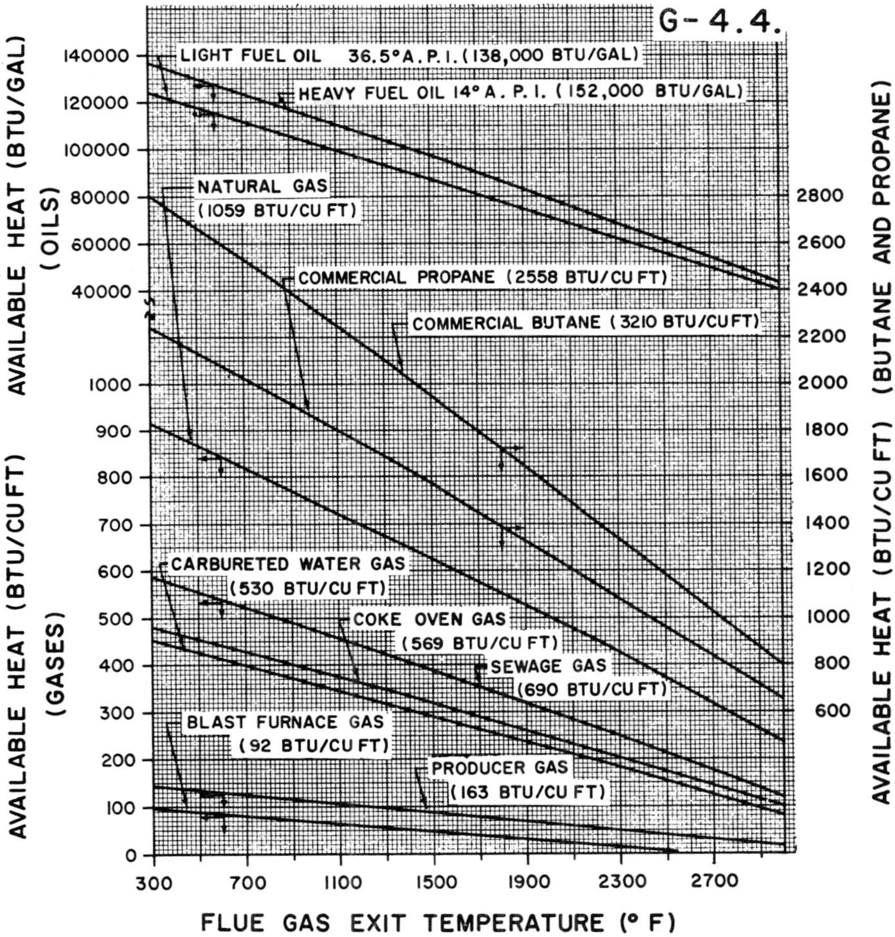

Available Heat vs Flue Gas Exit Temperature for Some Typical Fuels
Numbers in parentheses are the gross heating values for the particular fuels

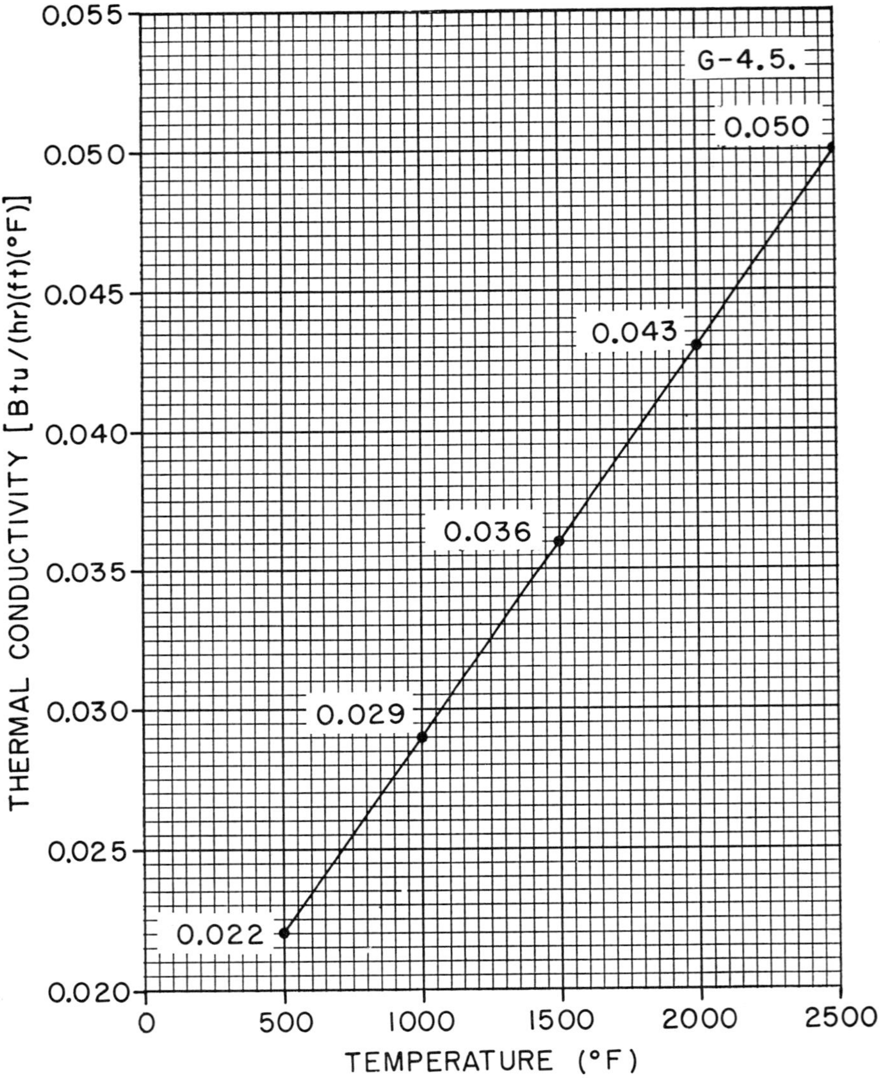

Thermal Conductivity vs Temperature

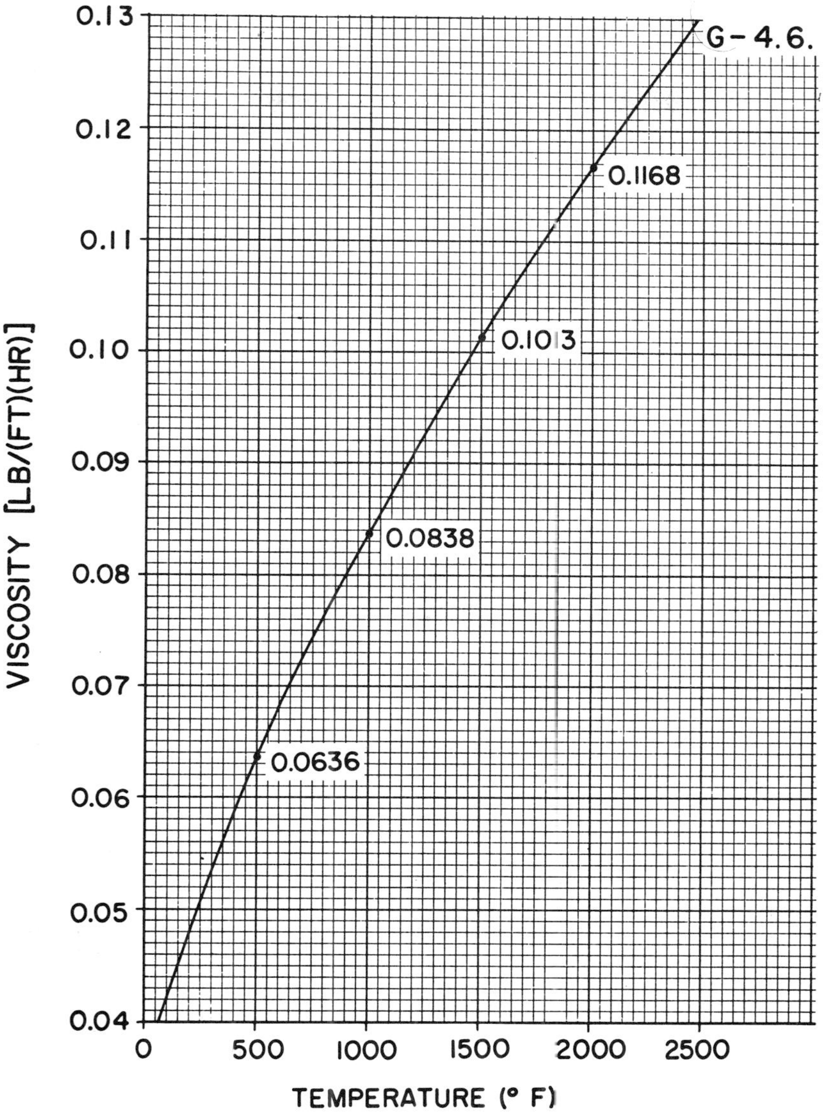

Viscosity vs Temperature

Data are for an average flue gas

G-5. HELIUM

Helium belongs to a family of normally gaseous elements known as the inert or noble gases. The two most singular features of this group are their lack of chemical activity and their monatomic nature. Helium has been used as a coolant in nuclear reactors at pressures of 50–100 atmospheres and temperatures up to 1500°F. Considerable attention has been devoted to the design, construction, and maintenance of systems employing helium at elevated temperatures.

Previously, the limited availability of helium was regarded as a major drawback. To alleviate this shortage, in 1959 the U.S. Bureau of Mines constructed a plant at Keyes, Oklahoma, where more than 290 million cu ft per year (1 atm–70°F) of high purity helium is extracted from natural gas by a combination of heat exchange and throttling. Existing helium reserves are depleted whenever combustion of natural gas takes place.

Properties of Importance

Density: Helium must be used at high pressures because of its low density (molecular weight of 4). The density at 932°F and atmospheric pressure is 0.0393 lb/ft^3.

Specific Heat: Helium has an approximately constant c_p of 1.25 Btu/(lb)(°F).

Thermal Conductivity: Helium is a good conductor of heat, having a thermal conductivity equal to that of hydrogen. See Figure G-5.1. Blaus and Mann[1] have proposed the following equation for the 1701° – 3321°F temperature range: $k \times 10^6 = 2.398 \times 10^5 + 6.719 \times 10^2 (T-1701)$.

Viscosity: See Figure G-5.2.

Physical Stability: Helium changes phase only in the vicinity of absolute zero.

Boiling Point: (1 atm) – 452.1°F

Melting Point: Helium does not solidify at 1 atm. The pressure calculated for solidification at absolute zero is 25 atm.

Critical Temperature: –450.4°F

Critical Pressure: 2.26 atm.

Chemical Stability: Since it is totally inert, helium possesses no chemical properties.

Thermal Stability: Having no intermolecular chemical bonds, helium is among the most stable substances known to man, as evidenced by its presence in the stars and its evolution in thermonuclear reactions.

Surface Tension: Allowable economic helium diffusion is 1 per cent of system volume per day. A high degree of helium leak-tightness can be secured with only a small increase in cost. In fact, all-welded systems can be constructed without having to develop a new handling technology.

Corrosion: The use of helium precludes many possible corrosion problems of fuel cladding and other elements of reactor structure encountered when gases such as carbon dioxide are employed. The non-oxidizing helium environment permits utilization of the outstanding high temperature strength of niobium for nuclear applications.

Physiological Effects: Nontoxic.

Flammability: None.

Electrical Properties: Good conductivity.

Appearance: Colorless and odorless.

Price Class: *C* per 100 cu ft at 2100 psi.

References

1. Mann, J. P., and Blaus, N. C., Atomic Energy Commission, LA–2316, (April 1959).
2. *Chemical Engineering*, No. 15, 67, (July 25, 1960).
3. Linde Engineering Corp., "Physical Property Equivalents of Some Cryogenic Fluids."
4. "Gas-Cooled Reactor Coolant Choice," ORNL–2699, June 15, 1959.
5. General Dynamics Corp., "Evaluation of Coolants and Moderators for the Maritime Gas-Cooled Reactor," December 10, 1958.

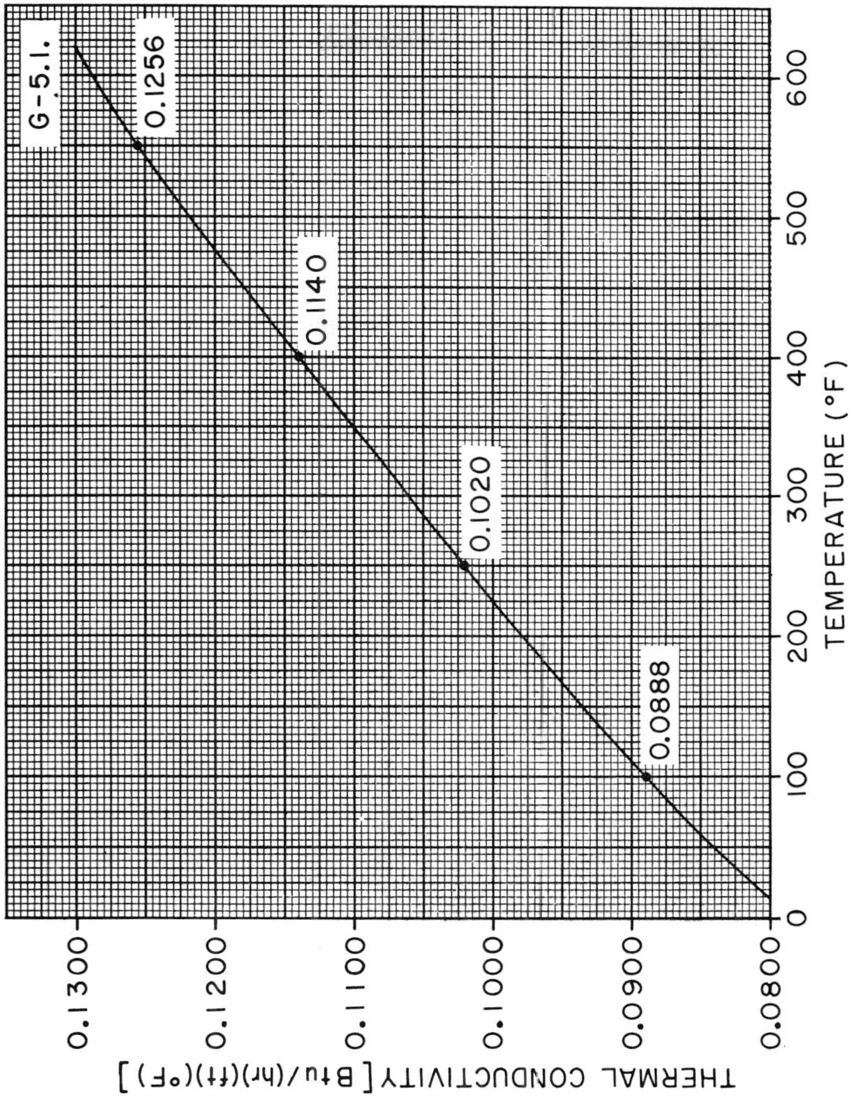

Thermal Conductivity vs Temperature (1 atm)

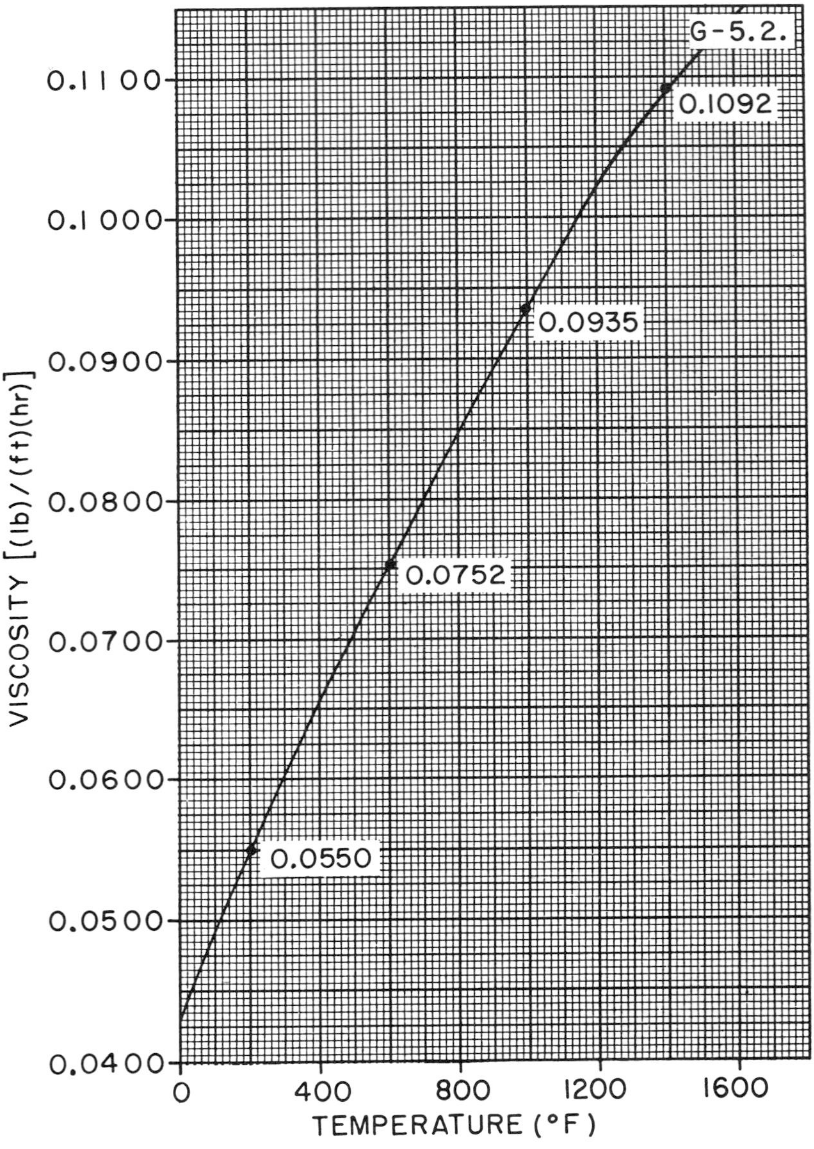

Viscosity vs Temperature (1 atm)

G-6. HYDROGEN

Hydrogen, a diatomic gas, is the lightest gas found in nature. Commercially, hydrogen is produced by the electrolysis of water, reactions of steam with iron, coke, methanol, and hydrocarbons, thermal decomposition of hydrocarbons, and as a by-product of petroleum refining. Hydrogen has a fine set of heat transfer properties: high thermal conductivity and specific heat, and low viscosity. Because of its high specific heat and reduced windage losses, hydrogen is used as a coolant in electric generators.

Properties of Importance

Density: Since it has a molecular weight of two, hydrogen must be compressed to high pressures in order to concentrate a sufficient mass of material in one location. See Figure G-6.1.

Specific Heat: See Figure G-6.2.

Thermal Conductivity: Hydrogen has the highest thermal conductivity of any gas. See Figure G-6.3.a and G-6.3.b. Blaus and Mann[1] have proposed the following equation for the 1701°-3321°F temperature range: $k \times 10^6 = 3.47 \times 10^5 + 304.1\,(T - 1701)$

Viscosity: See Figure G-6.4.

Physical Stability:
Boiling Point: −423.2°F
Melting Point: −434.6°F
Critical Point: −400.4°F and 12.8 atm.

Thermal Stability: The covalent chemical bond existing in a hydrogen molecule is ruptured at high temperatures, and free hydrogen atoms result. The degree of dissociation at one atmosphere is as follows:

256×10^{-34}	300° K
122×10^{-3}	2000° K
0.9409	5000° K
0.996	10000° K

See Figure G-1 for dissociation constants. The heat of dissociation is approximately equal to 205.4 Btu/lb-mole.

Chemical Stability: Hydrogen has chemical properties similar to those of elements of Group I and Group VII of the periodic system.

Surface Tension: Hydrogen has the highest coefficient of diffusion of all gases. It leaks through metals of the palladium and platinum groups. In spite of its excellent properties, the question of employing hydrogen as a reactor coolant is purely an academic one, since at elevated temperatures hydrogen rapidly diffuses through steel.

Corrosion: Nearly all metals absorb hydrogen, and reactions of metals with hydrogen vary considerably. Alkali earth metals, with the exception of beryllium and magnesium, and certain rare earths, react to form hydrides which are decomposed by water. In other metals, hydrogen alloys of differing composition are formed, rather than chemical compounds.

Flammability: The minimum ignition temperature of hydrogen is 1055°F. Hydrogen reacts with oxygen or air to form water with violence, provided a source of energy with sufficiently high intensity initiates the reaction. The limits of hydrogen flammability are as follows:

	Lower (%)	Upper (%)
H_2 in O_2	4	94
H_2 in air	4	74

Appearance: Odorless, colorless and tasteless.

Price Class: B per 100 cu ft at 2100 psi.

References

1. Mann, J. P., and Blaus, N. C., Atomic Energy Commission, LA-2316, April 1959.
2. Clark, George L., Editor, The Encyclopedia of Chemistry, New York, Reinhold Book Publishing Corp., 1957.
3. Linde Engineering Corp., "Physical Property Equivalents of Some Cryogenic Fluids."

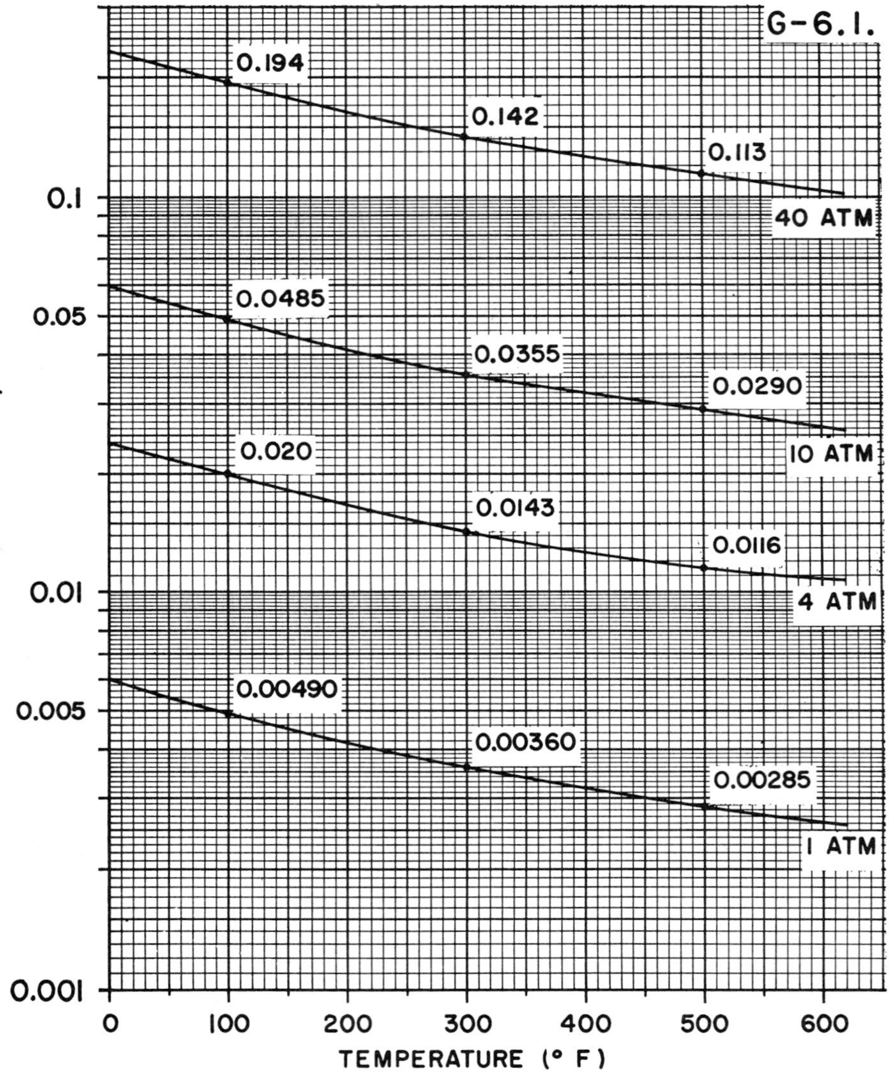

Density vs Temperature

G-6.2.

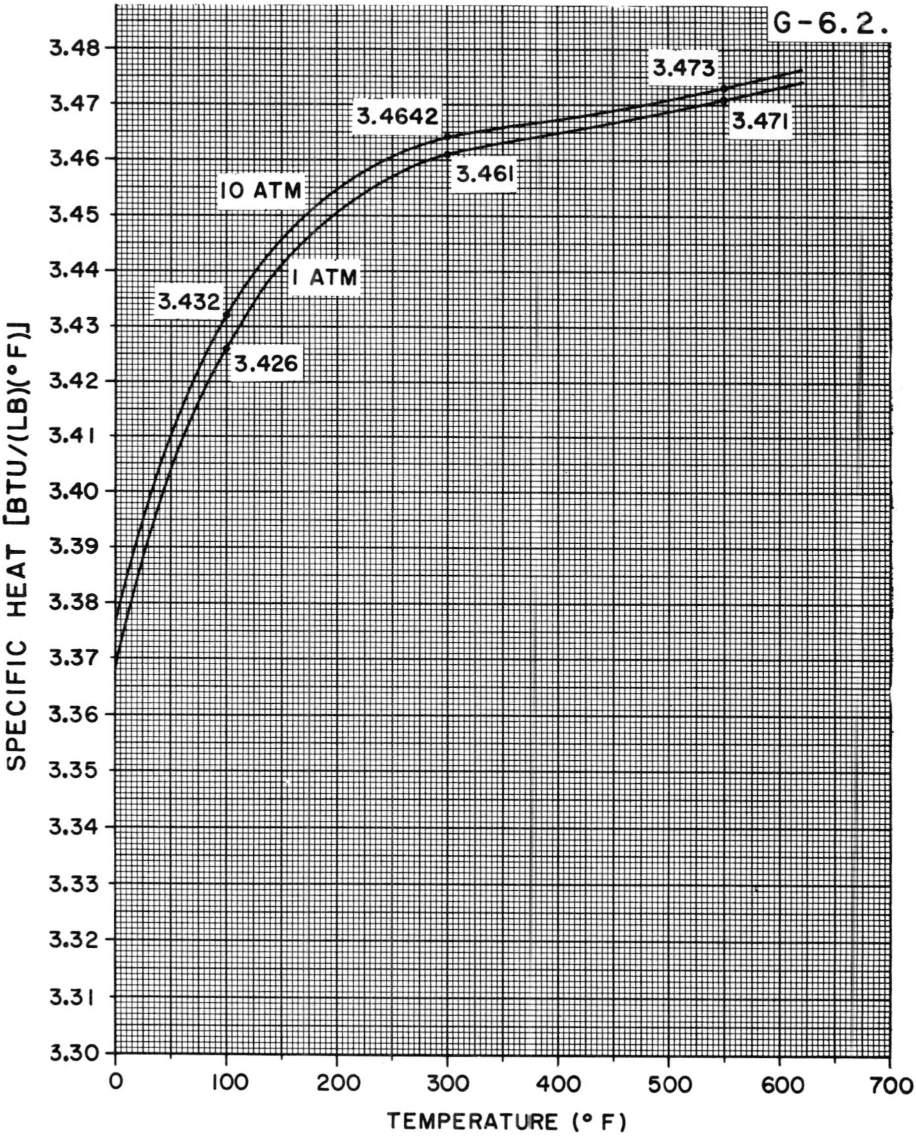

Specific Heat vs Temperature

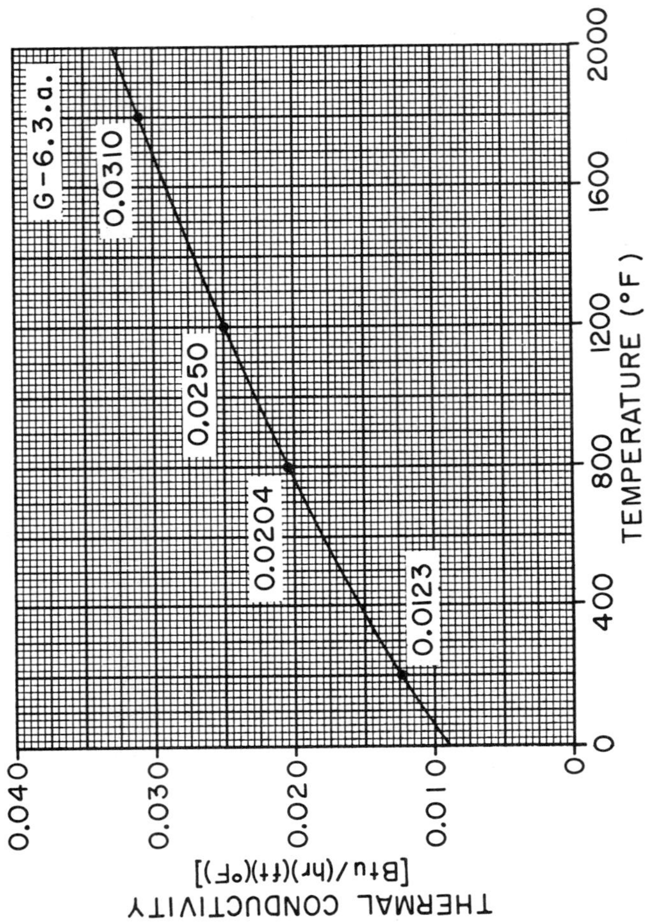

Thermal Conductivity vs Temperature (1 atm)

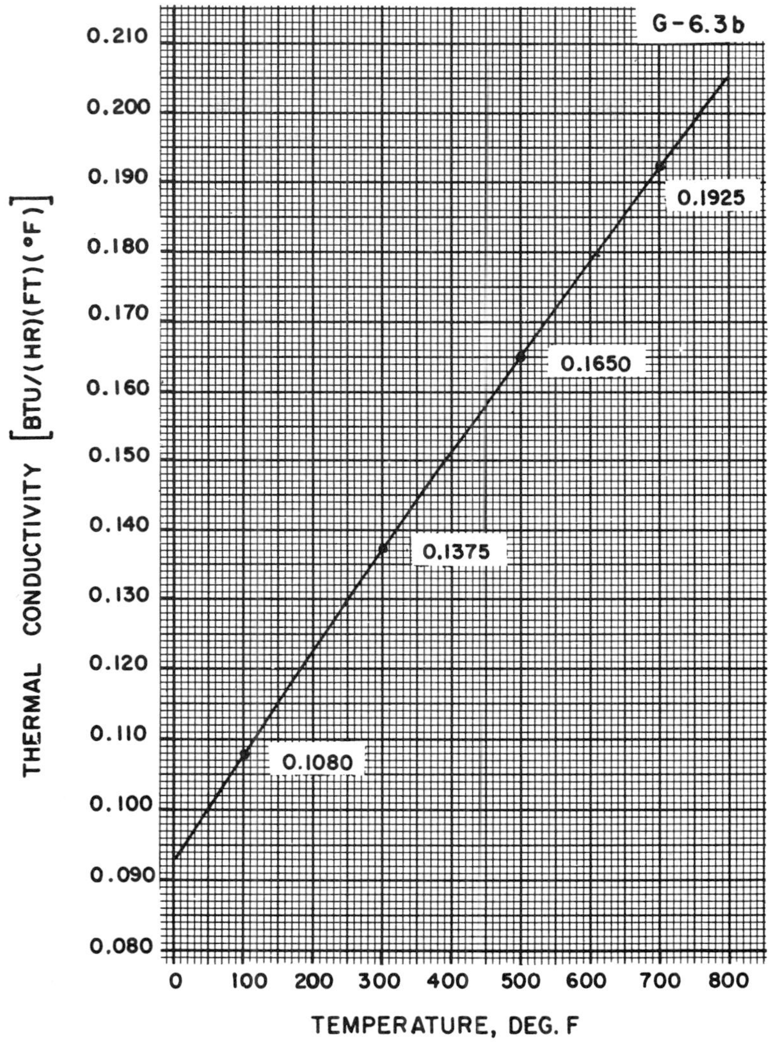

Thermal Conductivity vs Temperature (1 atm)

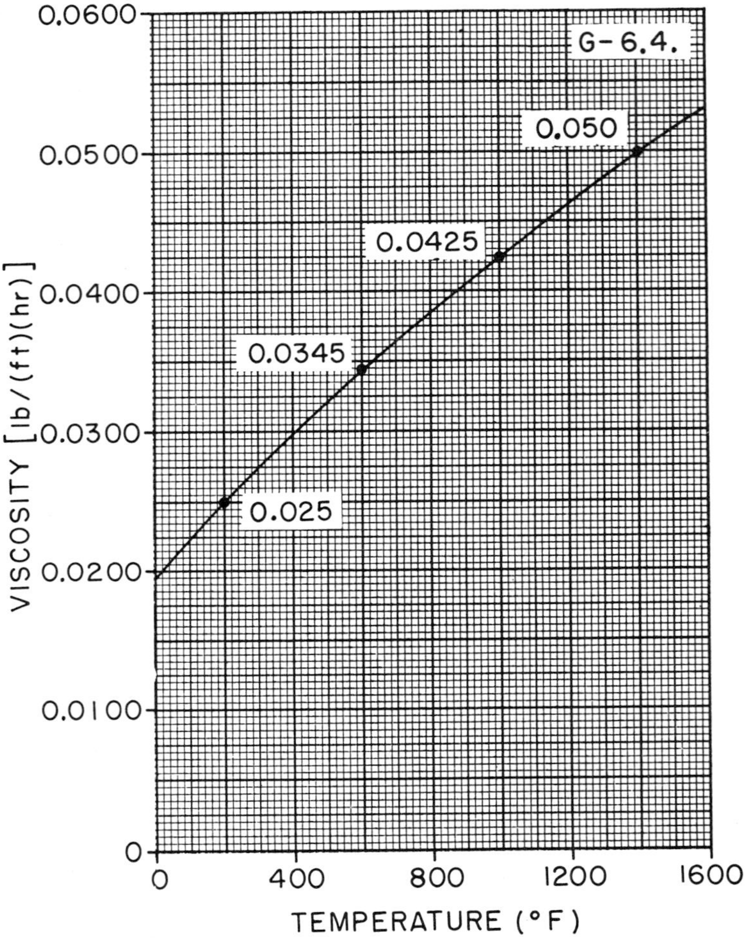

Viscosity vs Temperature (1 atm)

G-7. NITROGEN

Nitrogen is a colorless, odorless diatomic ga which constitutes 75.7 per cent of the atmosphere by weight and 78.06 per cent by volume. It is commercially produced either by fractional distillation of liquid air, with nitrogen volatilizing at −320°F, or by burning methane in air, dissolving under pressure the carbon dioxide formed in water, and then collecting the pure dry nitrogen. The relative inertness and noncorrosiveness of nitrogen are perhaps its foremost virtues.

Properties of Importance

Density: See Figure G-7.1.

Specific Heat: See Figure G-7.2.

Thermal Conductivity: See Figure G-7.3.

Viscosity: See Figure G-7.4.

Physical Stability:
Critical Point: $T_c = -232.85\,°F$, 33.5 atm
$P_c = 33.5$ atm
Melting Point: −345.75°F.
Boiling Point: −320.4°F.

Thermal Stability: The nitrogen triple bond has the highest bond energy of any diatomic molecule; 306,000 Btu/mole. Dissociation of diatomic molecules and ionization of the gas are the only reactions occurring in pure nitrogen.

Dissociation: $[N_2 \longrightarrow 2N]$

Ionization: $[N \longrightarrow N^+ + e^-]$
Values of the dissociation constant are given in Figure G-1, and the fraction of nitrogen dissociated and ionized at various pressures and temperatures is given in Table G-2.

Chemical Stability: In general, nitrogen is quite inert, except at high temperatures where it reacts with materials such as lithium, calcium, magnesium, and hydrogen to form nitrides.

TABLE 2. DISSOCIATION AND IONIZATION OF NITROGEN

Fraction Dissociated

$T \times 10^{-3}$ °K

p (atm)	5	6	8	10	12
1.0	.0065	.116	.822	.989	
10.0		.037	.417	.922	
20.0		.026	.308	.859	
50.0		.016	.201	.730	.995
100.0		.012	.143	.610	.980

Fraction Ionized

$T \times 10^{-3}$ °K

p (atm)	10	15	20	30
1.0	.0250	.544	.970	.998
10.0	.0079	.200	.775	.981
20.0		.143	.675	.962
50.0		.091	.444	.920
100.0	.0025	.063	.220	.855

Corrosion: Nitrogen is noncorrosive.

Flammability: Nitrogen is nonflammable.

Physiological Effects: Nontoxic. However, the discharge of waste nitrogen into unventilated locations dilutes the amount of oxygen available for breathing.

Price Class: B per 100 cu ft at 2100 psi.

References

1. Linde Corp., "Technical Data on Liquid Nitrogen," New York, 1957.
2. Martinek, F., "Thermodynamic and Transport Properties of Gases, Liquids, and Solids."
3. *Am. Soc. Mech. Engrs.*, p. 130. "Thermodynamic and Electrical Properties of Nitrogen at High Temperatures" (1959).

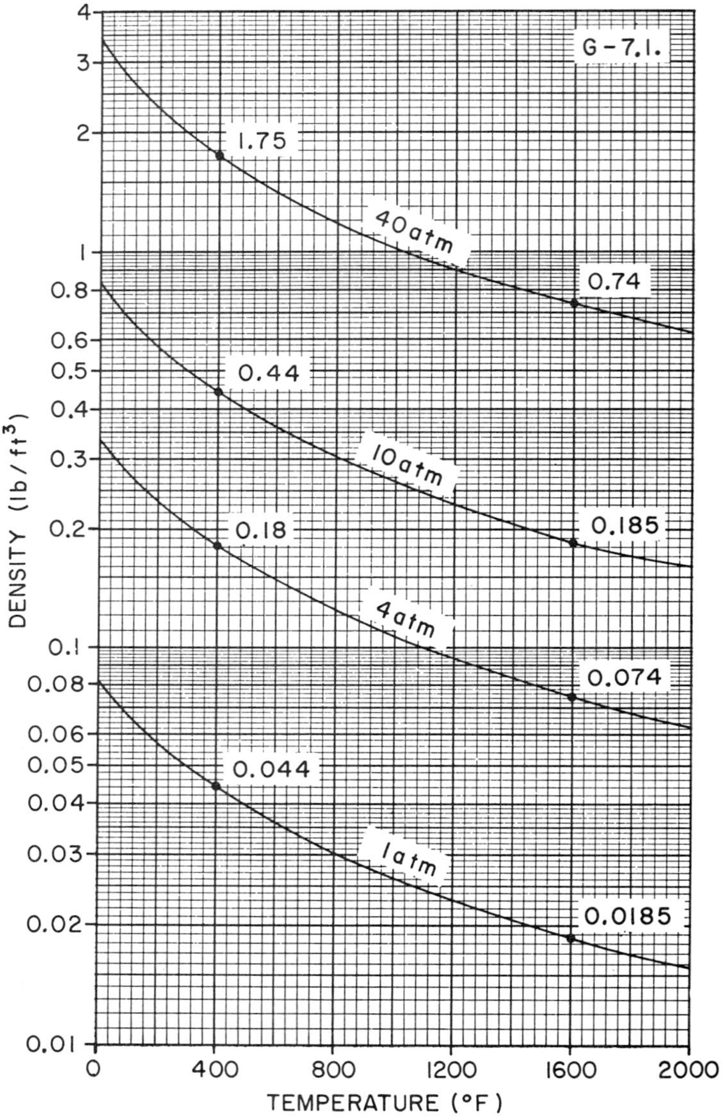

Density vs Temperature

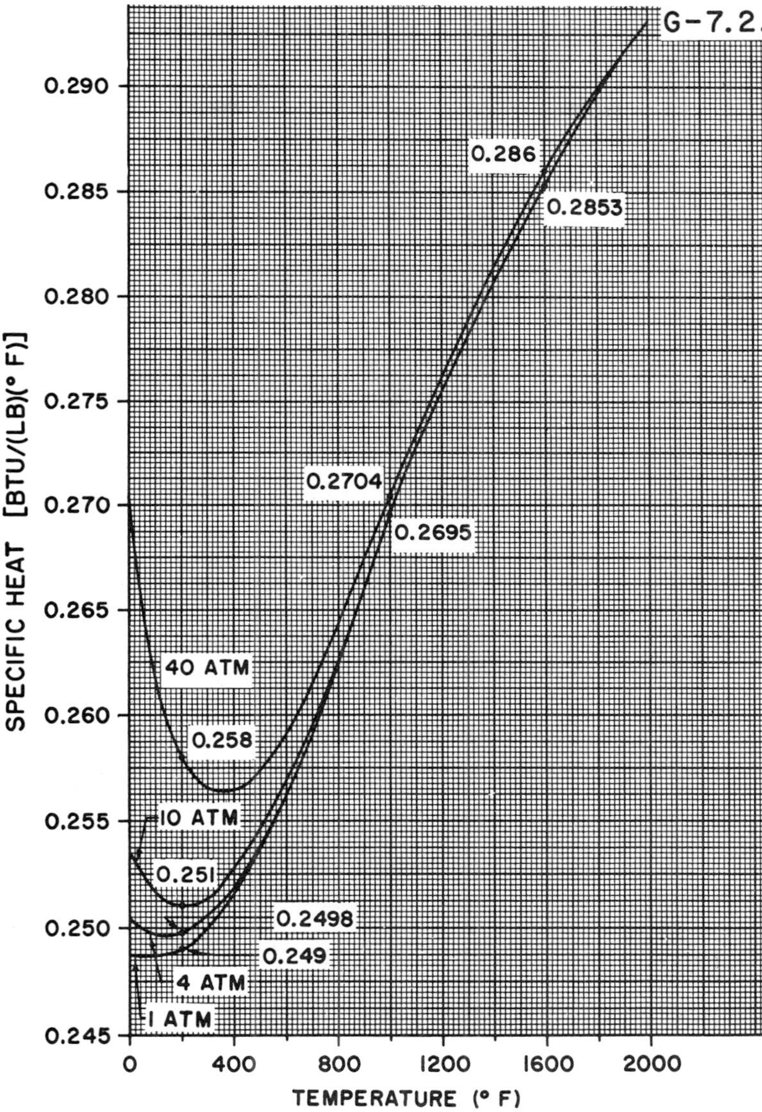

Specific Heat vs Temperature

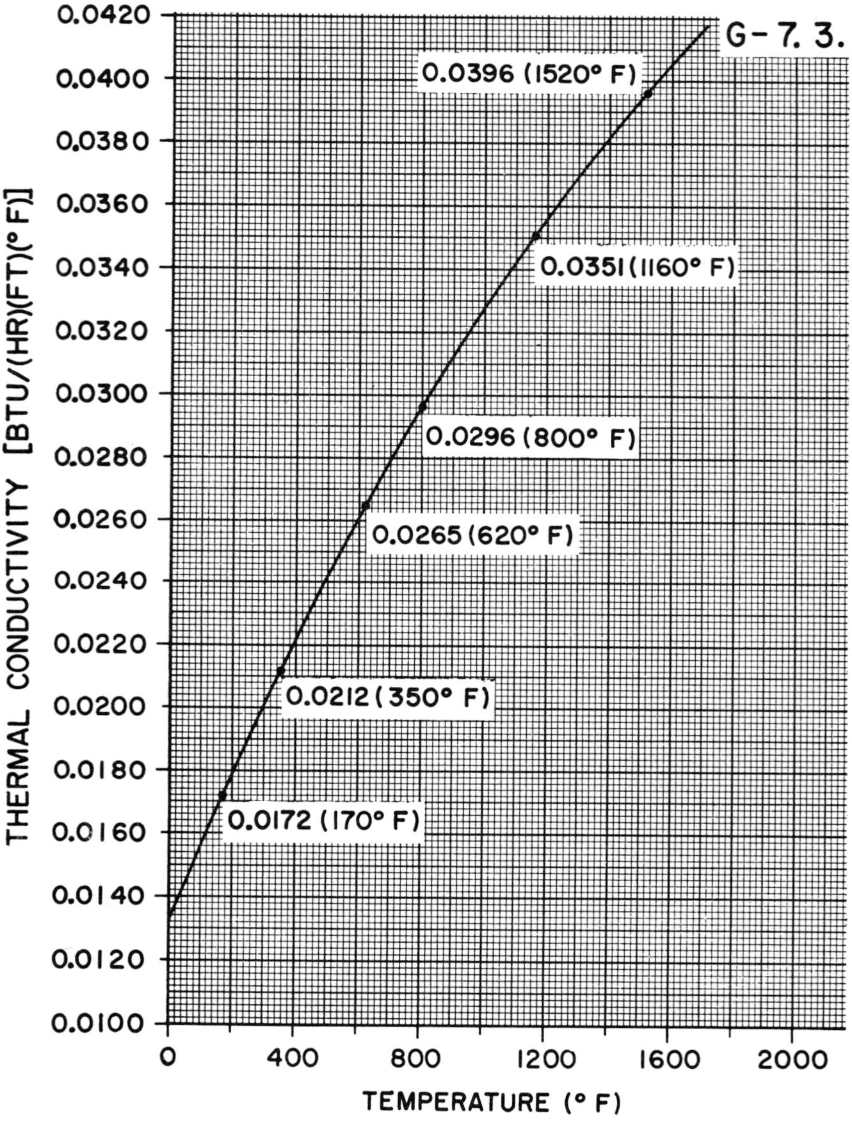

Thermal Conductivity vs Temperature (1 atm)

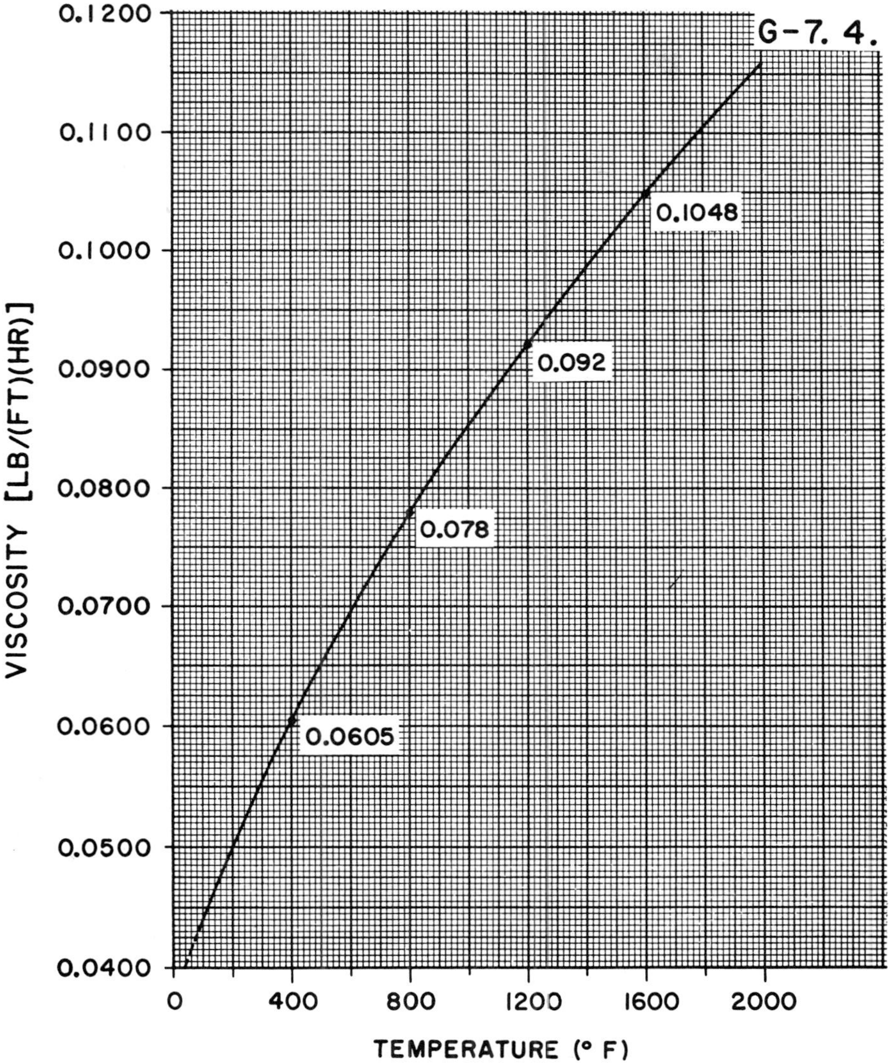

Viscosity vs Temperature (1 atm)

G-8. SULFUR HEXAFLUORIDE

Sulfur hexafluoride, SF_6, is a relatively inert gas used as an insulator and as a dielectric coolant. When SF_6 is used in place of oils in electronic equipment, design is simplified, lighter units are made possible and the units are easier to install and repair. The useful range of the gas is approximately from −40° to 300°F. This range can be extended with the proper equipment.

Properties of Importance

Density: See Figure G-8.1.

Specific Heat: See Figure G-8.2.

Enthalpy: The enthalpy, H, in Btu/lb is given in the following table. The temperatures are in °R.

T	H
536.69	0.
720.	319.83
900.	679.72
1080.	1169.19
1260.	1477.54
1440.	1896.82
1620.	2435.73
1800.	2763.27

Thermal Conductivity: The thermal conductivity of SF_6 is given by the equation,

$$k = 11.68 \times 10^{-5} T^{0.942} \quad T \text{ in } {}^\circ R, \quad k \text{ in } Btu/(hr)(ft)({}^\circ F)$$

Viscosity: See Figure G-8.3.

Physical Stability: See Figure G-8.4. The curve represents the pressure exerted by the vapors sulfur hexafluoride in equilibrium with liquid SF_6. However, in practical usage, the vapor is not in equilibrium with the liquid and its pressure may be determined by a calculation from an equation of state such as the Beattie-Bridgeman equation. The only time that the two phases are in equilibrium is when the material is in the cylinders in which it is shipped. The freezing point of SF_6 is −59.44°F.

Thermal Stability: SF_6 is very stable. However, certain metals seem to catalyze the decomposition of the material at temperatures above 300°F.

Chemical Stability: Sulfur hexafluoride is one of the least reactive substances known. The gas will slowly decompose when exposed to an open electric arc for a period of time.

Surface Tension: At −58°F, the surface tension is 11.63 dynes/cm; at −4°F, the surface tension is 8.02.

Corrosion: At temperatures below 300°F, SF_6 is not affected by any of the normal metals of construction. Above 300°F, only certain materials are applicable. The best ones are silver plate, aluminum, and stainless steel.

Flammability:
Flash Point: None.
Fire Point: None.
Ignition Point: None.

Physiological Effects: SF_6 is nontoxic. However, at very high temperatures, or in the presence of an electric arc, SF_6 is slowly decomposed to lower sulfur fluorides, some of which are toxic.

Electrical Properties: SF_6 has a higher dielectric strength than most gaseous dielectrics. At increased pressures, this discrepancy becomes even greater. The dielectric constant of SF_6 is 1 at zero pressure and increases to about 1.07 at 22 atmospheres. These values are for a temperature of 77°F and a frequency of 24240 megacycles/sec. The dielectric constant of SF_6 does not vary with changes in frequency because the material has no dipole moment. SF_6 also has the ability to extinguish any electric arc that may form because of high transient voltages.

Refractive Index: (32°F-1 atm) 1.000783.

Color: Colorless.

Price Class: B per lb.

Reference

1. Allied Chemical Corp., "Sulfur Hexafluoride," *Tech. Bull. TB-85602*, (1955).

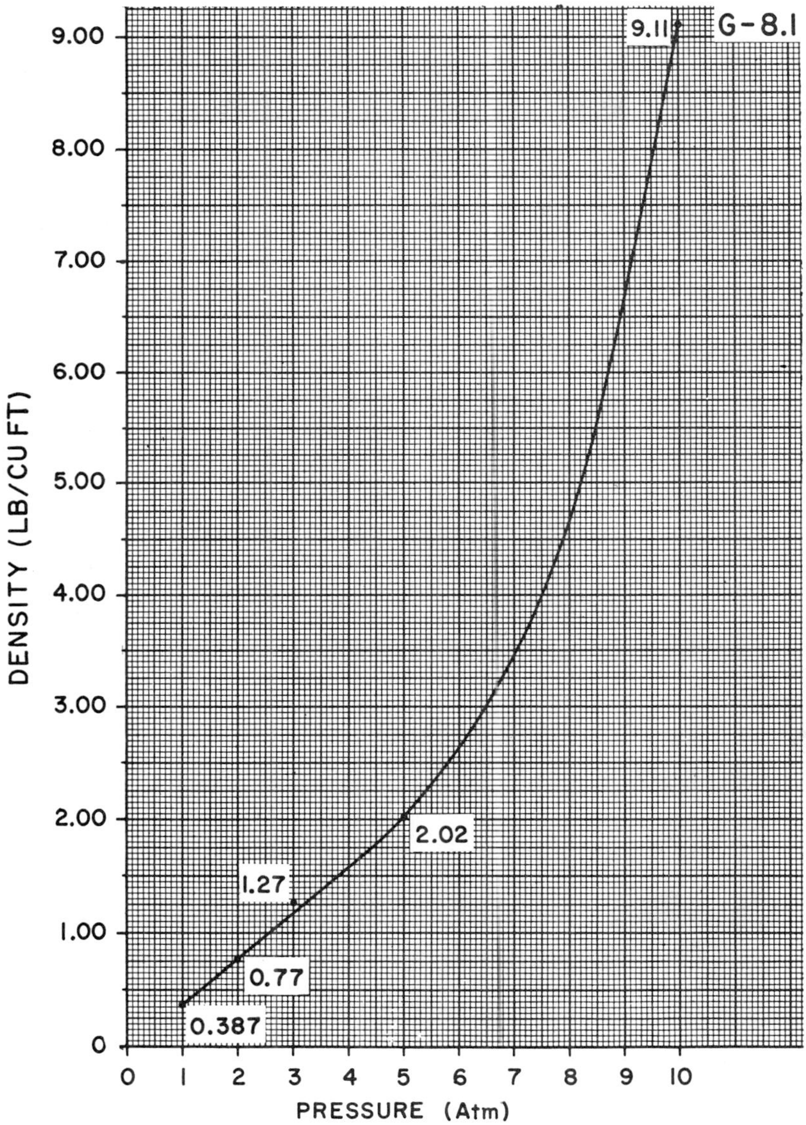

Vapor Density vs Pressure (at 70 °F)

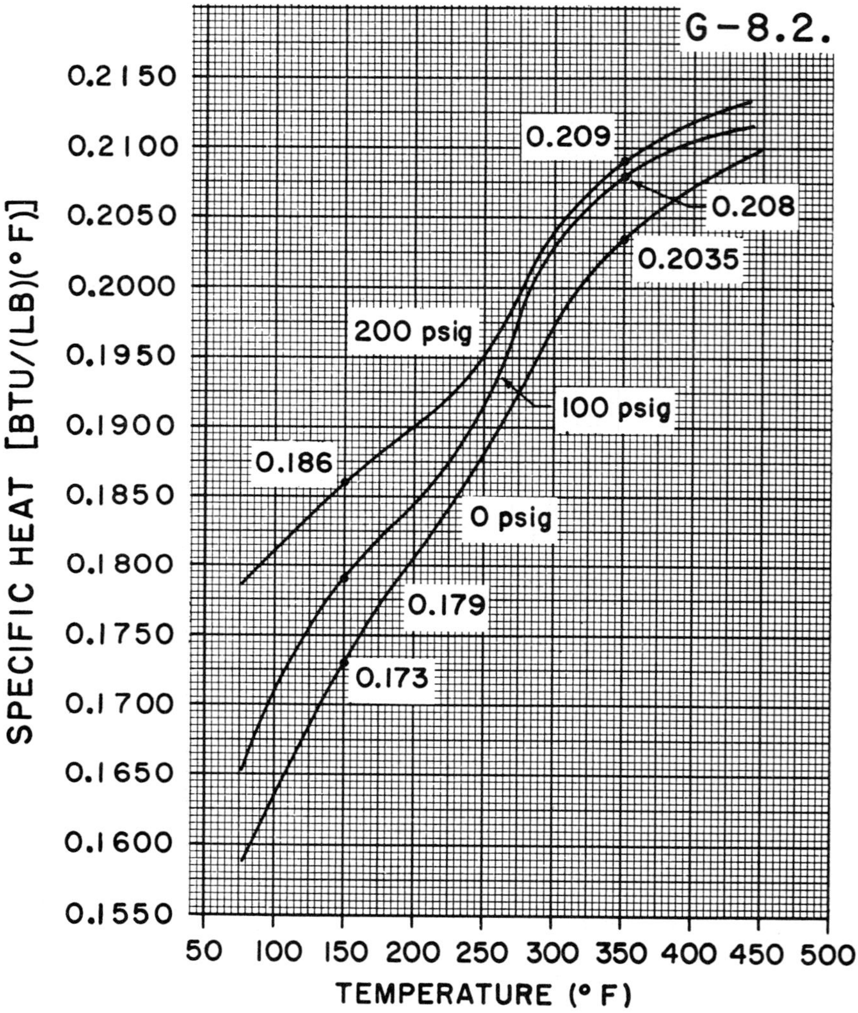

Specific Heat vs Temperature

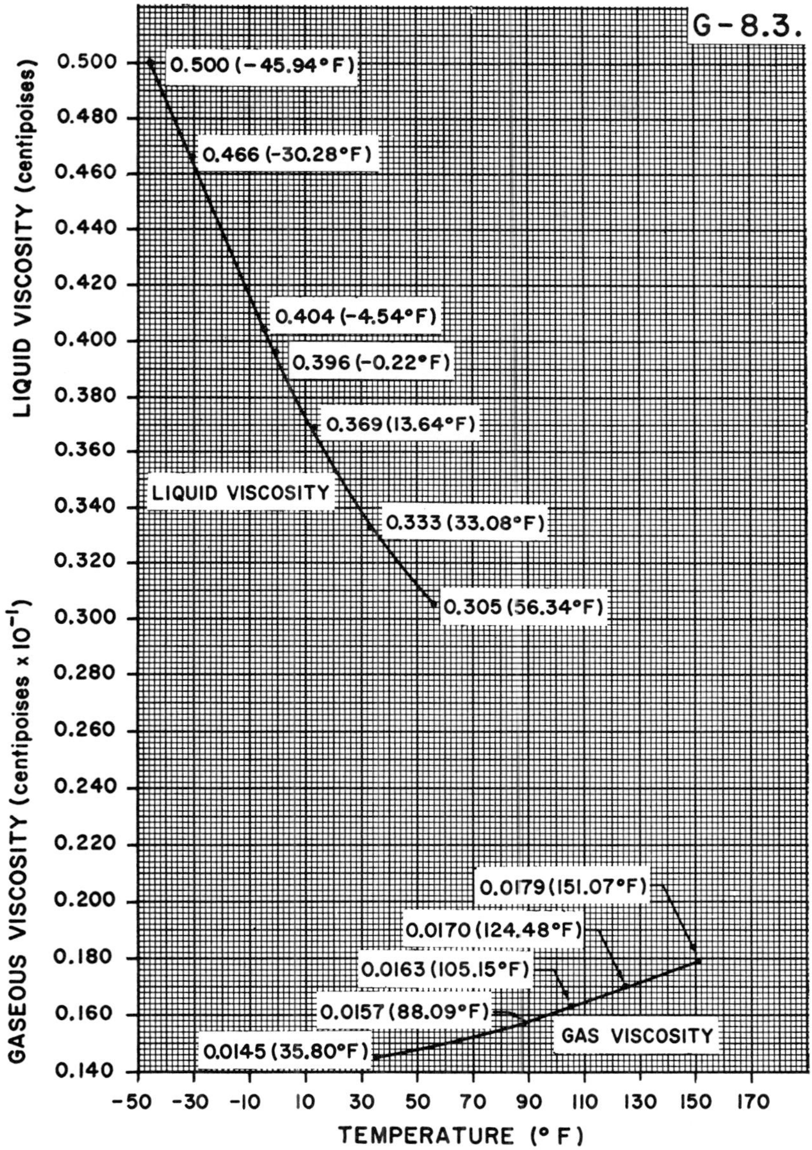

Vapor Pressure vs Temperature

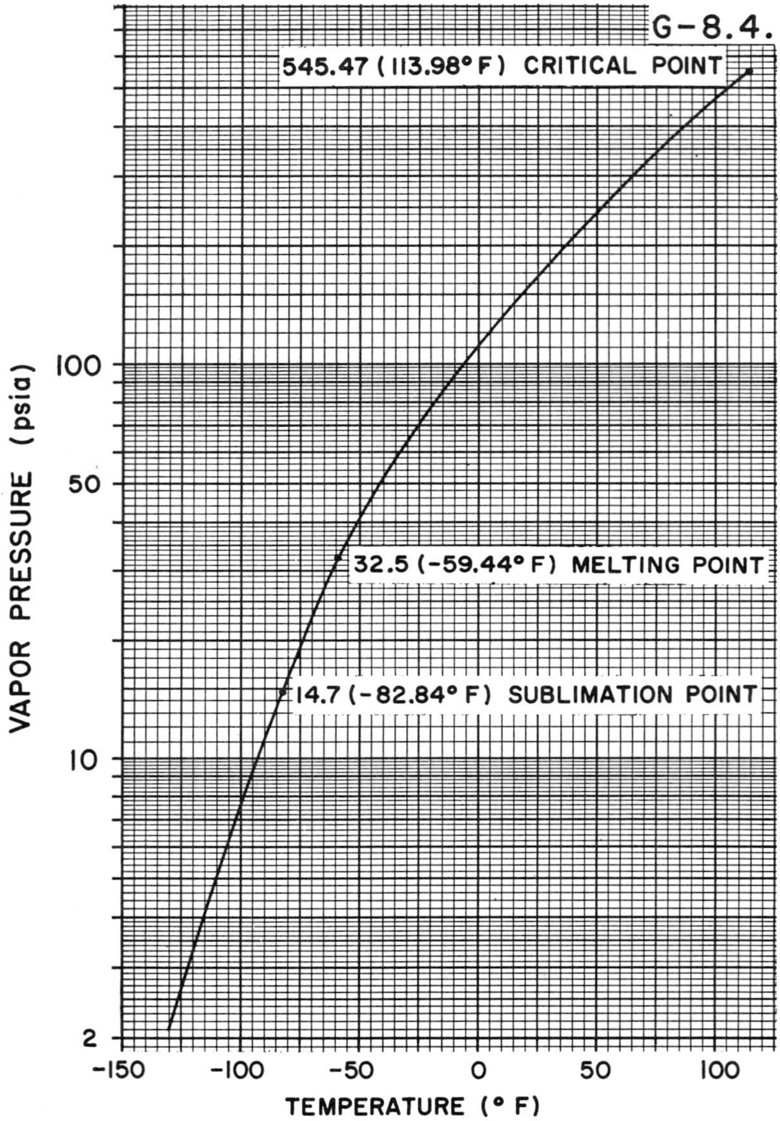

Liquid Viscosity vs Temperature
Gas Viscosity vs Temperature

V. VAPORS

V-1. ALKYL NAPHTHALENES

Alkyl naphthalenes were utilized as heat transfer media in Great Britain under a number of different trade names (Bectherm, Petrotherm, and Catarex). These products are not currently manufactured. Nevertheless, the chemical class is worthy of mention.

The idea of using alkyl naphthalene as heat transfer material has been introduced in Japan by Soken Chemical Engineering Company, Ltd., in competition with the different materials used in England and America. This material is marketed under the name of Neo SK-Oil. Its characteristics are quite favorable. There are presently three different molecular weights on the market—260, 240, and 170.

The solidifying point of −10°C of Nos. 260 and 240, or even lower, is one of the definite advantages of this material.

The application range is claimed to be between 340° to 350°C.

These materials are highly flammable, as are all alkyl naphthalenes.

Alkyl naphthalenes, aromatic in nature, are prepared from the liquid residues of petroleum cracking. They are also derived from coal tar, together with paraffins and hydrocarbons of similar volatility. Their P-T relationship permits their application as a liquid and as a vapor. At 600°F, alkyl naphthalenes possess a vapor pressure of 47 psig.

As alkyl naphthalenes must undergo continuous long-term use and repeated condensation and vaporization, thermal stability is of crucial significance. The maximum bulk operating temperature recommended is approximately 590°F. Reforming reactions, heralded by a drift in vapor pressure and sludging, are purported not to take place in the 550°–600°F range. Experience with alkyl naphthalene mixtures is quite limited and no comprehensive studies have been conducted. It has been reported that an industrial fatty alcohol distillation unit operated at 540°F over three years without noticeable deterioration of the heat carrier.[3] A few of the principal heat transfer properties of alkyl naphthalenes are given in Figures V-1.1 and V-1.2.

Alkyl naphthalenes do not attack normal structural materials. Copper or its alloys should not be employed, because they promote cracking, and small quantities of dissolved sulfur compounds produce corrosion. Further advantages of alkyl naphthalenes are their low freezing points (below 14°F) and surface tension values higher than those for similar oil mixtures.

The major reason for their utilization in England was an economic one. Considerable savings result by employing heat transfer agents produced at home, rather than imported fluids. "Spent" fluid, which has seen long service, can be readily rejuvenated for continued use.

References

1. Brearley, N., *Proc. Brit. Inst. Mech. Engrs.* **166A**, 350, (1952).
2. Rottenburg, P., *Trans. Brit. Inst. Chem. Engrs.* **35**, (February 1957).
3. Tripp, G. W., *The Engineer*, 453, (October 21, 1949).
4. Neo SK-OIL 260, 240 and 170 Bulletin, Soken Chemical Engineering Co., Ltd., Tokyo, Japan.

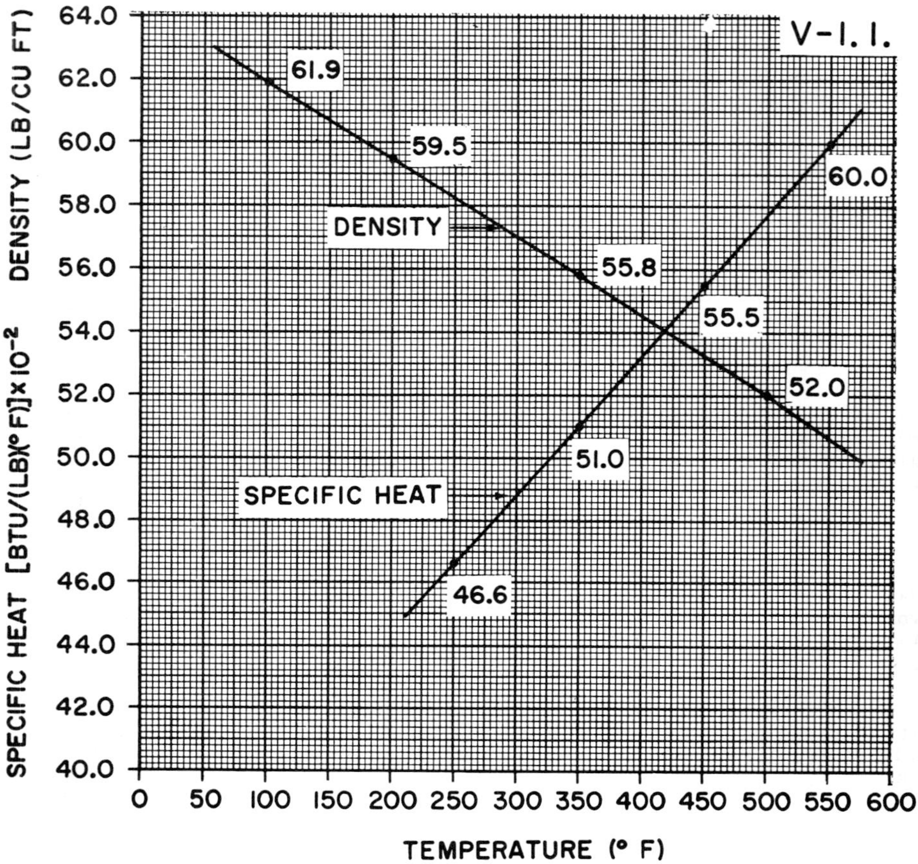

Specific Heat vs Temperature
Density vs Temperature

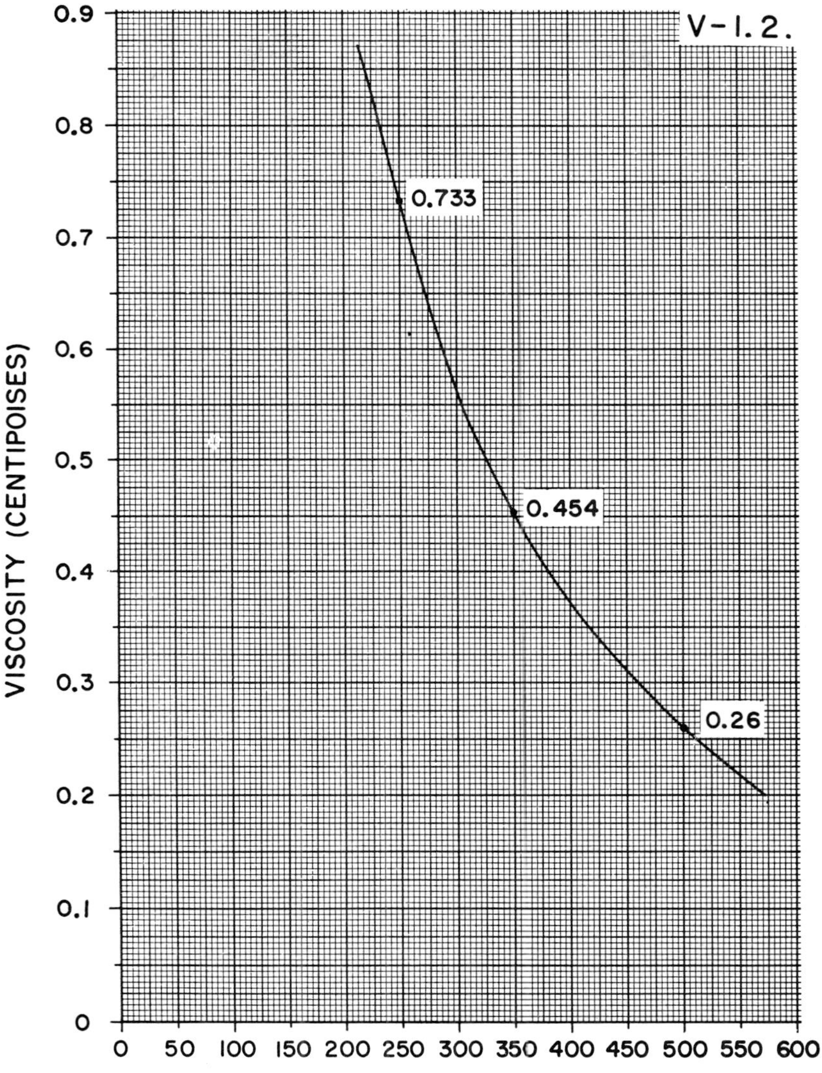

Viscosity vs Temperature

V-2. ANISOLE (PHENYL METHYL ETHER)

Anisole, a high boiling aryl alkyl ether, serves as a heat carrier in the −30° to 500°F range. The fluid may be applied in either the vapor or liquid phase.

Properties of Importance

Density: 62.7 lb/ft^3 (60°F)

Specific Heat: 75.2°F-0.422
88.9°F-0.462

Thermal Conductivity: 0.09 Btu/(hr)(ft)(°F)

Viscosity: See Figure V-2.1.

Latent Heat of Vaporization (b.p.): 154 Btu/lb

Physical Stability: See Figure V-2.1.
Boiling Point: 308.8°F
Critical Temperature: 723°F
Melting Point: −35.1°F

Thermal Stability: In one example after two weeks of holding the liquid at 482°F, no permanent gas was formed. Low boilers were below 0.04 per cent and high boilers below 0.02 per cent. Recommended operating range is −30° to 300°F. As a vapor, Anisole should not be employed at temperatures exceeding 500°F, or at vapor pressures greater than 121 psig.

Chemical Stability: The fluid is immiscible with water but dissolves completely in all other common solvents, with the exception of glycols.

Surface Tension: 36 dynes/cm (60°F)

Corrosion: The fluid is noncorrosive.

Flammability: The flash point of Anisole is 125°F.

Physiological Effects: Anisole is relatively nontoxic. It appears to lack the acute corrosive action of phenol and the blood cell damage of benzene. This may be attributed to the presence of the low toxicity methoxy group.

Color: Colorless.

Refractive Index: 1.5165 (68°F)

Dielectric Constant: 4.41 (68°F)

Price Class: C

References

1. Du Pont De Nemours, E. I., & Co., Inc., "Anisole (Phenyl Methyl Ether)," Grasselli Chemical Dept.
2. Factory Insurance Association, "Recommended Good Practice for Heat Transfer Mediums in Closed Systems," June 1954.

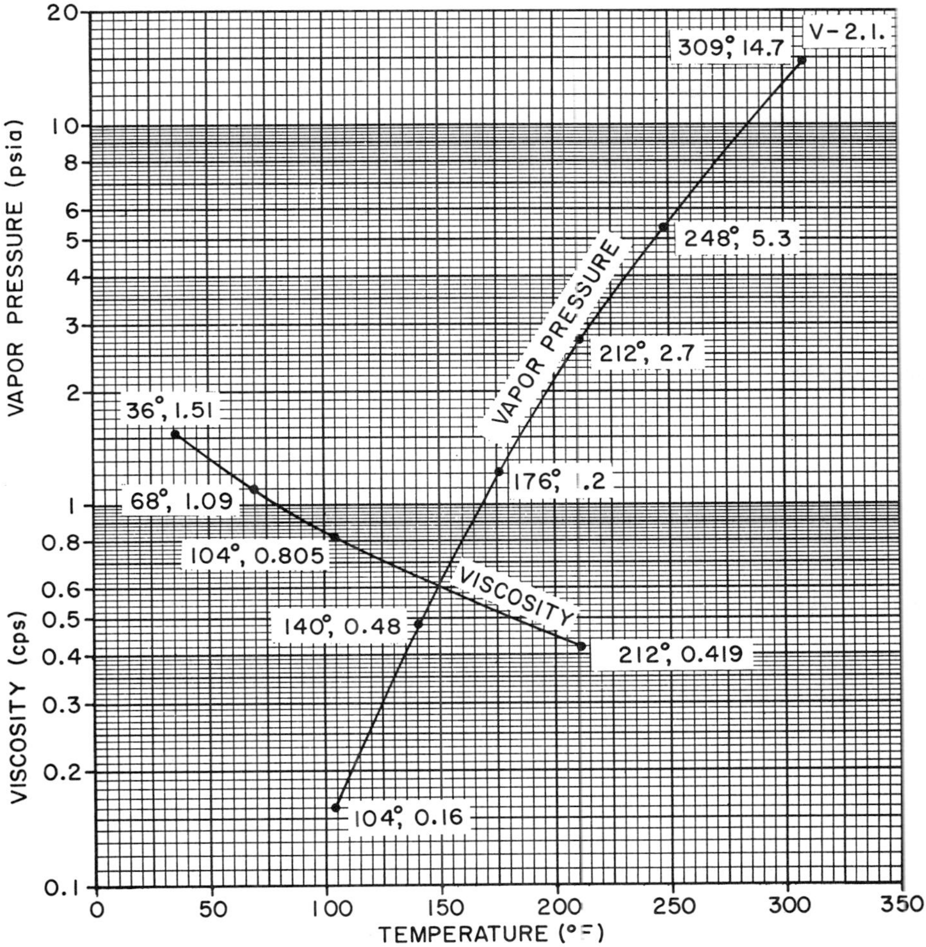

Vapor Pressure and Viscosity vs Temperature

V-3. o-DICHLOROBENZENE

(Dowtherm E)

Dowtherm E is a grade of o-Dichlorobenzene which has been purified and stabilized for heat transfer use. It may be used in either the liquid or the vapor state; however, the vapor is more commonly used for heating applications. The recommended operating temperature range is 300°-500°F. Decomposition may occur when Dowtherm E is heated above 500°F.

Properties of Importance

Density: See Fig. V-3.1.

Specific Heat: See Fig. V-3.2.

Heat of Vaporization: See Fig. V-3.2.

Thermal Conductivity: See Fig. V-3.3.

Viscosity: See Fig. V-3-3.

Physical Stability: See Fig. V-3.4 for the vapor pressure curve. The freezing point of Dowtherm E is −7°F.

Chemical Stability: Dowtherm E decomposes at high temperatures to form hydrochloric acid. It is believed that aluminum catalyzes thermal breakdown. For this reason, it should not be used in Dowtherm E systems. Otherwise, Dowtherm E is a very stable material within the recommended operating range. The system must be kept free of water or moisture.

Surface Tension: The surface tension of Dowtherm E is 37 dynes/cm at 77°F.

Corrosion: Aluminum pipes and fittings should not be used with Dowtherm E. All other common structural materials are acceptable.

Flammability:
Flash Point: 155°F
Fire Point: 285°F
Self Ignition Point: Above 932°F

Physiological Effects: Dowtherm E has a low oral toxicity. Inhalation of the vapor is hazardous; however, the vapor has such a pungent odor that it is easily detected before reaching a dangerous concentration.

Electrical Properties: Dowtherm E has a specific resistivity of 1×10^7 ohm-cm. No data on the dielectric constant is available.

Color: Dowtherm E is a clear, colorless to light straw-colored liquid.

Price Class: B.

References

1. Dow Chemical Co., "Dowtherm for Industrial Process Heating."
2. Dow Chemical Co., "Dowtherm Handbook."

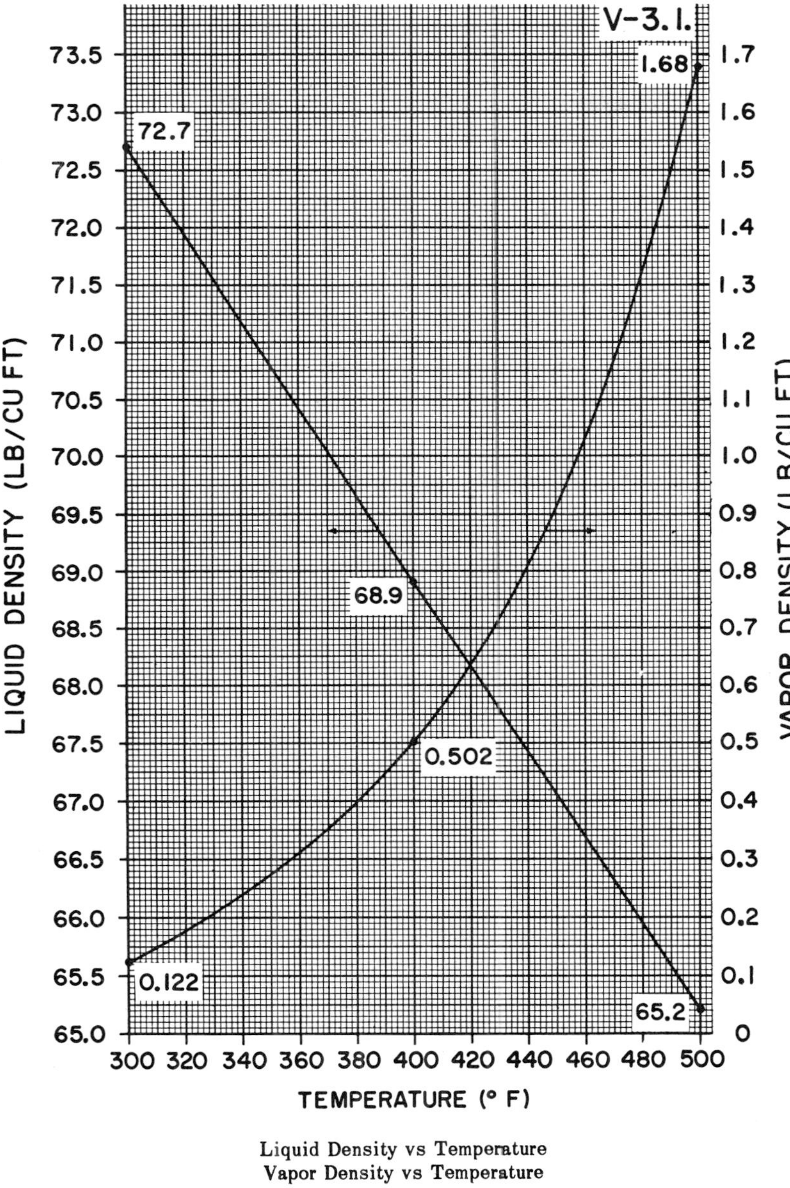

Liquid Density vs Temperature
Vapor Density vs Temperature

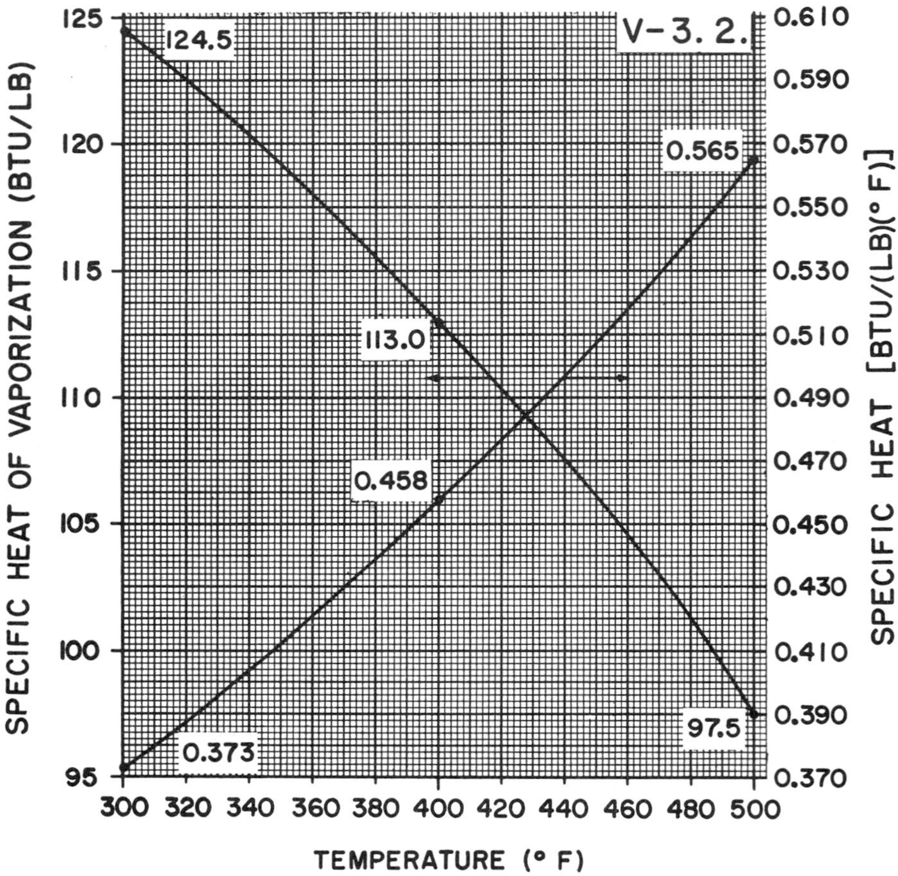

Specific Heat of Vaporization vs Temperature
Specific Heat of Liquid vs Temperature

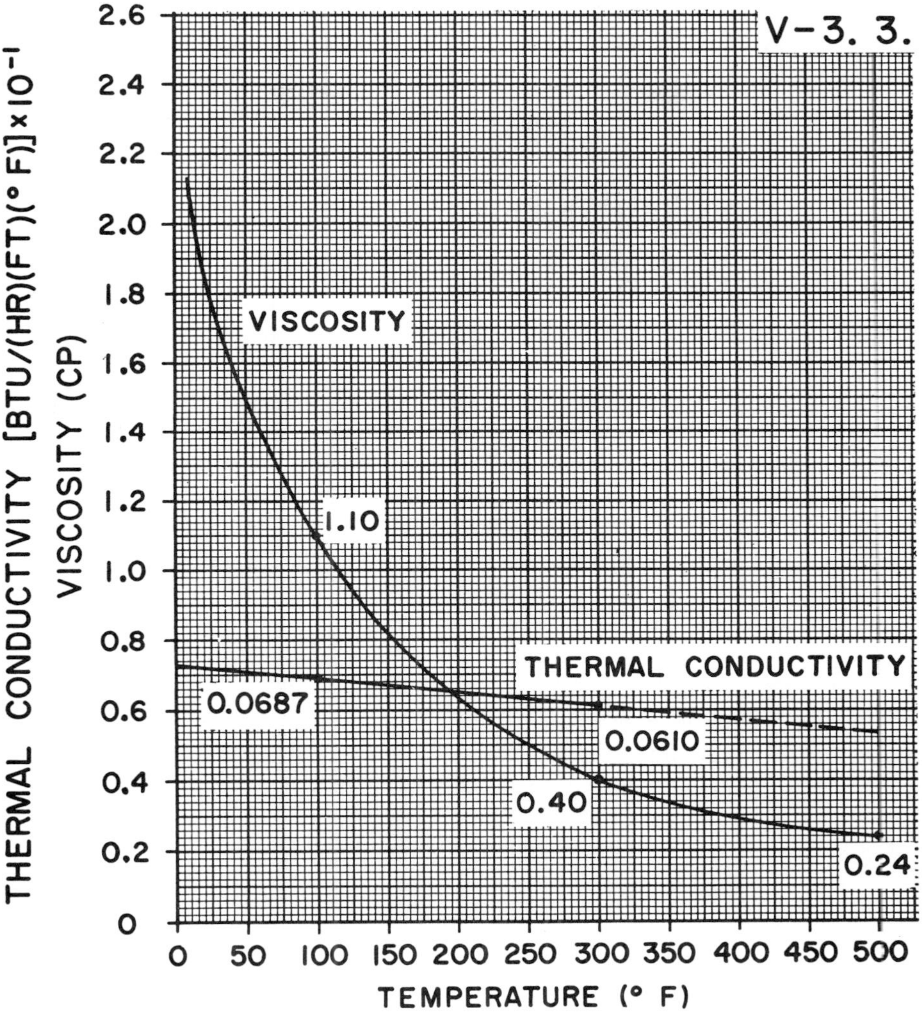

Viscosity of Liquid vs Temperature
Thermal Conductivity of Liquid vs Temperature

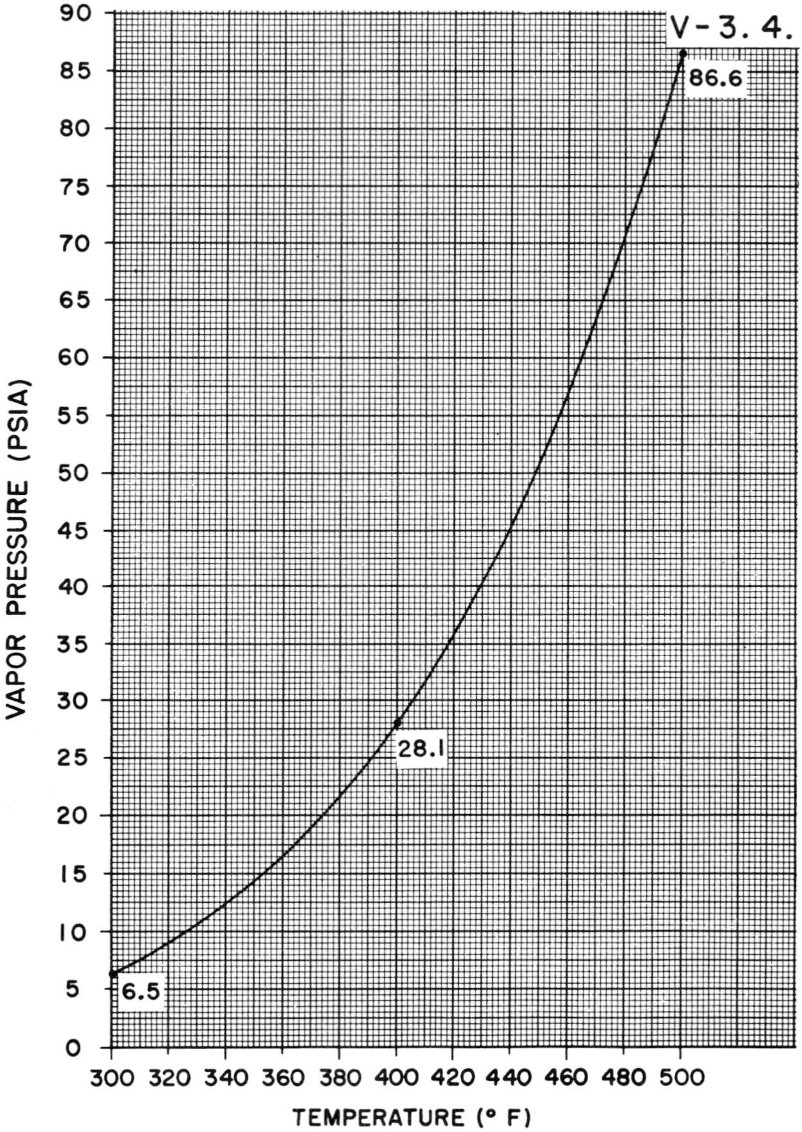

Vapor Pressure vs Temperature

V-4. DIPHENYL-DIPHENYL OXIDE

Dowtherm A in U.S.A.
Thermex in United Kingdom
Diphyl in Germany

Dowtherm A is a eutectic mixture containing 73.5 per cent diphenyl oxide, and 26.5 per cent diphenyl. It may be employed for heat transfer application either in the liquid or the vapor state. Vapor phase heating is more efficient. Since the vapor pressure curves for diphenyl and diphenyl oxide are almost identical, Dowtherm A may be considered to be a pure compound for heat transfer purposes. Dow Chemical Company suggests use of the fluid in the temperature range between 450°–750°F. Dowtherm A is one of the oldest and widely used organic heat carriers in the chemical process industries.

Properties of Importance

Density: Both the liquid and the vapor densities are shown in Figure V-4.1.

Specific Heat: See Figure V-4.2.

Enthalpy: The heat of vaporization is shown in Figure V-4.2.

Thermal Conductivity: See Figure V-4.2.

Viscosity: The liquid viscosity of Dowtherm A is shown in Figure V-4.3. The vapor viscosity has been calculated to be in the range of 0.006 to 0.0125 centipoises.

Physical Stability: See Figure V-4.4. Note that Dowtherm A exerts a pressure of only 109.6 psia at 700°F, while steam at the same temperature would exert a pressure in the neighborhood of 3000 psia. The freezing point of Dowtherm A is 53.6°F.

Thermal Stability: When heated near the upper limit of its recommended temperature range, Dowtherm A tends to slowly decompose in the manner described under *Chemical Stability.*

Chemical Stability: As soon as Dowtherm A is heated, it darkens because of the precipitation of fine particles which remain suspended in the liquid. However, these particles have no effect on the operation of the system. At higher temperatures, polymerization to higher weight materials which remain in solution may take place. This will not affect the system either, as long as dissociation into carbon and hydrogen does not occur. When carbon is deposited on the pipe walls, heat transmission is impeded; the fluid temperature rises rapidly, hastening decomposition. Dowtherm A may be recovered from its dissociation products by distillation. Dow Chemical Company recommends this when there is more than 10 per cent of high boiling materials present, as measured by a standard ASTM distillation.

Surface Tension: 43 dynes/cm at 77°F. Dowtherm A is an excellent wetting agent and seeps through any imperfections in joints, welds, stuffing boxes, etc.

Corrosion: Dowtherm A does not react with any of the common metals used in heat transfer systems.

Flammability:
Flash Point: 255°F
Fire Point: 275°F
Self-Ignition Point: 1150°F

Physiological Effects: Dowtherm A does not present any toxicity problems. Its pungent odor is detected quite easily, and there is little danger of excess of fumes accumulating. In the liquid phase, Dowtherm A may be a skin irritant and should be handled carefully.

Electrical Properties: Dowtherm A has a specific resistivity of 3.36×10^{11} ohm-cm. The dielectric constant at 75°F is 3.3 for frequencies of 1, 10, or 100 kilocycles.

Color: The material is originally a clear, light, straw-colored liquid. However, it darkens rapidly with use.

Price Class: B.

References

1. Dow Chemical Co., "Dowtherm for Industrial Process Heating."
2. Dow Chemical Co., "Dowtherm Handbook."

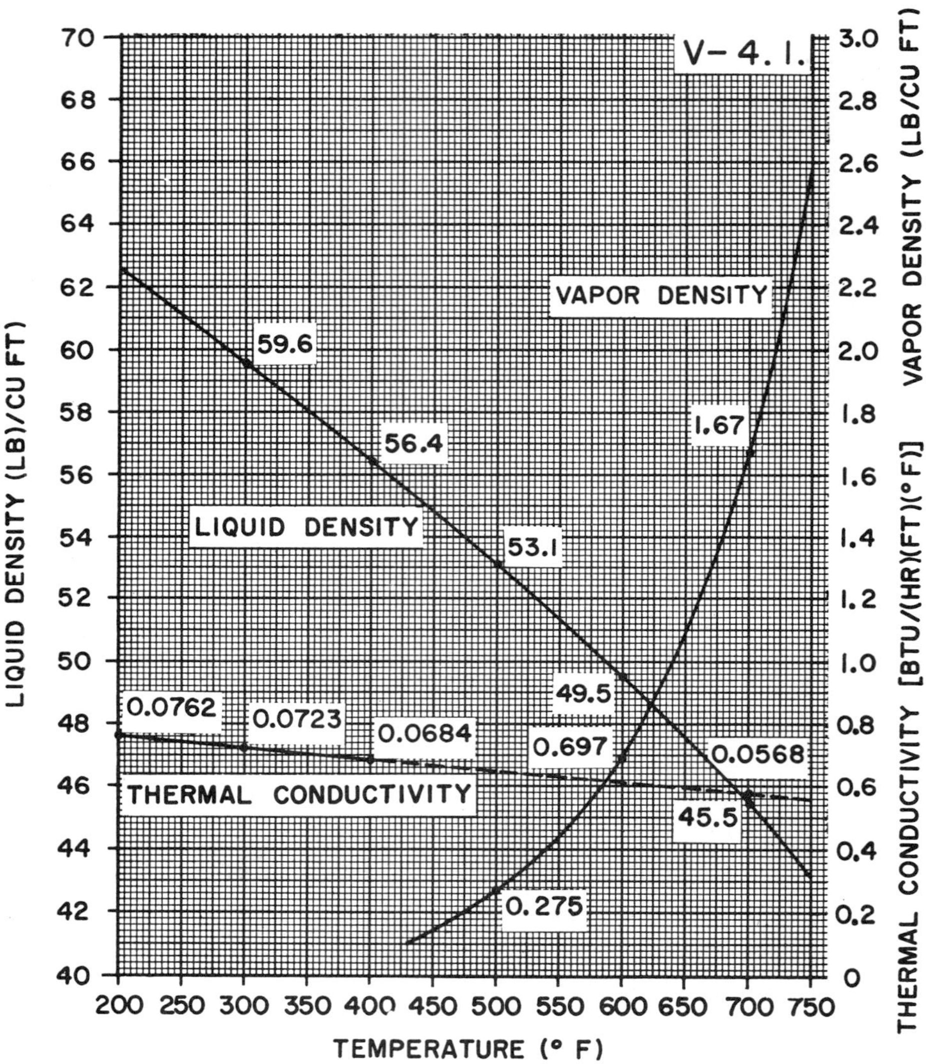

Liquid Density vs Temperature
Vapor Density vs Temperature
Thermal Conductivity vs Temperature

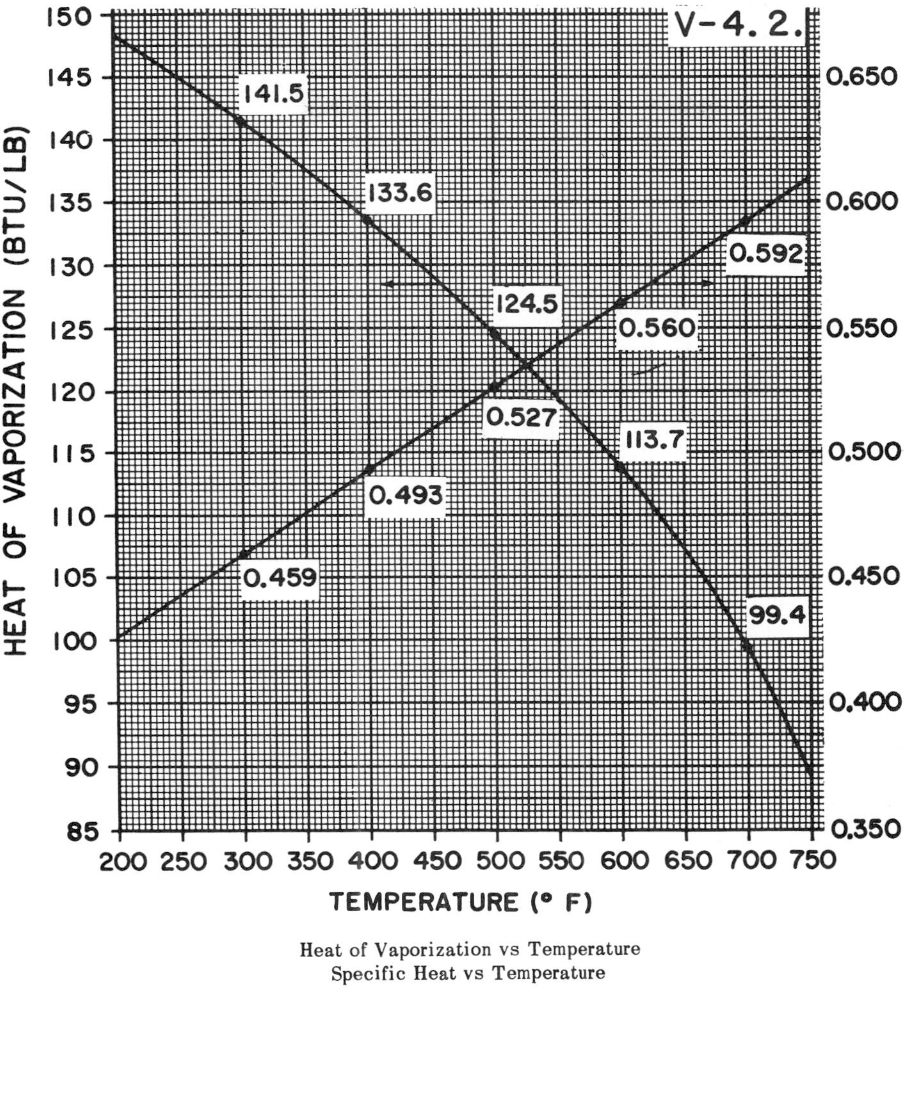

Heat of Vaporization vs Temperature
Specific Heat vs Temperature

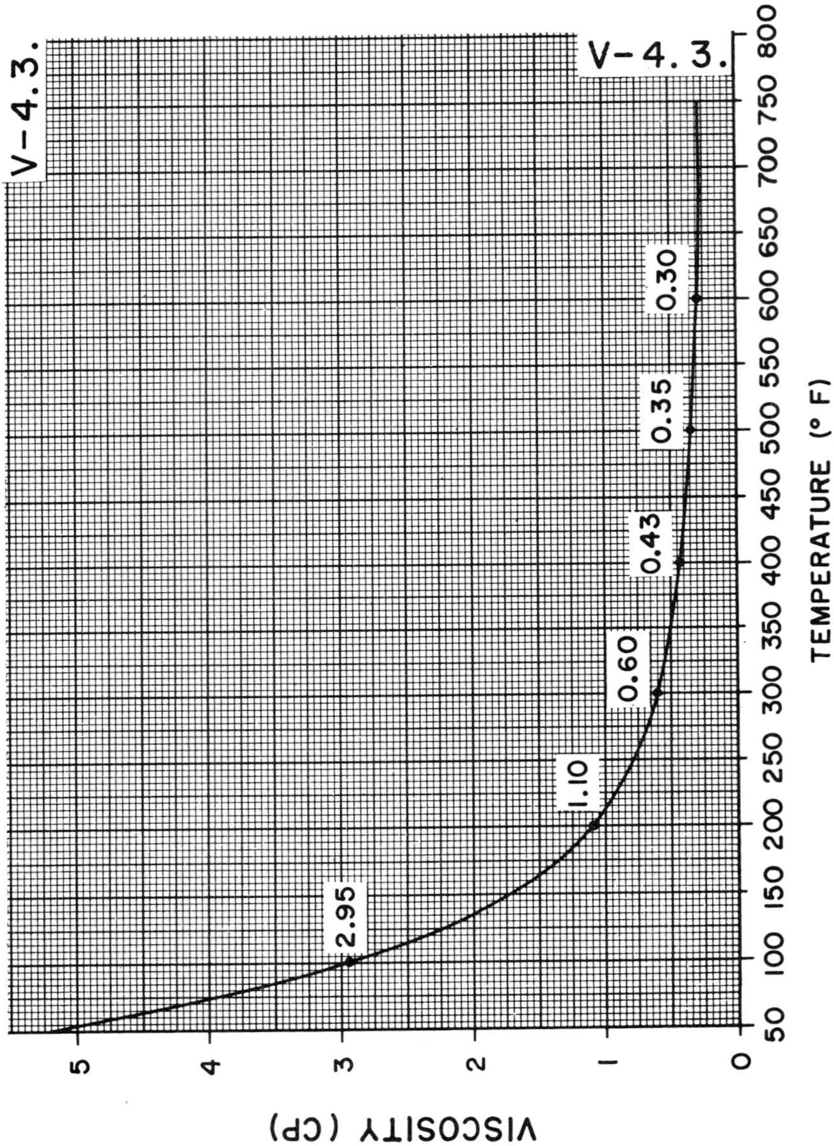

Viscosity of Liquid vs Temperature

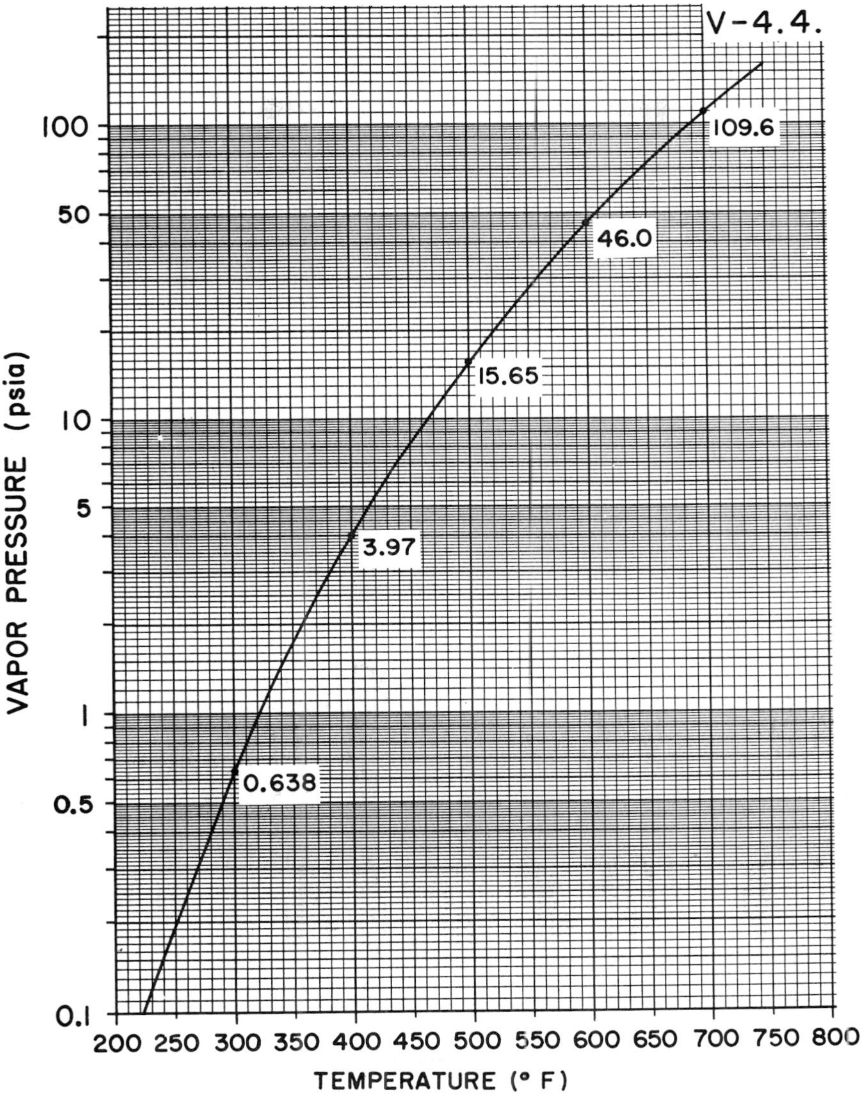

Vapor Pressure vs Temperature

V-5, V-6. FLUOROCARBONS

The properties of fluorocarbons are ideally suited to their use as dielectric coolants. Fluorocarbons possess low surface tension, low viscosity, high density, and a boiling point low enough to permit highly effective evaporative cooling and/or convective cooling with either the liquid or its vapors. Electrical properties are outstanding. The dielectric constants and dissipation factors are low, dielectric strengths and resistivities are high. The inertness of fluorocarbons to thermal and chemical change allows them to be used under extremely rigorous conditions. The fact that fluorocarbons are nonflammable and do not attack common construction materials adds to their value.

V-5. FC-43

Chemically speaking, FC-43 is a fully fluorinated product composed primarily of an isomeric mixture of the perfluoro amine, $(C_4F_9)_3N$. Although it has the structure of a tertiary amine, it has none of the properties usually associated with these compounds.

Properties of Importance

Density: See Figure V-5.1 for Specific Gravity.

Coefficient of Volume Expansion:

$77°-104°F - 2.04 \times 10^{-3}\ °F^{-1}$
$284°-320°F - 3.78 \times 10^{-3}\ °F^{-1}$

Viscosity: See Figure V-5.1 for kinematic viscosity.

Specific Heat: FC-43 has a specific heat of 0.27 Btu/lb at 77°F.

Heat of Vaporization: 29.8 Btu/lb at b.p.

Thermal Conductivity: See Figure V-5.2.

Physical Stability: See Figure V-5.3 for the vapor pressure curve. The maximum value for pour point is -40°F (ASTM D-97).

Thermal Stability: FC-43 has excellent thermal stability during prolonged exposure to temperatures up to 617°F. No degradation has been observed after prolonged exposure up to 570°F in either copper or stainless steel containers. Liquid FC-43 has remained in contact with a hot wire at 1100°-1200°F for considerable lengths of time with no observed breakdown of the fluid or burnout of the wire.

Chemical Stability: FC-43 is not attacked by strong acids, strong alkalis, strong reducing agents, or strong oxidizing agents such as fuming nitric acid, liquid oxygen, ozone, hydrogen peroxide, or gaseous fluorine. At elevated temperatures, FC-43 is a poor solvent for most materials with the exception of other fluorinated liquids.

Surface Tension: FC-43 has a very low surface tension of 15.0–18.0 dynes/cm at 77°F, therefore it easily penetrates through openings.

Corrosion: FC-43 undergoes no electrical or chemical change in contact with various materials of construction such as metals, plastics, elastomers, even above maximum operating temperatures for dielectric coolants.

Flammability: FC-43 is nonflammable and nonexplosive.

Physiological Effects: Clinical studies revealed no harmful effects from limited exposure to the vapors or from extensive contact with the skin. In many years of handling FC-43 in laboratories and plants, no harmful effects have been observed. However, pyrolysis products formed at elevated temperatures may be harmful, and these products should be removed by an adequate ventilation system.

Appearance: Colorless and odorless.

Refractive Index: 1.291

Electrical Properties:

	Resistivity	ohm/cc
-85°F		8×10^{15}
75°F		3×10^{14}
202°F		2×10^{14}
302°F		8×10^{13}

Dielectric Constant:

t	1.0–40 KC	1 KMC	3 KMC	8.5 KMC
-85°F	2.00	–	–	–
75°F	1.90	1.92	1.92	1.90
302°F	1.71	–	–	–

Price Class: D

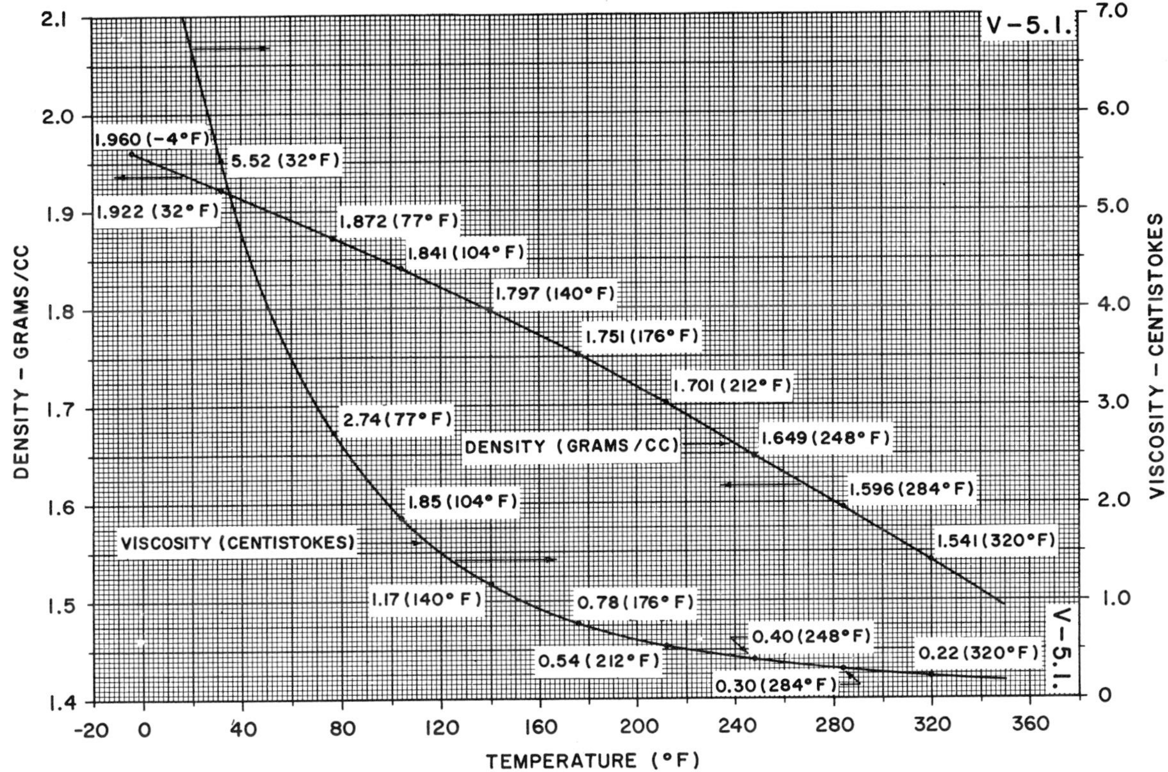

Density and Kinematic Viscosity vs Temperature

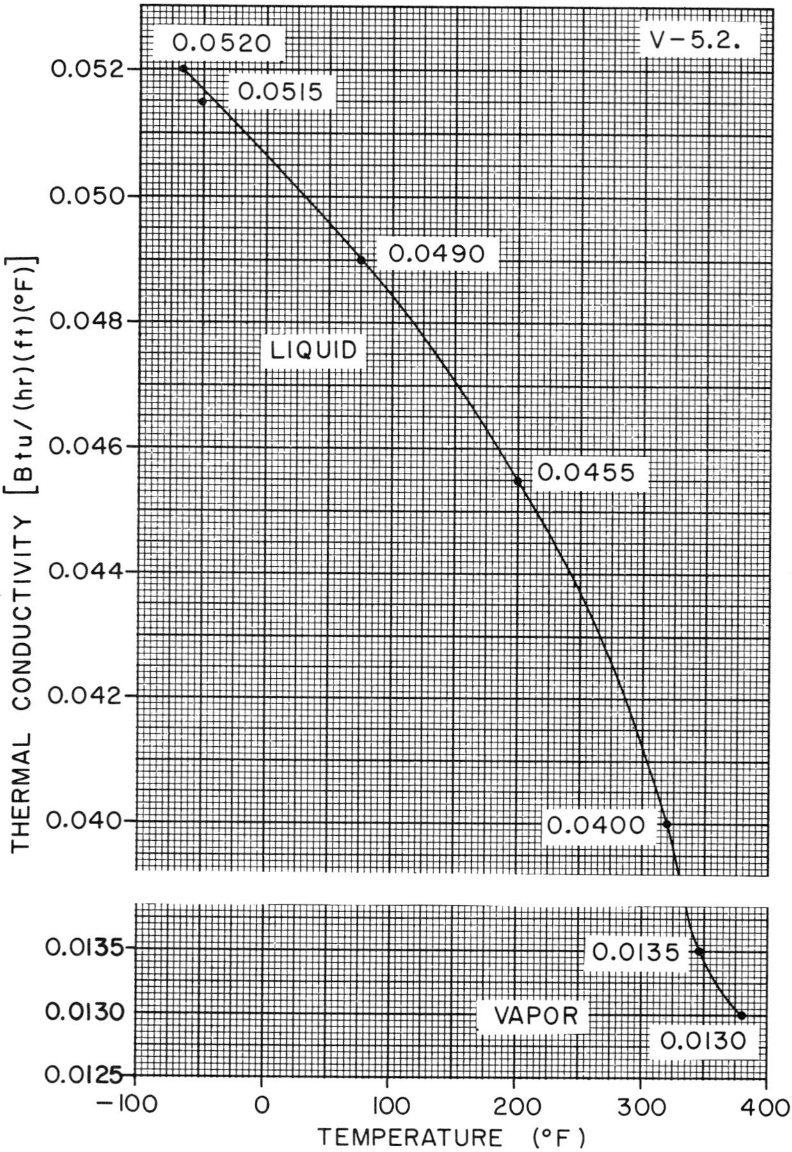

Thermal Conductivity of Liquid and Vapor (1 atm) vs Temperature

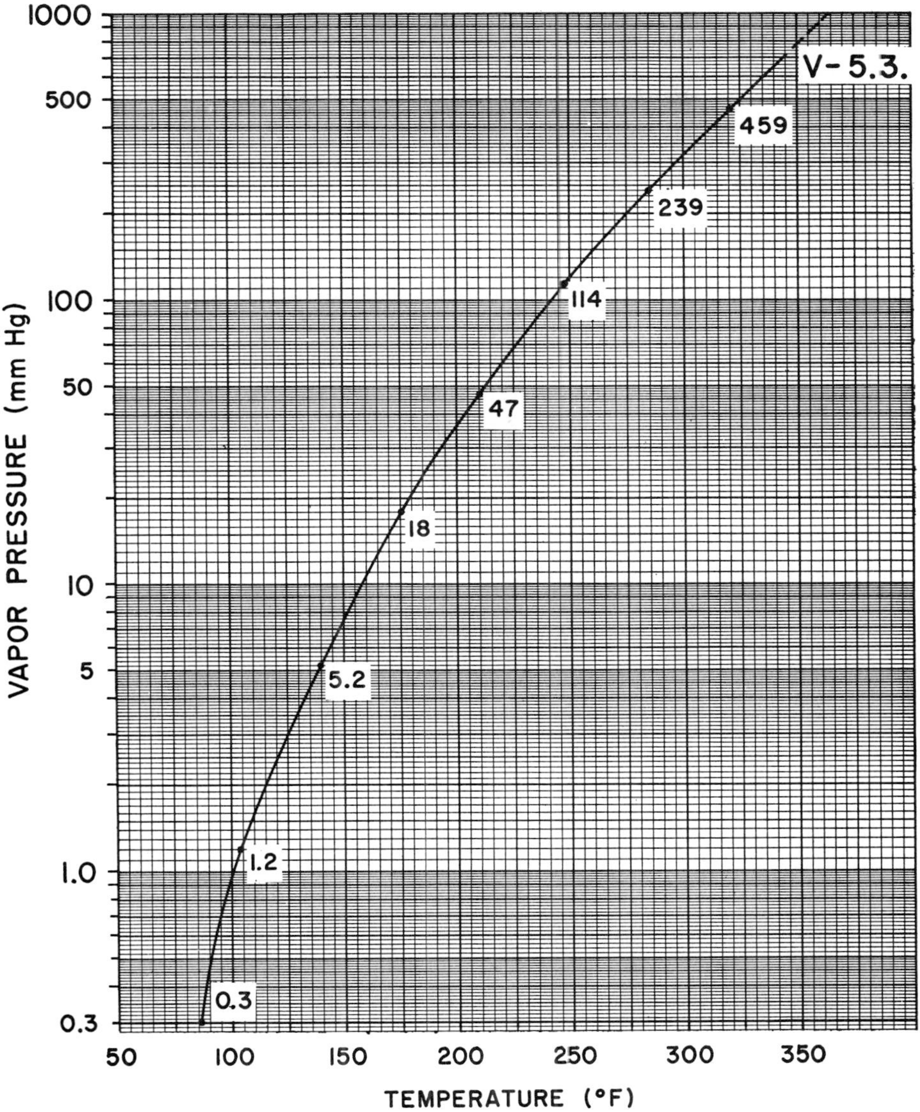

Vapor Pressure vs Temperature

V-6. FC-75

FC-75 is a fully fluorinated product composed primarily of a mixture of eight carbon materials. An isomeric mixture of the perfluoro cyclic ether, $c-C_8F_{16}O$, is present in a high percentage. This class of compounds has high thermal stability and electrical insulating properties far superior to hydrocarbon oils and silicon esters.

Properties of Importance

Density: See Figure V-6.1.

Coefficient of Volume Expansion:
66°-104°F – 2.9×10^{-3} per °F
104°-176°F – 3.6×10^{-3} per °F

Specific Heat: The liquid has a specific heat of 0.248 Btu/(lb)(°F) at 77°F. The vapor has a specific heat of 0.236 Btu/(lb)(°F) at 230°F.

Thermal Conductivity: See Figure V-6.2.

Viscosity: See Figure V-6.3.

Latent Heat: At boiling point, FC-75 has a heat of vaporization of 37.8 Btu/lb.

Physical Stability: See Figure V-6.4. The maximum value of the pour point is –80°F (ASTM D-97).

Surface Tension: –40°F – 23.0 dynes/cm
32°F – 18.1 dynes/cm
77°F – 15.1 dynes/cm
122°F – 12.0 dynes/cm

Thermal Stability: The upper limit of thermal stability to prolonged exposure is 750°F. No degradation has been found after prolonged exposure up to 570°F. Less than one percent degradation has been noted after 240 hours at 750°F in either copper or stainless steel containers. Exposure at 850° to 930°F for 60 hours in stainless steel containers results in 15 per cent decomposition.

Chemical Stability: FC-75 is chemically inert and thus is not attacked by strong acids, bases, amines, oxidizing agents, or reducing agents. However, at elevated temperatures, violent reactions with molten alkali metals, alkaline-earth metals, powdered aluminum or magnesium are possible.

Corrosion: FC-75 does not attack common materials of construction. Nearly all metals, plastics, and elastomers are not affected by liquid or vapor FC-75.

Flammability: FC-75 is nonflammable.

Physiological Effects: Identical to those of FC-43.

Radiation Stability: A report prepared by the American Chemical Society states that fluorocarbon liquids are as stable as, or more stable than, some aromatic hydrocarbons toward ionizing radiation. Variations of viscosity and density with gamma radiation, along with the formation of gaseous products on irradiation, were investigated for FC-75.

Appearance: Colorless and odorless.

Refractive Index: 77°F – 1.277

Electrical Properties:
Resistivity:
–85°F – 2×10^{16} ohm/cc
75°F – 6×10^{14} ohm/cc
212°F – 2×10^{15} ohm/cc

Dielectric constant:

	1.0 – 40 KC	1 KMC	3 KMC	8.5 KMC
–85°F	2.05	...	...	...
75°F	1.86	1.87	1.86	1.80
212°F	1.75	...	...	...

Price Class: D.

References

1. Minnesota Mining and Manufacturing Co., "Technical Data Inert Liquid FC-43," St. Paul, Minn., April 1, 1957.
2. Minnesota Mining and Manufacturing Co., "Technical Data, Inert Fluorochemical Liquid FC-75," 1959.
3. Yarrington, R. M., and Kay, W. B., *J. Chem. Eng.*, Data 5, 24-29) "Thermodynamic Properties of Perfluoro-2-butyl tetra-hydrofuran," (1940).
4. Spilhaus, A. F., and Troeltzsch, L. A., "Heat Transfer Properties of FC-75 and FC-43," Project 459-5059, Minnesota Mining and Manufacturing Co., October 1959.
5. LaZerte, J. D., Smith, J. E., and Olyphant, M., "Fluorocarbon FC-75—Properties and Applications."

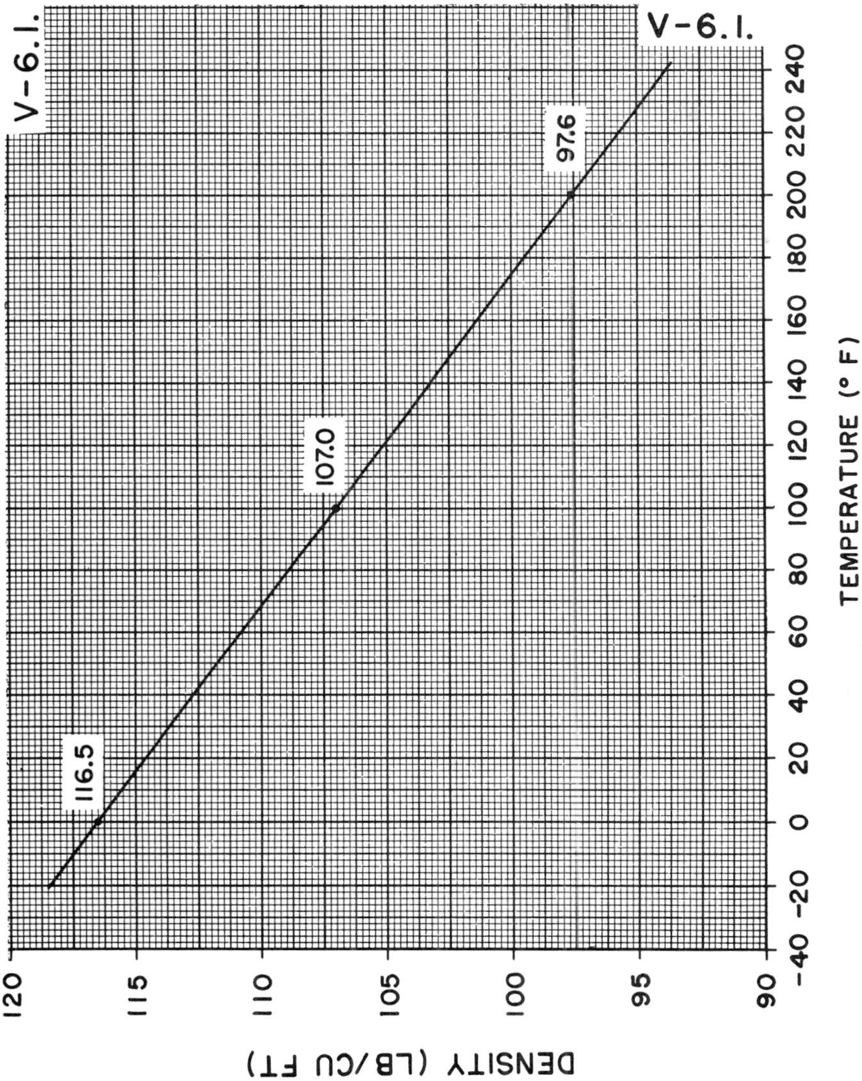

Density vs Temperature

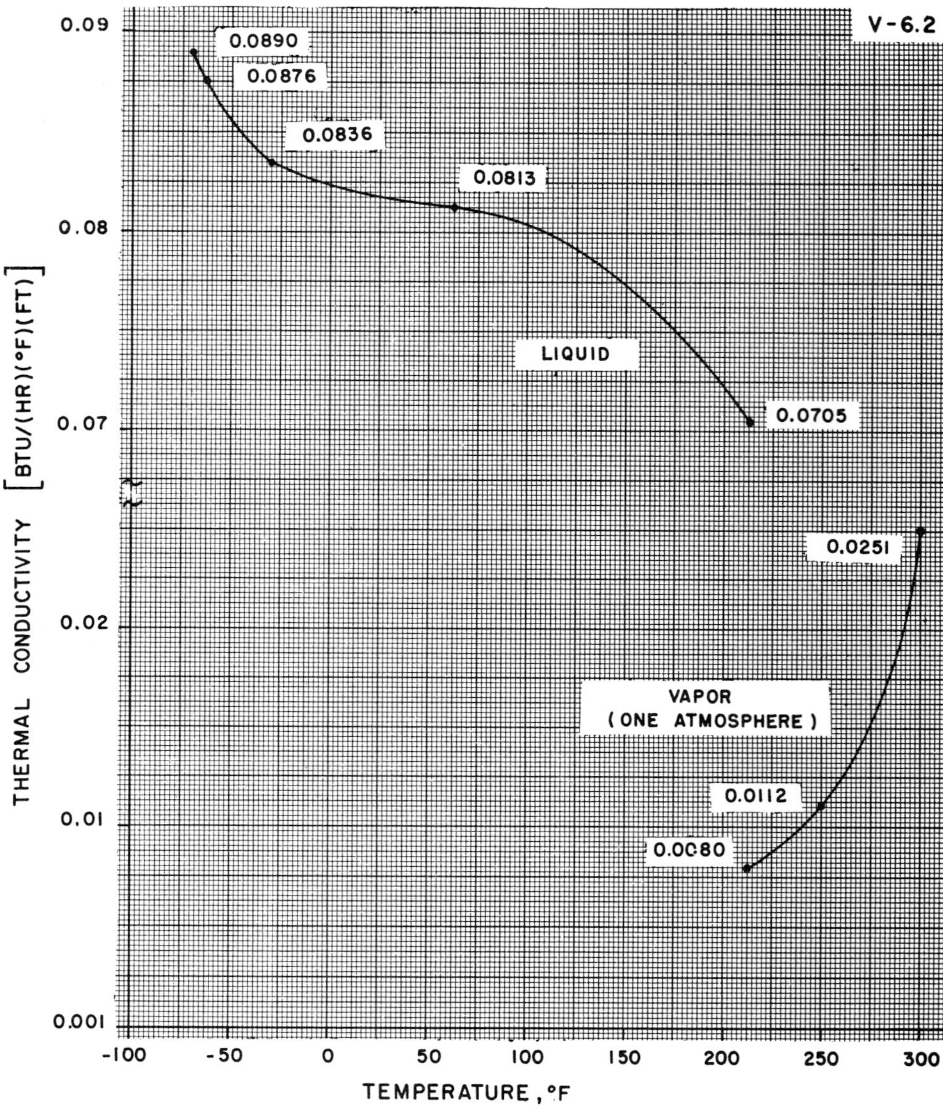

Thermal Conductivity vs Temperature

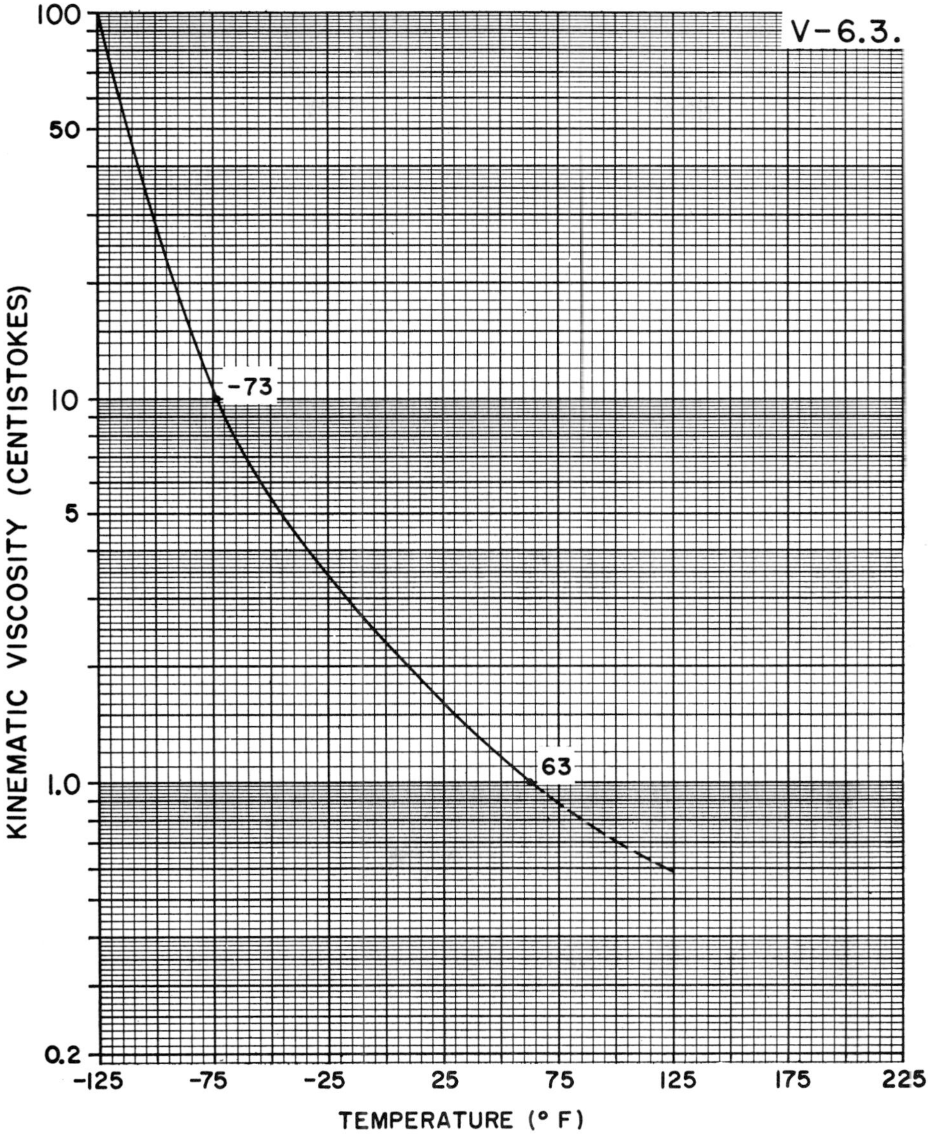

Kinematic Viscosity vs Temperature

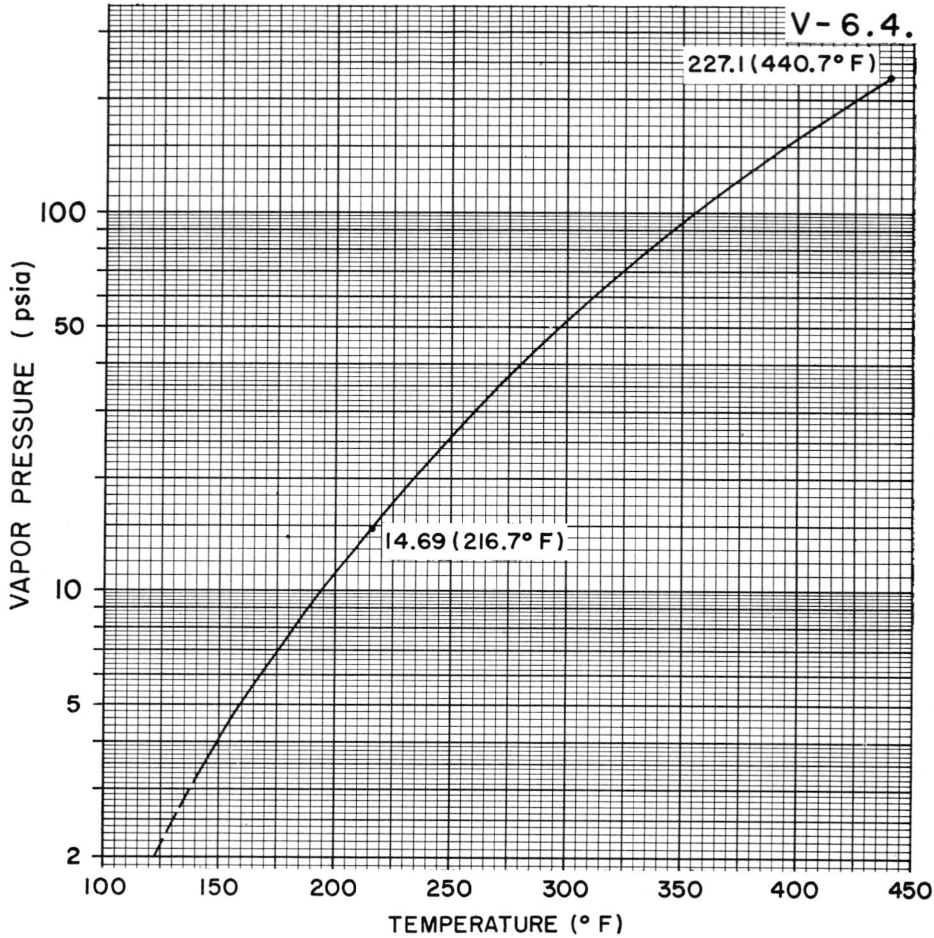

Vapor Pressure vs Temperature

V-7. PARA-CYMENE

Para-cymene, or *p*-methyl isopropyl benzene, can be used as a heat transfer medium in either the liquid or the vapor phase. The liquid is used primarily for cooling in closed systems at temperatures up to 300°F. The vapor phase is recommended for heating systems in the temperature range of 350°–550°F. It is one of the few materials available for this range.

Properties of Importance:

Density: Figure V-7.1 shows the liquid density. Figure V-7.4 shows the vapor density.

Specific Heat: See Figure V-7.1.

Enthalpy: See Figure V-7.2 for the heat of vaporization.

Thermal Conductivity: The thermal conductivity expressed in Btu/(hr)(ft)(°F) has been estimated to be as follows:

400°F-0.0808
300°F-0.08
200°F-0.0824
100°F-0.088
0°F-0.0956

Viscosity: See Figure V-7.3.

Physical Stability: See Figure V-7.4. Note that the vapor pressure at the maximum recommended operating temperature is only about 140 psia. The freezing point of *p*-cymene is -100.3°F.

Surface Tension: The surface tension is high enough to avoid the problem of leakage at the joints in most cases. At 77°F, it is 28 dynes/cm.

Thermal Stability: Para-cymene is very stable within its operating range. The liquid has even been applied to systems in which the temperature reaches 575°F without any great increase in the rate of decomposition.

Chemical Stability: The fluid is insoluble in water.

Corrosion: Para-cymene is noncorrosive.

Boiling Point: 350°F at 760 mm pressure.

Flammability:
Flash Point: (Tag Open Cup) 138°F.
Fire Point: 152°F.

Physiological Effects: The toxicity limit in air is 10 ppm.

Refractive Index: 1.490 at 68°F.

Appearance: Para-cymene is a colorless liquid.

Price Class: B.

References

1. Newport Industries Co., "Para-cymene as a Heat Transfer Medium," 1960.
2. *Chem. Eng.*, "Media for Heat Transport," 225 June 1953.

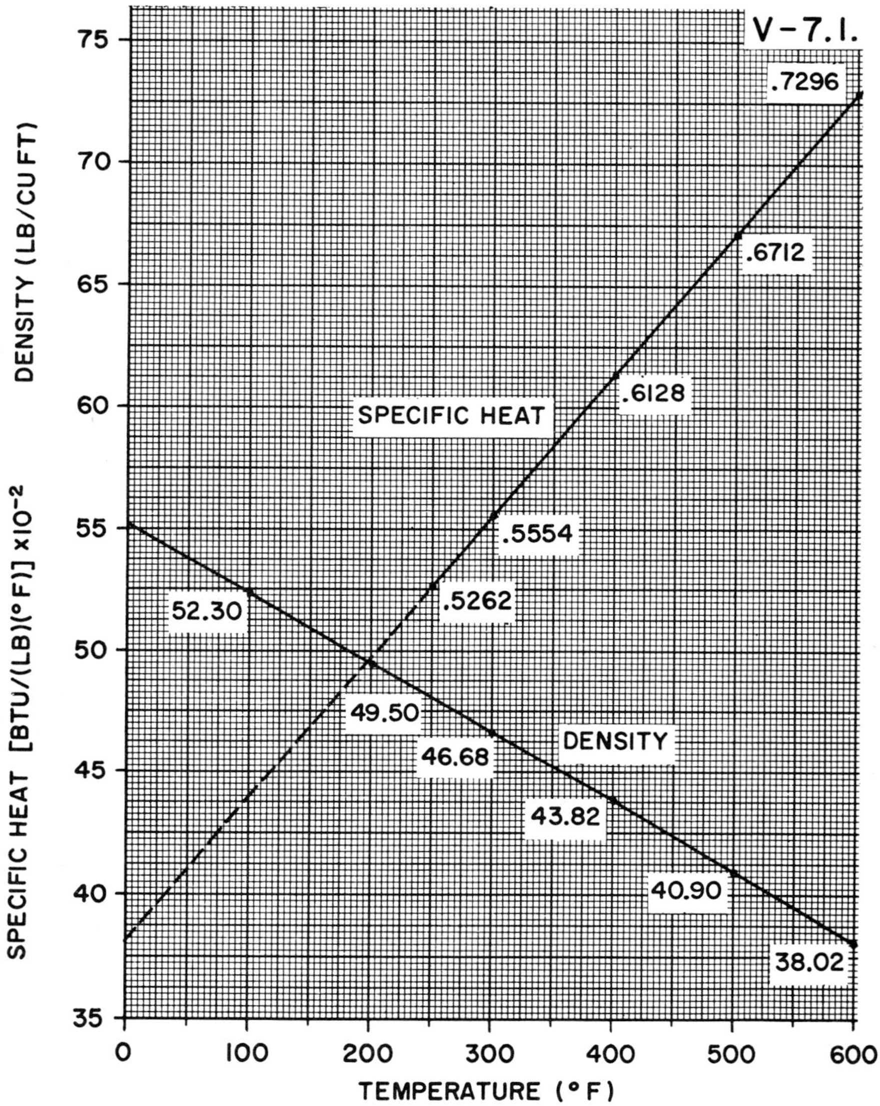

Specific Heat vs Temperature
Density vs Temperature

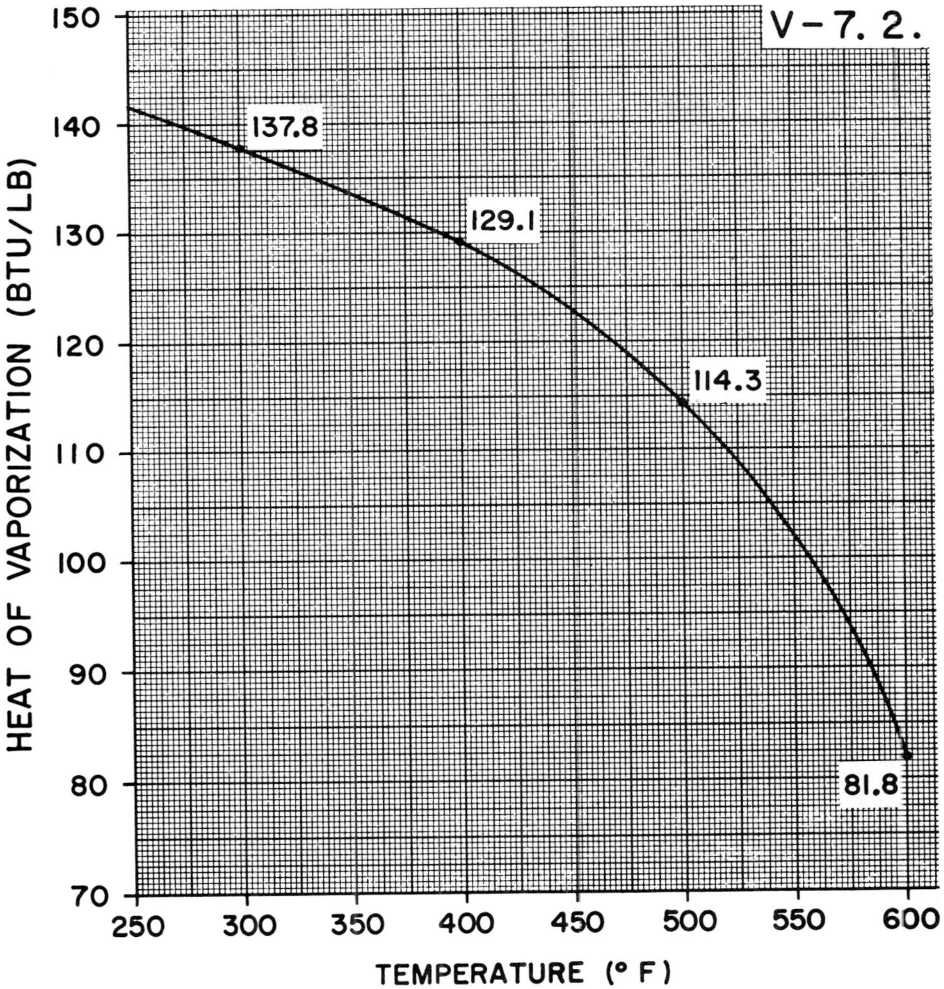

Heat of Vaporization vs Temperature

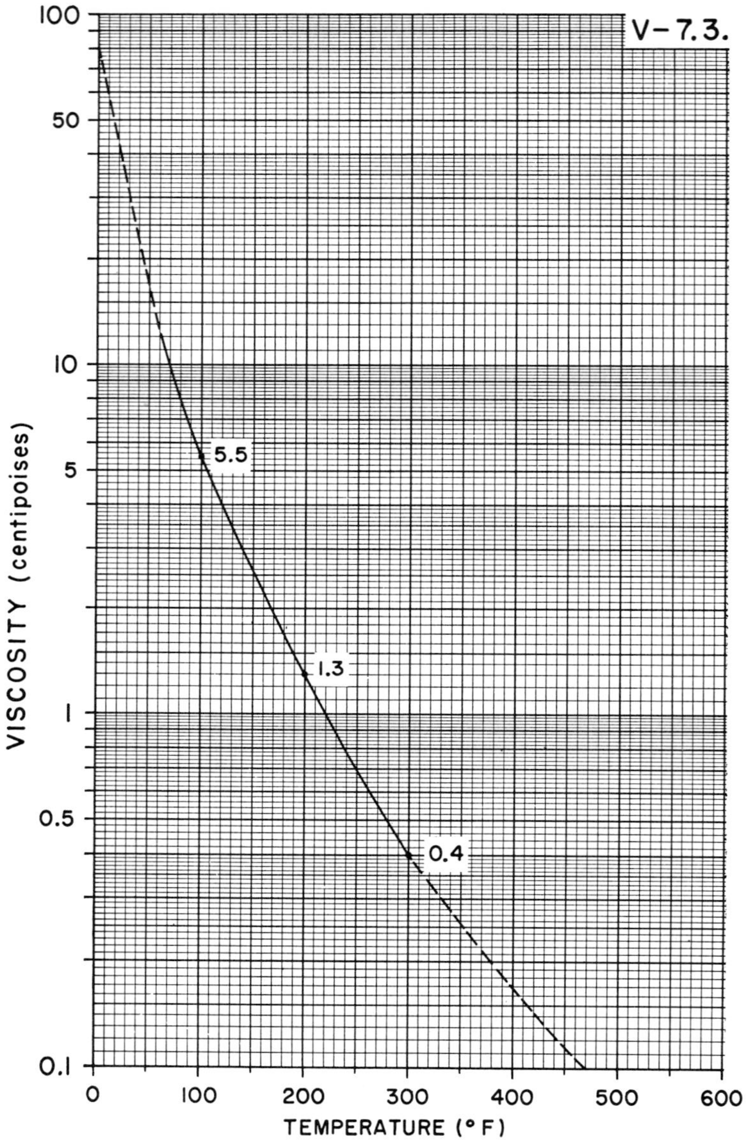

Viscosity vs Temperature

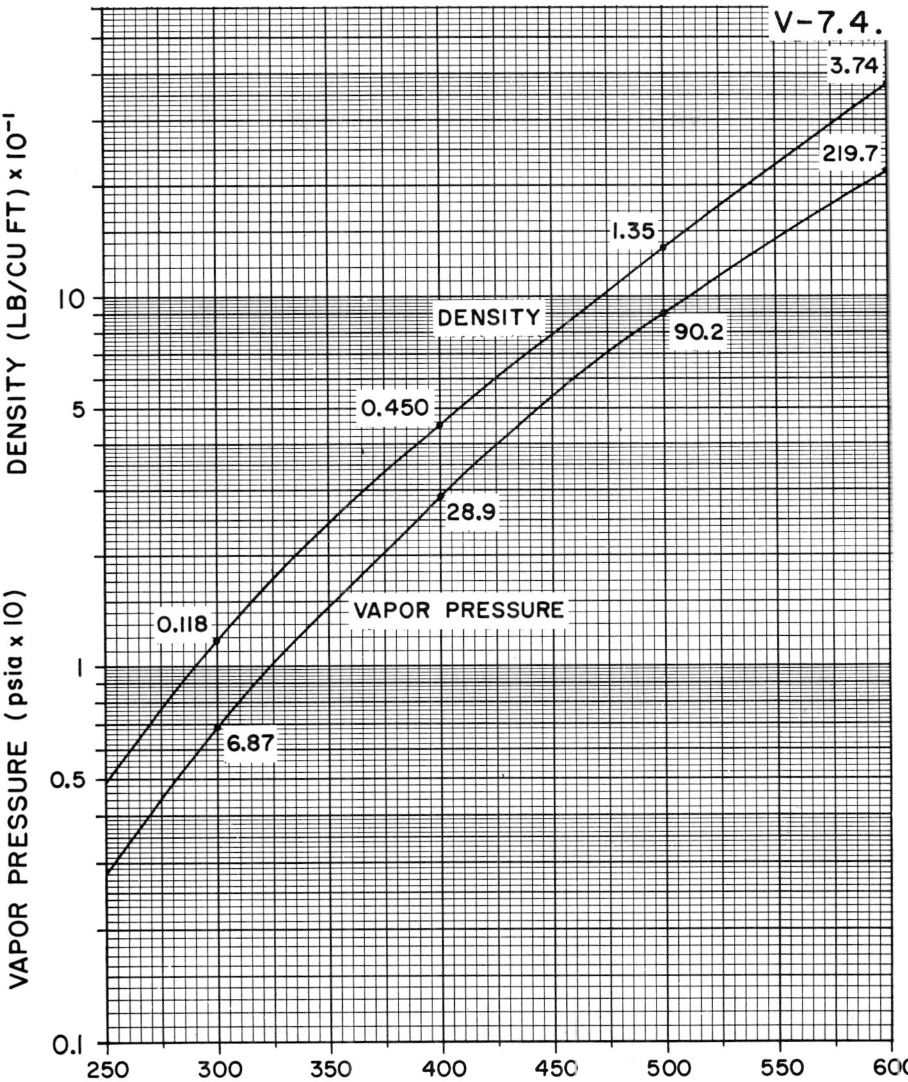

Vapor Pressure vs Temperature
Density vs Temperature

V-8. STEAM

Steam is generated in boilers for the dual purpose of acting as the operating fluid for thermodynamic power cycles and as a process heating medium. The power field is exclusively reserved for steam, thus assuring its ready availability at every plant. The physical properties of steam have undoubtedly been more thoroughly investigated than those of any other fluid.

As a superheated vapor, steam can serve as a heat transfer medium from 600°-1200°F. As a saturated vapor, steam is applied in either one of two ways: (1) direct mixing with the material; (2) indirect heat transfer across a heat exchange surface. For direct mixing, a dispersed cloud of steam droplets has tremendous surface area per unit volume and, coupled with a vast store of latent heat, provides excellent heat fluxes. For indirect heat transfer, steam releases large quantities of thermal energy per pound at a relatively constant temperature which can be regulated by pressure control. Extraordinarily high film coefficients up to 30000 Btu/(hr)(ft^2)(°F) are obtained for the "drop-wise" condensation of steam on clean surfaces. When surfaces are covered with condensing water, the coefficient drops rapidly below 1000 Btu/(hr)(ft^2)(°F).

Within the 200°-450°F range, high temperature water has proved to be superior to steam. Considerable literature is available on the merits of steam versus high temperature water.

Properties of Importance

The saturation region for steam is found at relatively elevated temperatures compared to other permanent gases. Properties such as specific heat, thermal conductivity, density, and viscosity increase rapidly when saturation is approached. Ideality vanishes, and intermolecular van der Waals' forces predominate. As the critical region is reached, accurate prediction of rapidly changing values becomes quite difficult.

Density: See Figure V-8.1.

Specific Heat: See Figure V-8.2.

Thermal Conductivity: See Figure V-8.3.

Viscosity: See Figure V-8.4.

Physical Stability: Above 450°F, the accelerated increase of vapor pressure with temperature caused by the logarithmic nature of the P-T characteristic, makes the use of saturated steam questionable.

Thermal Stability: Thermal dissociation of water vapor proceeds by two mechanisms: decomposition into uncharged molecules of hydrogen and oxygen, and formation of charged hydrogen and hydroxyl ions. At a film temperature of 2200°F the amount of vapor dissociating with H^+ plus OH^- is slightly greater than that into $H_2 + O_2$. Figure G.1 shows the equilibrium constants for both reactions, while Table V-1 presents the effect of temperature and pressure on degree of ionization and degree of dissociation.

Chemical Stability: Steam undergoes no major chemical reactions which influence its use as a heat carrier up to a film temperature of 2200°F.

Corrosion: Rigid alkalinity control, removal of dissolved oxygen, silica, and sludge are all important elements of water treatment. Steam purity can be set quite close to perfection. Ultimately only 1 ppm of foreign material which can attack structural materials, remains.

Flammability: None.

Physiological Effects: Nontoxic.

TABLE V-1

Dissociation					Ionization		
Temp. (°F)	Pressure (atm) 0.1	1.0	10	100	Temp. (°F)	Pressure (atm) 0.1	1.0
2730	0.00043	0.0002	0.00009	0.00004	3272	0.021	0.010
3630	0.0125	0.0058	0.0027	0.00125	3632	0.053	0.023
4530	0.0884	0.0421	0.0198	0.000927	3992	0.113	0.058
5430	0.284	0.144	0.0704	0.0333	4352	0.200	0.103
					4712	0.320	0.173
					4892	0.393	0.220

Appearance: Colorless, odorless, and tasteless.

Dielectric Constant: The dielectric constant of water vapor is 1.005-1.006 in the 350°-240°F range.

References

1. Geiringer, P. L., "High Temperature Water," New York, John Wiley and Sons, Inc., 1963.
2. Dorsey, E. N., "Properties of Ordinary Water Substance," p. 152, New York, Reinhold Publishing Corp., 1940.
3. Ledinegg, M., "Dampferzeugung Dampfkessel, Feuerungen Theorie," p. 141, Springer Verlag, Wien, 1952.
4. Marks, L. S., "Mechanical Engineers Handbook," p. 368, New York, McGraw Hill Book Co., 1941.
5. The Babcock & Wilcox Co., "Steam—Its Generation and Use," pp. 21-26, 21-30, 1955.
6. Keenan, J. H., and Keyes, F. G., "Thermodynamic Properties of Steam," New York, John Wiley and Sons, Inc., 1936.

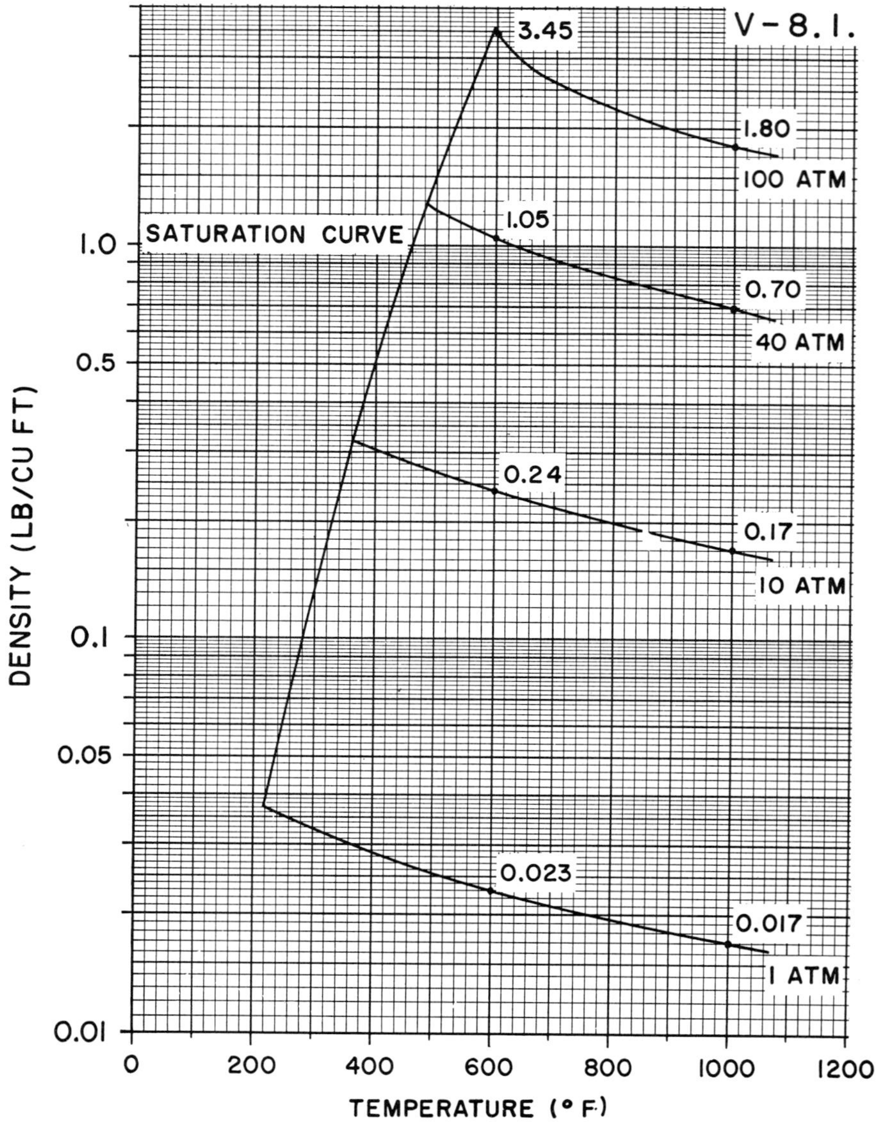

Density vs Temperature

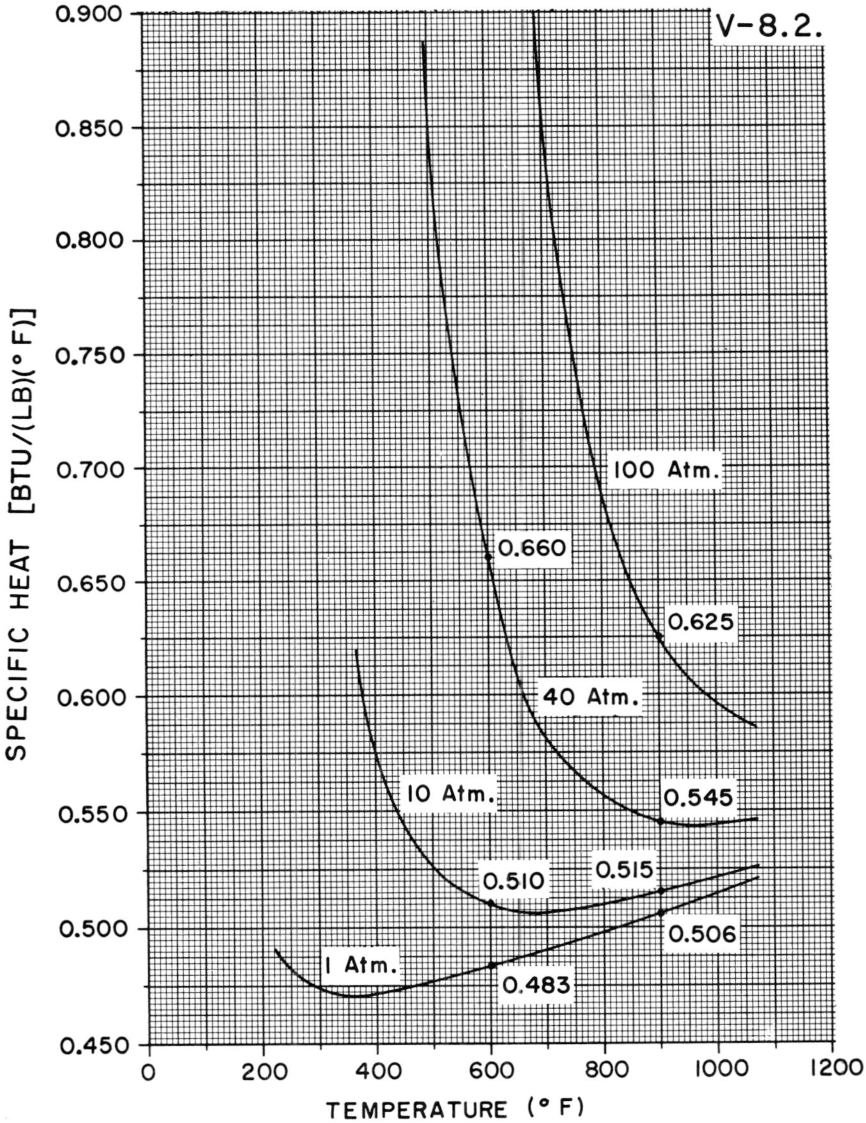

Specific Heat vs Temperature

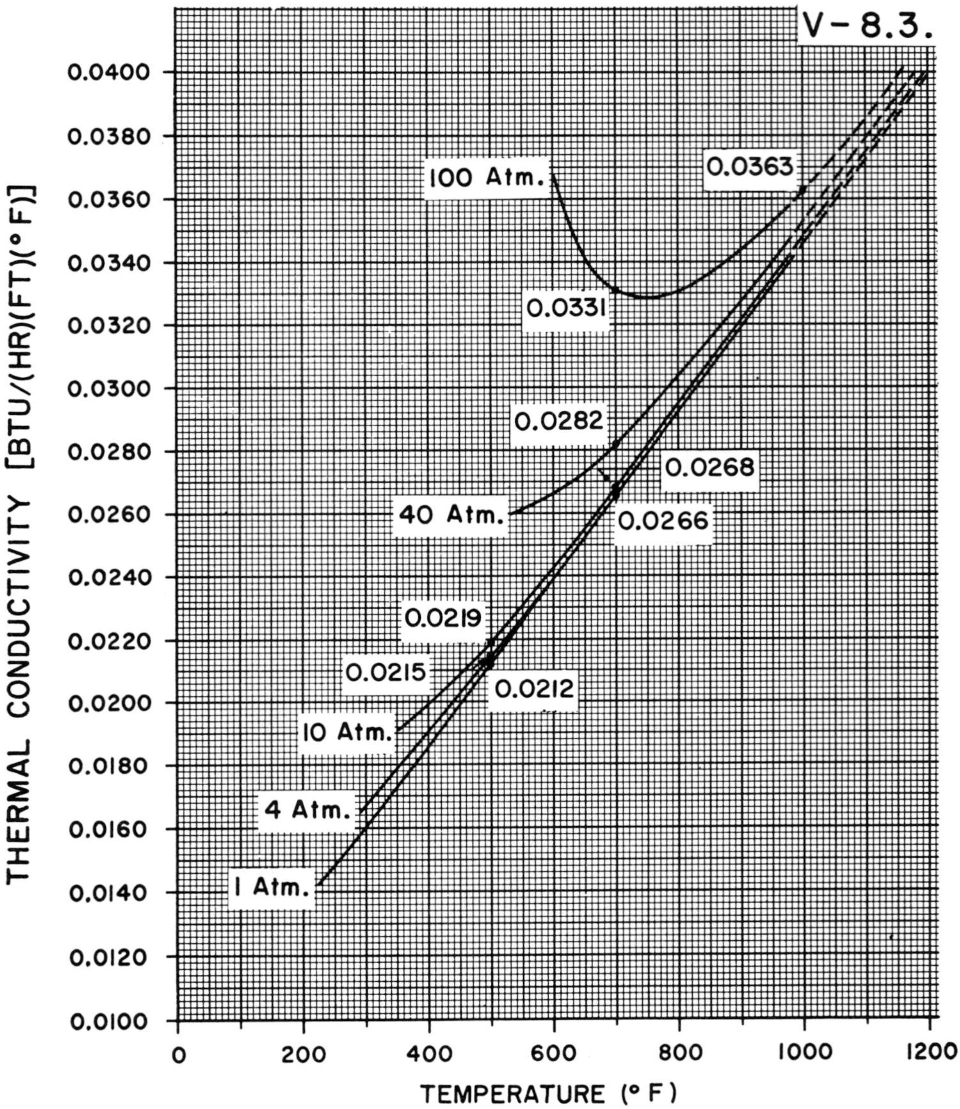

Thermal Conductivity vs Temperature

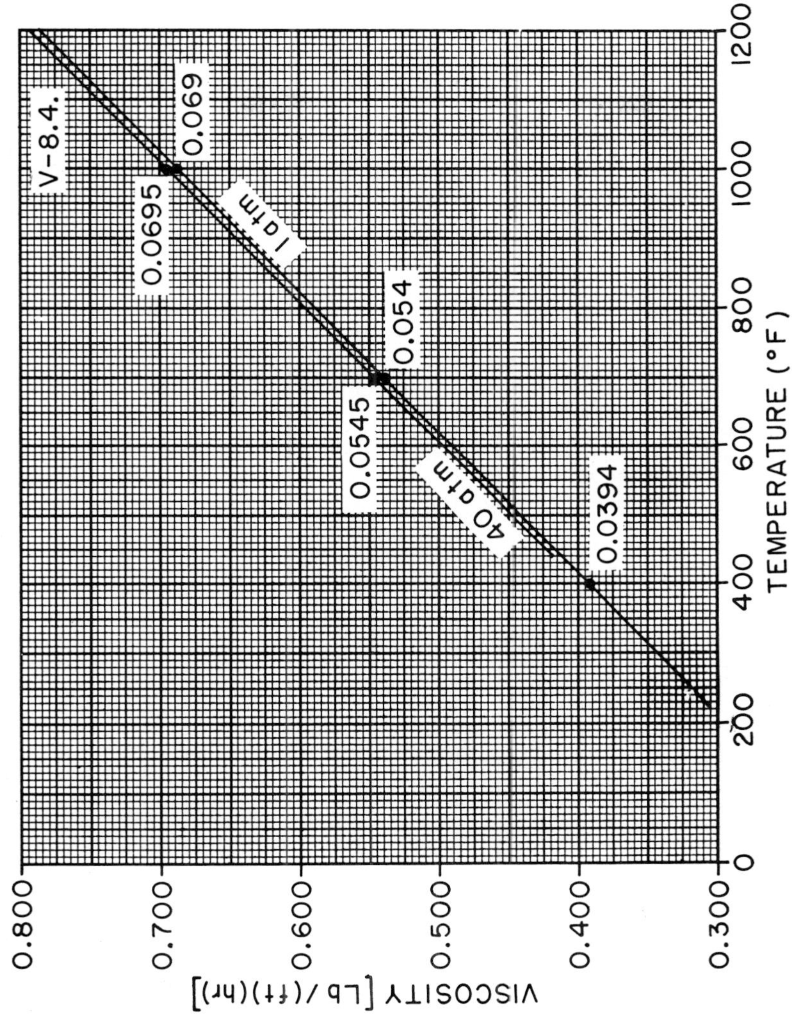

Viscosity vs Temperature

L. LIQUIDS

L-1. WATER

Water is an unequaled heat transfer medium. It is the most abundant liquid on earth and can be easily obtained in almost unlimited amounts. Surprisingly enough, as a consequence of its ready availability, water is often overlooked in favor of more exotic and chemically complex materials. As a heat transfer fluid, the exceptionally auspicious combination of the properties of water is unmatched by any substance known to man. In fact, water is often chosen as a reference fluid or standard, in terms of which other less favorable heat carriers are evaluated. The high temperature properties of water are even more noteworthy. At the present time, the practical temperature limit for the use of water is about 450°F. (In the following paragraphs, the properties of water are discussed in more detail.)

Properties of Importance

Density: The property of liquid density plays a dual role in heat transfer. Materials of high density with tight molecular packing transfer heat more readily than diffuse substances. Density also reflects the degree of expansion and contraction which a fluid undergoes as a result of a temperature change within the system. The expansion caused by a temperature increase is appreciable. At 450°F, a given weight of water occupies a volume 20 per cent larger than that at ambient conditions. Also, the rate of change of volume increases as the temperature increases. Figure L-1.1 shows the density-temperature relationship of saturated water. See also Figure L-1.7.

Specific Heat: Specific heat is one of the most propitious properties of water, as shown in Figure L-1.2. Water has a specific heat over 1 Btu/(lb)(°F) at high temperatures. Very few liquids even approach a value as high as 1.

Enthalpy: One of the most commendable properties of high temperature water is its extraordinary ability to collect and store large quantities of heat energy. The high specific heat of water, combined with its relatively high density, leads to the largest values of ρc_p, volumetric heat capacity, known in nature. Figure L-1.2 shows the enthalpy of water, and Figure L-1.3 shows the increase of the heat accumulation capacity of water with increasing temperature.

Thermal Conductivity: The thermal conductivity of water (Figure L-1.4), is higher than that of most liquids and much higher than that of steam. However, it is low enough so that there is only a small heat flow between adjacent strata of water. Cold water at the bottom of a container is heated only very slowly by the warmer layers above. This property can be utilized in the design of vertical heat accumulators. As shown in Figure L-1.4, the conductivity rises to a maximum of 0.396 Btu/(hr)(ft)(°F) at 260°F. The conductivity decreases slightly between 260° and 450°F.

Viscosity: Water has a moderate absolute viscosity (μ), in comparison to most organic liquids. The viscosity decreases rapidly from 0.84 lb/(ft)(hr) at 200°F to 0.33 lb/(ft)(hr) at 450°F, a 61 per cent decrease. This fact is quite significant for both heat transfer and fluid flow. The frictional head loss sustained by a heat carrier flowing through a long straight pipe is approximately proportional to $\mu^{0.18}$–$\mu^{0.25}$ Obviously, a viscosity decrease at high temperatures will produce a marked drop in fluid resistance. Thus, as the bulk temperature of water increases, viscosity diminishes and optimum conditions are closely approached.

The kinematic viscosity (ν) is the quotient of absolute viscosity divided by density. As stated in Section I, the kinematic viscosity is important in the computation of fluid friction losses and heat transfer coefficients, as evidenced by its presence in the Reynolds number, $\frac{DV}{\nu}$. Kinematic viscosity does not decrease with a rise in temperature, as sharply as the absolute viscosity, because of the corresponding drop in density. One can see that decreasing the kinematic viscosity will increase the Reynolds number and therefore lower the friction factor. The use of water at higher temperatures will then result in lowering of frictional resistance for a given pipe diameter, and flow velocity because of increased Reynolds number. See Figures L-1.4, and L-1.5.

Film Coefficient: Since water is such an important heat transfer material, a discussion of the film coefficient of heat transfer might be valuable. One common equation for the film coefficient in turbulent forced convection flow is:

$$h = (a)\left(\frac{k}{D}\right)\left(\frac{DVe}{\mu}\right)^{0.8}\left(\frac{c_p\,\mu}{k}\right)^{0.4} \qquad (3\text{-}2)$$

However, for our purpose, it is more convenient to express the equation in the following form:

$$h = (a)\left(\frac{V^{0.8}}{D^{0.2}}\right)\left(\frac{c_p}{\mu}\right)^{0.4}(k)^{0.6}(\rho)^{0.8} \qquad (3\text{-}3)$$

where h is expressed in Btu/(hr)(ft^2)(°F). The influence of each temperature-dependent property in the equation is shown in Table L-1. The table clearly shows that it is more advantageous to use high temperature water for maximum heat transfer. If the velocity is doubled from 5 ft/sec to 10 ft/sec, maintaining all other conditions constant, the coefficient is increased by 74 per cent. Decreasing the pipe diameter from 4 inches to 2 inches, again holding all other conditions constant, increases the coefficient by 14.8 per cent. Controlling the film coefficient by adjusting the temperature, velocity, and pipe diameter properly will result in higher process efficiency.

Physical Stability: It has been shown that the high temperature properties of water are those which are most desirable. In order to maintain the liquid state at high temperatures, high pressures must be employed. High pressures require the use of costly equipment and thick-walled vessels. Therefore, a temperature of about 450°F is very close to the practical limit of economic applications at the present time, except for industrial processes of special nature. The slope of the vapor pressure curve of water increases rapidly above 300°F, as shown in Figure L-1.6. There is more than a tenfold increase in vapor pressure from 300°-500°F. The freezing point and pour point of water are both at 32°F.

Thermal Stability: Water does not undergo any thermal degradation until it reaches the dissociation point, which is discussed in the description of steam. At this point, the water splits up into atomic hydrogen and oxygen.

TABLE L-1. INCREASE OF HEAT TRANSFER COEFFICIENT h

Property	From 100° to 300°F: Increase in Property		% Increase in h	From 100° to 400°F: Increase in Property		% Increase in h
Thermal Conductivity	k	0.031		k	0.020	
	$k^{0.6}$	0.026	+4.77	$k^{0.6}$	0.018	+3.30
Specific Heat	c_p	0.033		c_p	0.070	
	$c_p^{0.4}$	0.013	+1.30	$c_p^{0.4}$	0.033	+3.30
Density	ρ	−4.68		ρ	−8.34	
	$\rho^{0.8}$	−1.78	−6.5	$\rho^{0.8}$	−3.0	−11.0
Viscosity	μ^{-1}	+1.625		μ^{-1}	+2.455	
	$\mu^{-0.4}$	+0.397	+65.5	$\mu^{-0.4}$	+0.617	+102
TOTAL:	(1.0477)(1.013)(0.935)(1.655) =	1.64	+64%	(1.033)(1.033)(0.89)(2.02) =	1.90	+90%

Chemical Stability: Water has excellent chemical stability.

Surface Tension:

Temperature (°F)	Surface Tension (dynes/cm)	Temperature (°F)	Surface Tension (lb/ft)
215	58	32	0.0520
300	49	68	0.0500
394	37	104	0.0455
436	32	176	0.0430
		212	0.0404

Corrosion: The liquid must be conditioned to remove corrosion-causing impurities. Brass and bronze are usually not used in high temperature water systems.

Flammability: Water is completely nonflammable.

Physiological Effects: There are no toxicity problems encountered in the use of water.

Electrical Properties: The dielectric constant of water is 80 at 10^8 cycles/sec.

Refractive Index: The refractive index of water is 1.334 at 32°F and 1.187 at 212°F.

Appearance: Water is colorless and odorless.

References

1. Brown and Marco, "Introduction to Heat Transfer," New York, McGraw-Hill Book Co., 1942.
2. Dorsey, E. N., "Properties of Ordinary Water Substance," p. 152, New York, Reinhold Publishing Corp., 1940.
3. Geiringer, P. L., "High Temperature Water," New York, John Wiley & Sons, Inc., 1963.
4. Keenan, J. H., and Keyes, F. G., "Thermodynamic Properties of Steam," New York, John Wiley and Sons, Inc., 1936.
5. Lottes, P. A., "The Reactor Handbook" 2nd Edition.

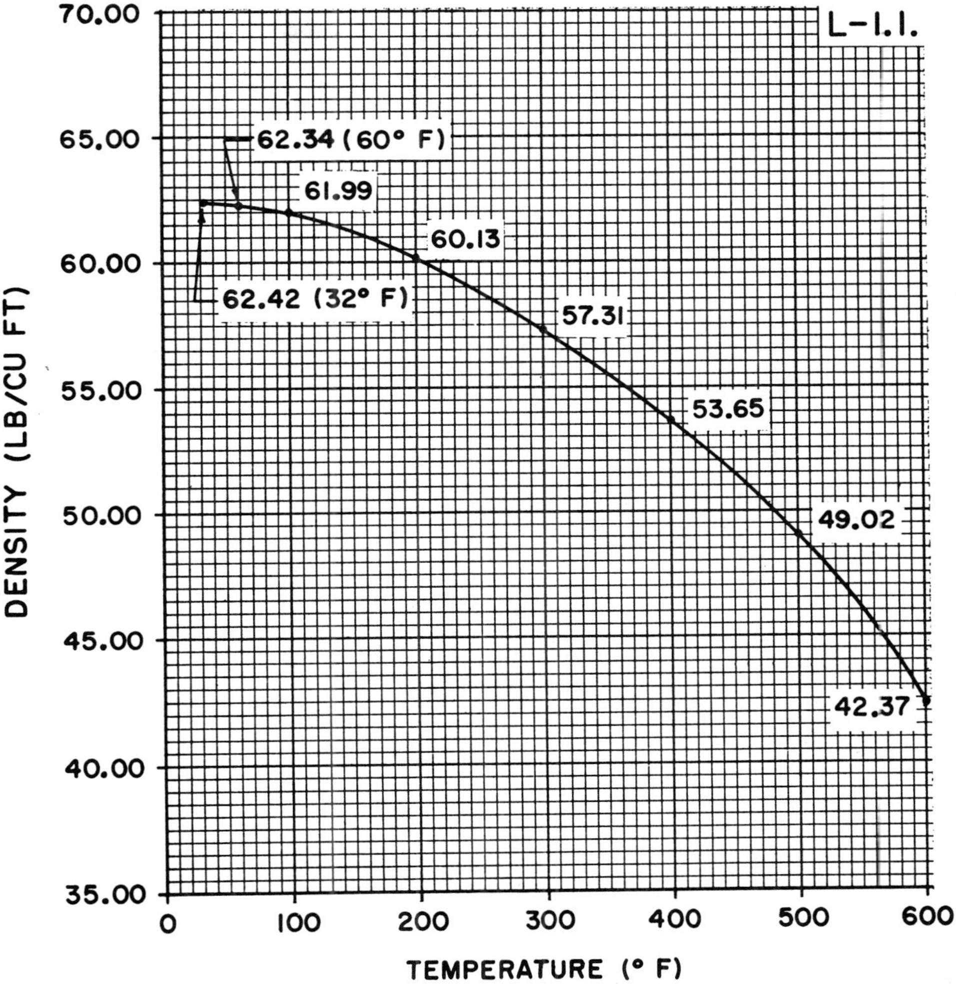

Density vs Temperature

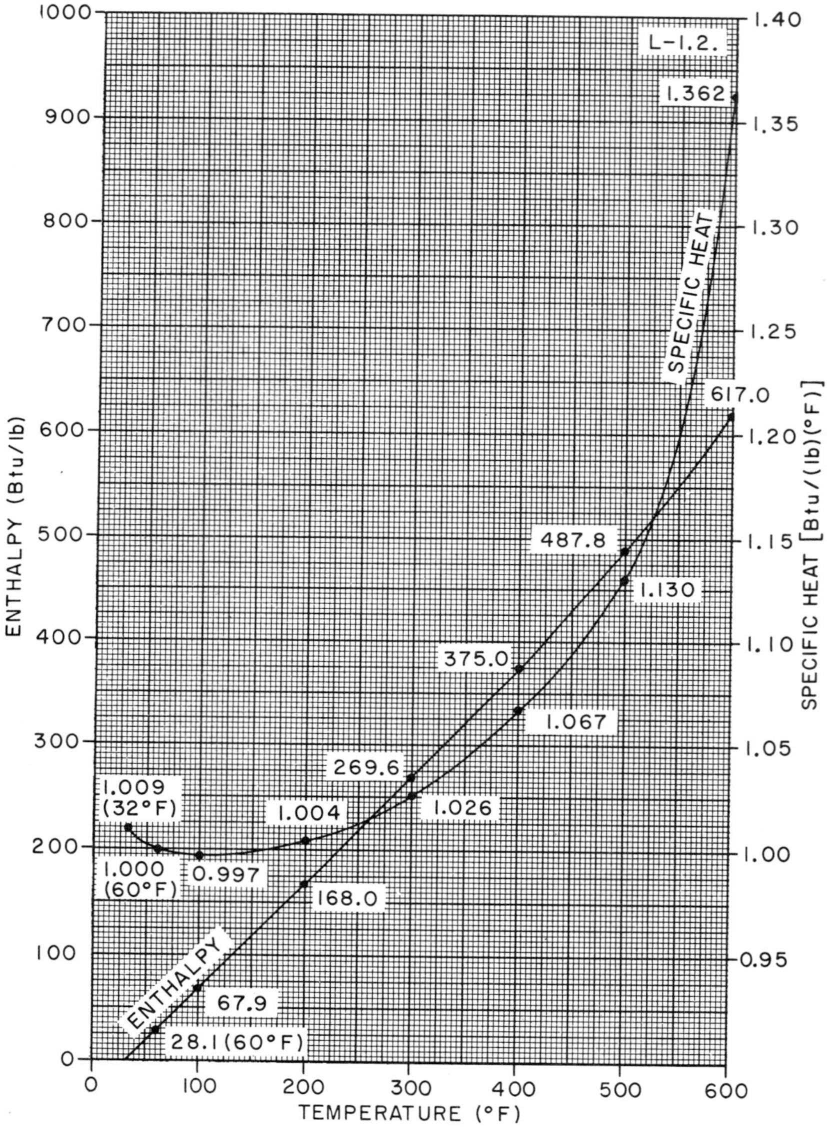

Enthalpy vs Temperature
Specific Heat vs Temperature

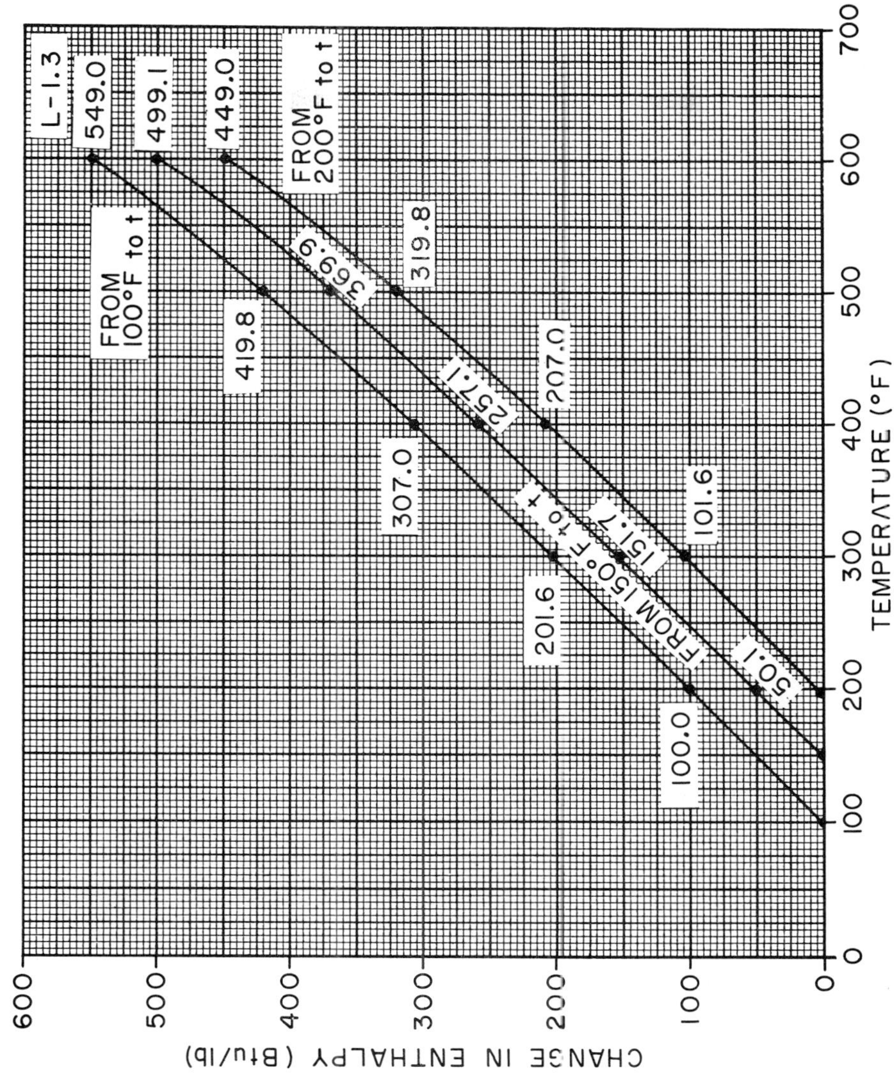

Change in Enthalpy vs Temperature

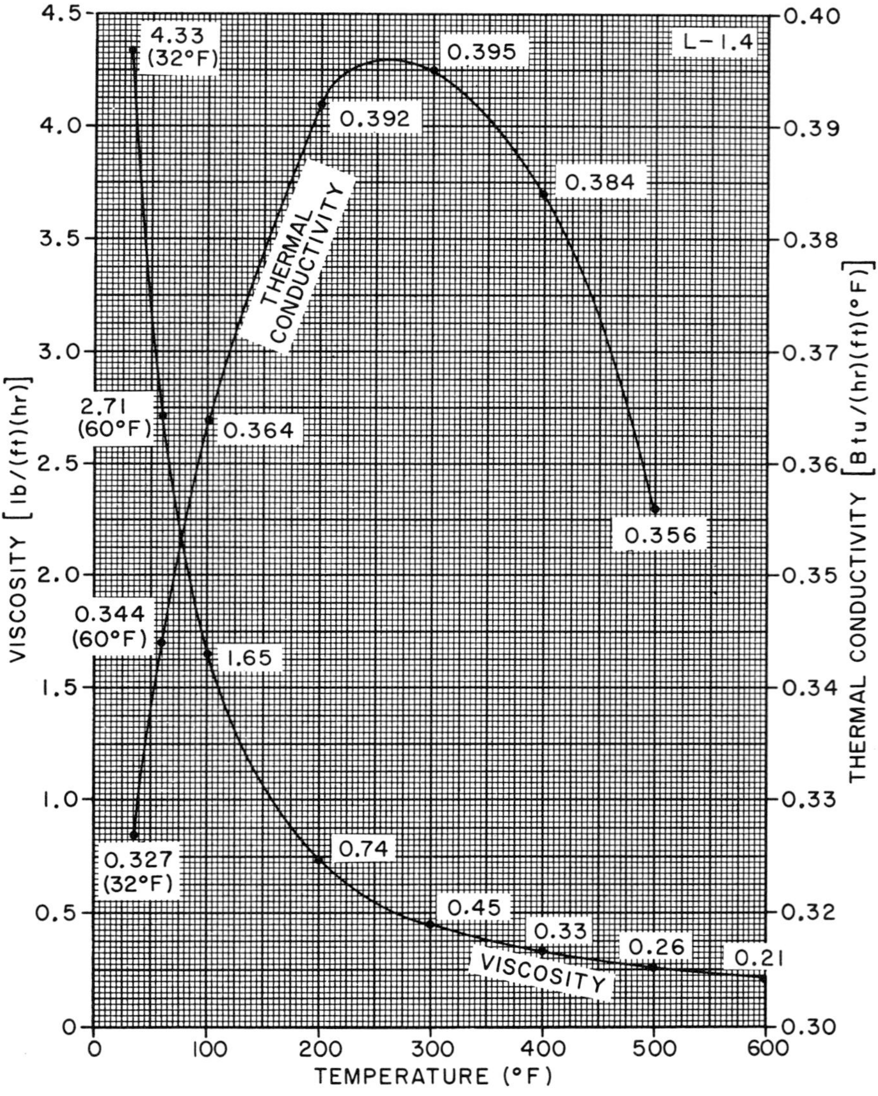

Viscosity vs Temperature
Thermal Conductivity vs Temperature

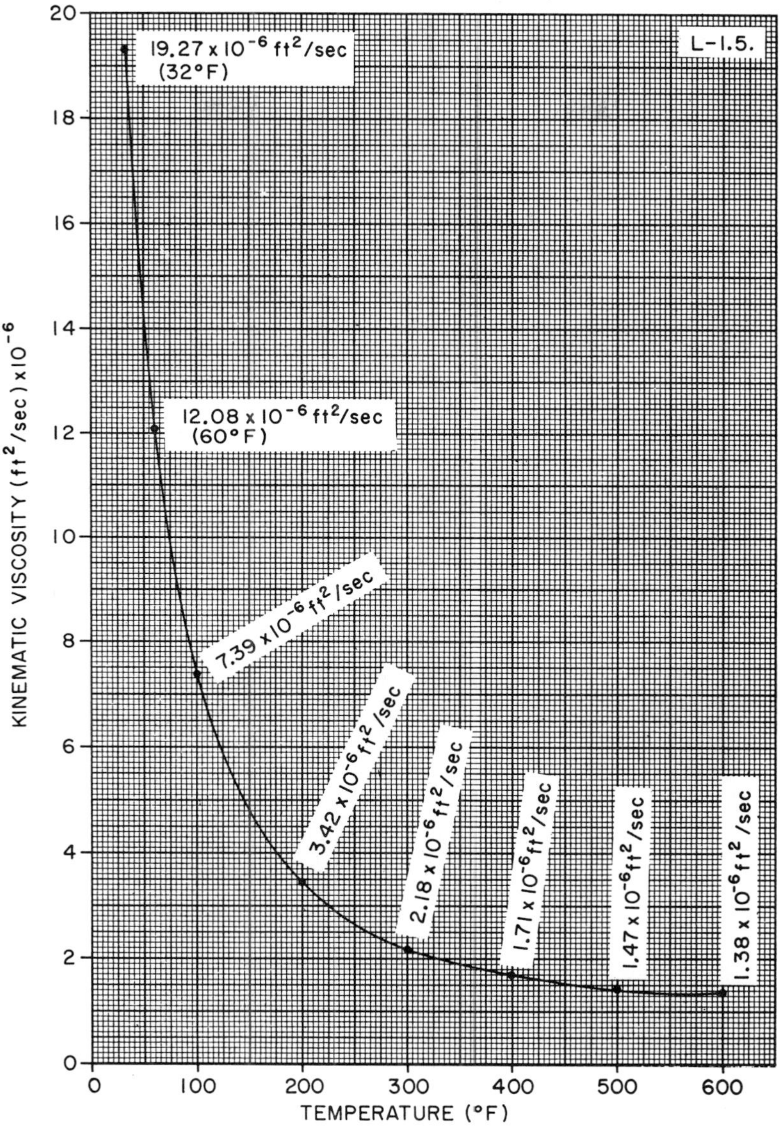

Kinematic Viscosity vs Temperature

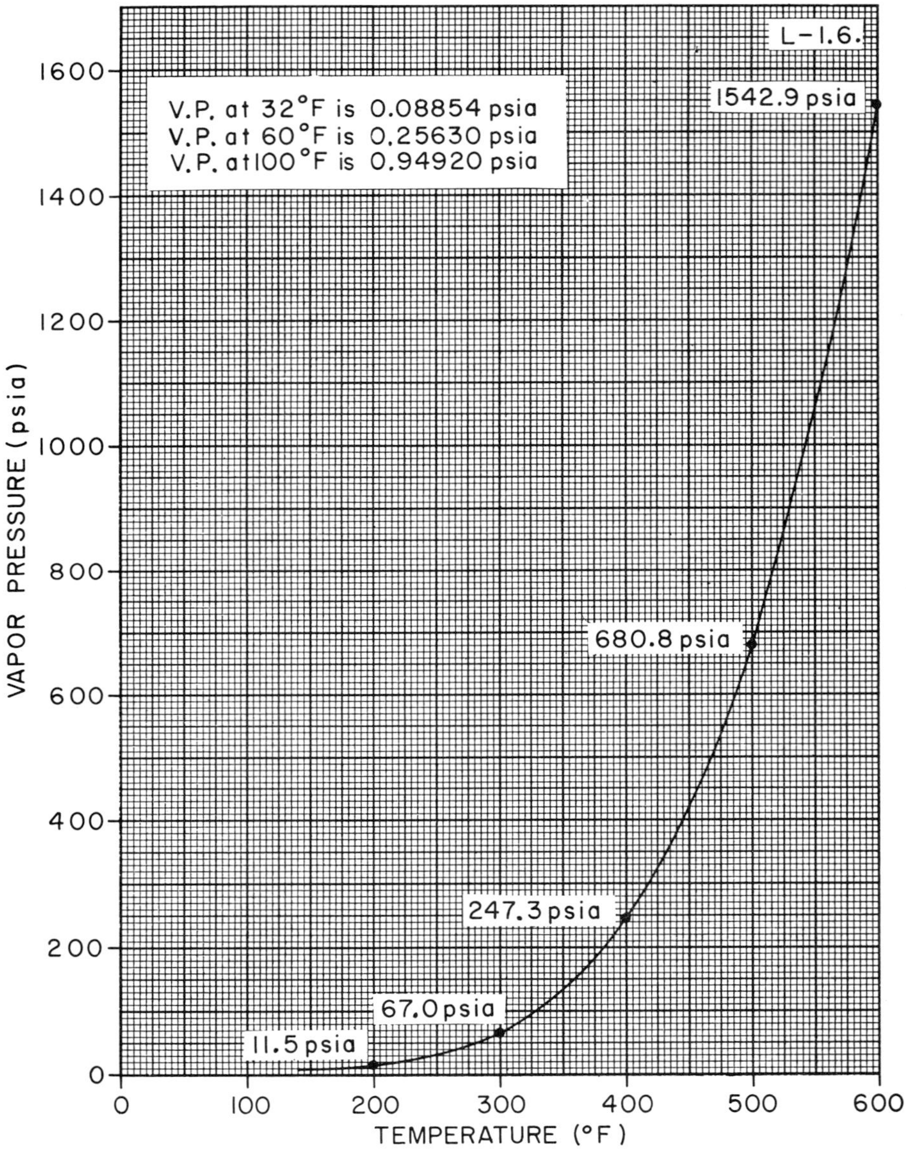

Vapor Pressure vs Temperature

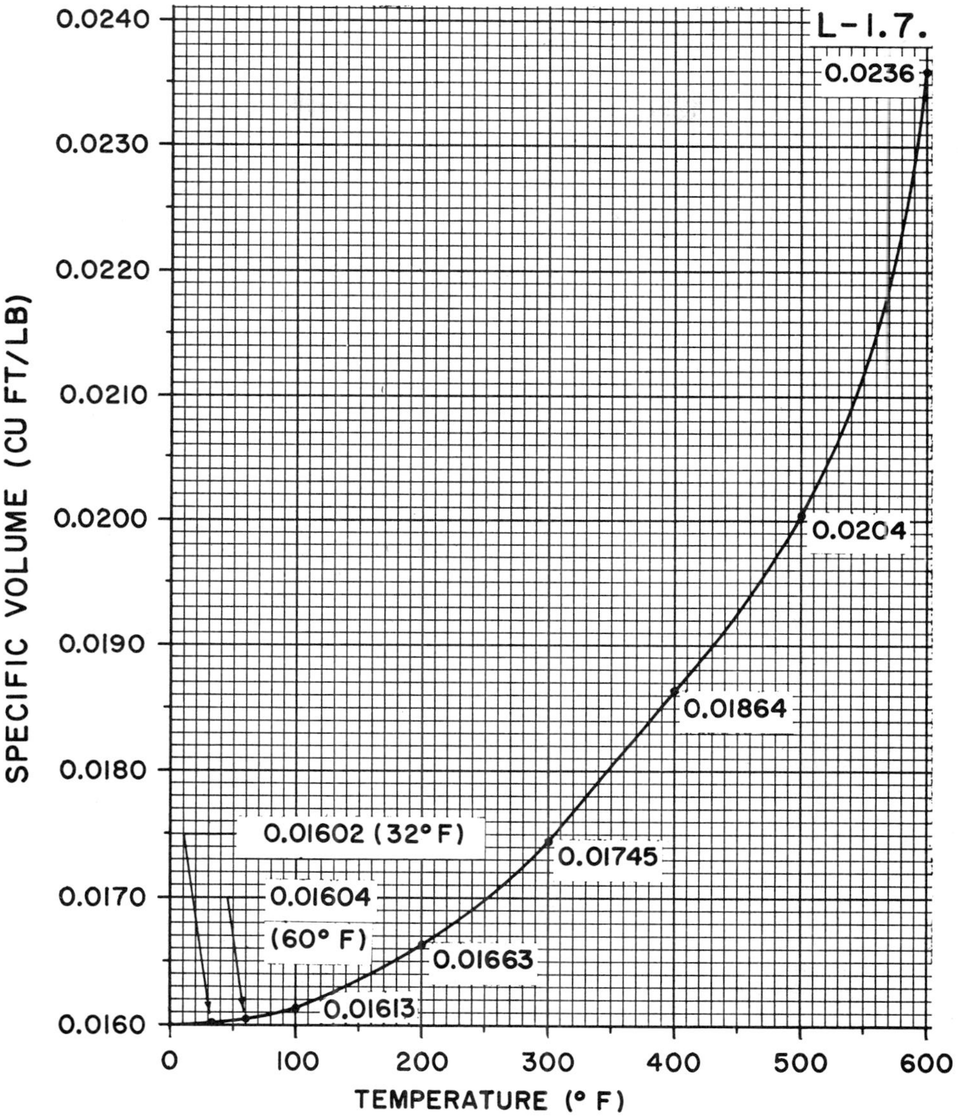

Specific Volume vs Temperature

L-2. HEAVY WATER (D_2O)

A thousand gallons of natural water, weighing approximately four tons, contain about one pound of heavy water. Despite this sparse concentration, heavy water is extracted from natural water in sizable quantities to respond to the growing demands of nuclear technology. Several methods are used for this purpose. The U. S. Atomic Energy Commission Savannah River plant uses a combination of three processes: (1) dual temperature exchange between H_2S and H_2O; (2) vacuum distillation, and (3) electrolysis.

Properties of Importance:

Density:

t(°F)	ρ (lb/ft³)
100	68.7
200	66.7
300	63.5
400	59.3
500	54.0

Specific Heat: See Figure L-2.1.

Viscosity:

t(°F)	μ(lb/(ft)(hr))
100	1.99
200	0.849
300	0.508
400	0.358
500	0.289

Thermal Conductivity: The thermal conductivity of heavy water, like that of water, increases with rising temperature to a shallow maximum at about 230°F, and then decreases at still higher temperatures. The thermal conductivity of heavy water is slightly inferior to that of water. $\frac{k_{H_2O}}{k_{D_2O}}$ increases in a linear fashion from 1.045 at 167°F to reach a maximum of 1.12 near 482°F, and then drops steeply with further temperature rise. At the critical temperature the ratio attains a value of unity.

Physical Stability: Figure L-2.2 gives a comparison of the vapor pressure of heavy water with that of water. Heavy water has a critical temperature of 700°F, a freezing point of 39.2°F, and a boiling point of 214.5°F.

Chemical Stability: Definite chemical differences exist between the two kinds of water. Acid solutions of heavy water have greater acidity than corresponding water solutions. Also, numerous salts and some gases such as carbon dioxide are less soluble in heavy water.

Thermal Stability: Chemical bonds between deuterium and other atoms have slightly greater stability than similar bonds with hydrogen. In fact, kinetic considerations indicate that a carbon-hydrogen bond should react at seven times the speed of a carbon-deuterium bond.

Handling Properties: Because of the fairly close resemblance between water and deuterium oxide, all handling problems are practically identical.

Electrical Properties:

Dielectric Constant: The following equation has been proposed for the range of the freezing point up to 212°F:

$$\varepsilon = 87.48 - 0.40509t + 9.638 \times 10^{-4}t^2 - 1.333 \times 10^{-6}t^3$$

Price Class: D.

References

1. *Nuclear Science Abstract*, **12**, No. 2, p. 15615.
2. *Nuclear Science Abstract*, **14**, No. 13, p. 10374.
3. *Nuclear Science Abstract*, **13**, No. 18A, p. 16291.
4. Katz, J. J., "Sci. American" **203**, No. 1, 106.
5. *Nucleonics* **16**, No. 9, 109.
6. *Inst. Mech. Engrs.* (*London*), "Proceedings of Joint Conference on Thermodynamics and Transport Properties of Fluids," 1042, (July, 1957).
7. Lottes, P. A., "The Reactor Handbook," 2nd Edition.

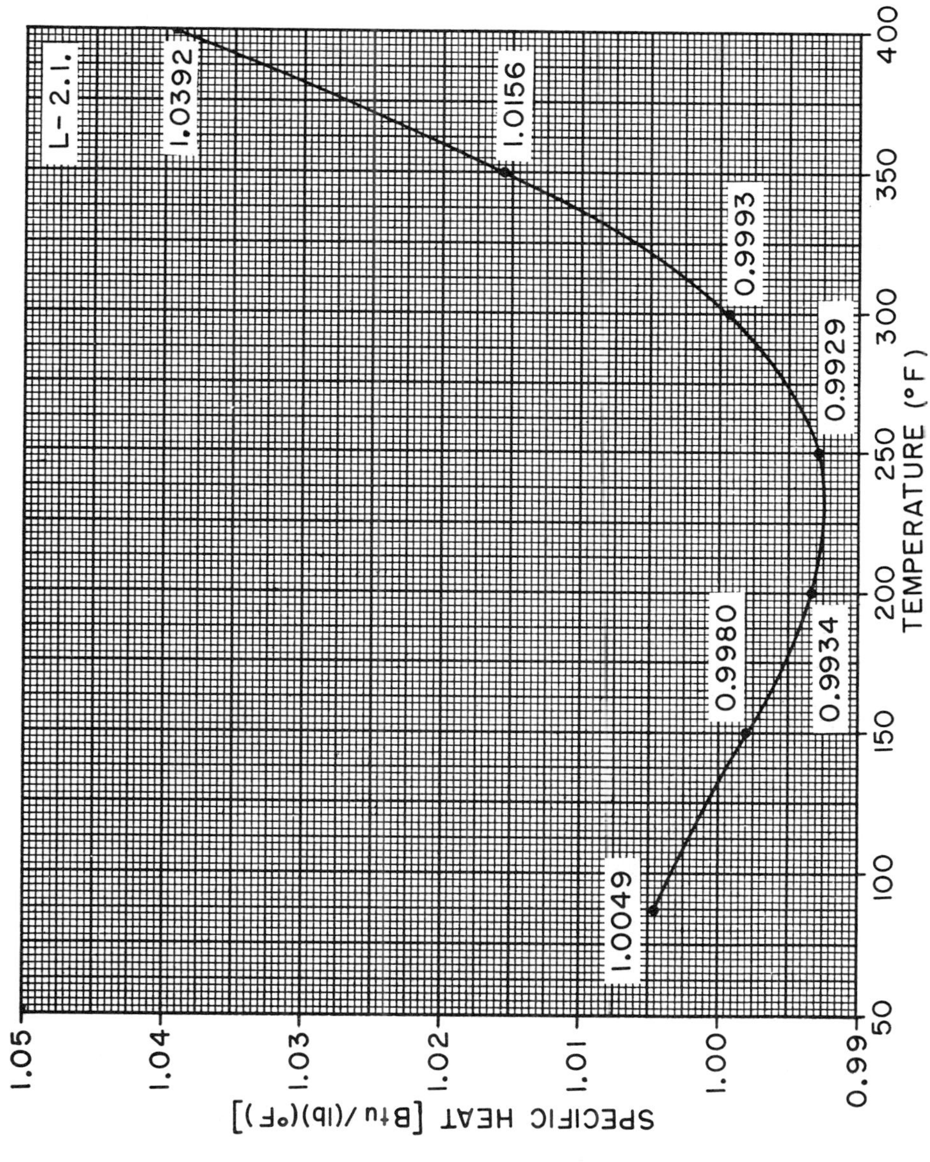

Specific Heat vs Temperature

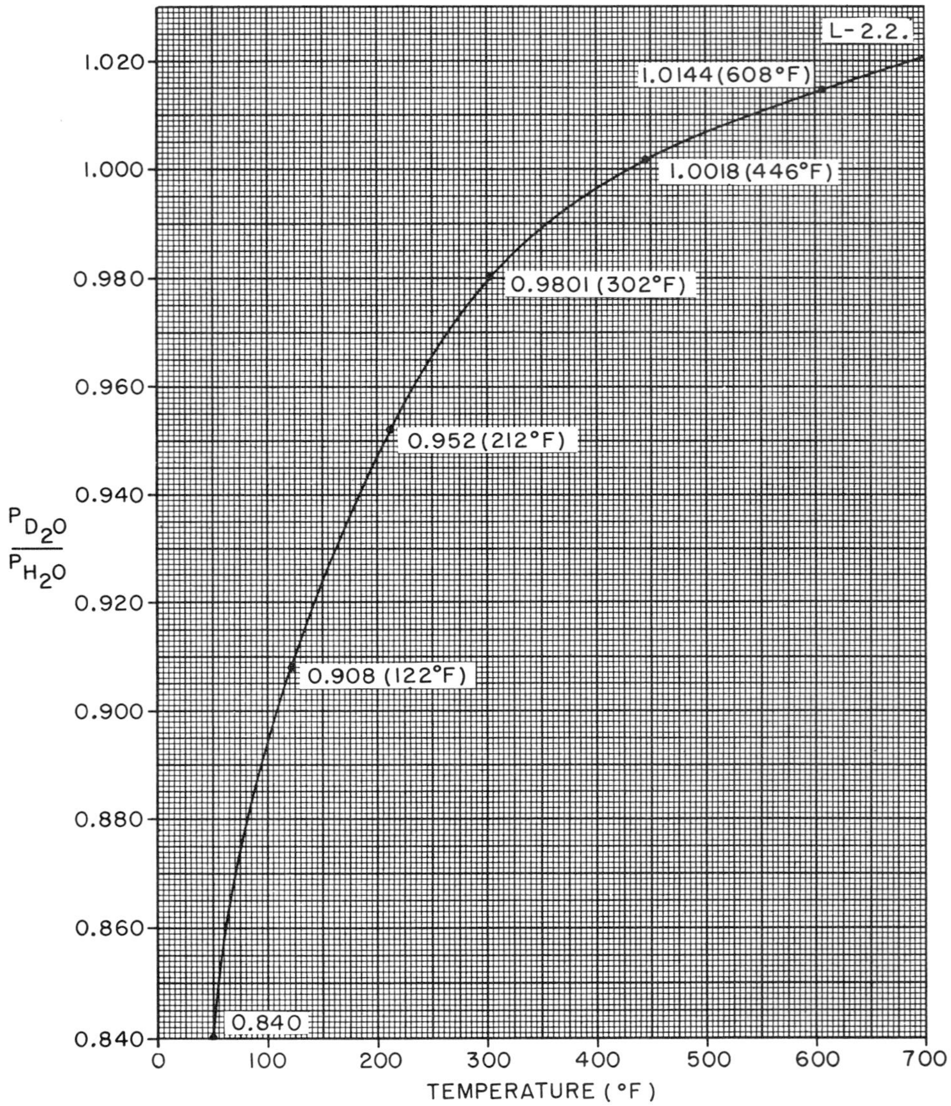

Vapor Pressure vs Temperature

L-3. ALCOHOLS

Methanol, ethanol, and isopropanol are the three alcohols used in anti-freezes. They are all soluble in water, and their function is to lower the freezing point of water, but at the same time they lower the boiling point. Methanol is the most efficient of the alcohols for depressing the freezing point of water and is usually the one employed for this purpose.

Properties of Importance

Specific Gravity: The following table lists the specific gravity of various alcohol solutions at 68°F.

Weight %	Methanol	Ethanol	Isopropanol
20	0.968	0.971	0.970
40	0.937	0.937	0.931
60	0.897	0.894	0.835
80	0.848	0.844	0.836
100	0.793	0.791	0.786

Specific Heat: At 68°F the specific heat of methanol, anhydrous ethanol, and 99 per cent isopropanol, is 0.594, 0.573, and 0.599 Btu/(lb)(°F), respectively. It should be noted that the specific heat of isopropanol becomes greater than 1.00 Btu/(lb)(°F) at temperatures above 330°F.

Thermal Conductivity: The thermal conductivity of several methanol-water solutions is given in the following table.

Methanol By Weight	68°F	104°F	140°F
0	0.341	0.360	0.377
20	0.278	0.295	0.307
40	0.232	0.237	0.242
60	0.191	0.189	0.186
80	0.155	0.148	—
100	0.128	0.116	—

Viscosity: The viscosity of alcohol-water solutions at −4° is given in centipoises in the following table.

Weight % Alcohol	Methanol	Ethanol	Isopropanol
40	9.0	—	—
60	6.2	21.	—
80	3.2	10.1	33.0
100	1.2	3.7	14.0

Physical Stability: Figure L-3.1.1 shows the vapor pressure of methanol-water solutions at different temperatures. In Figures L-3.1.2 and L-3.1.3, the effect of the alcohols on the freezing point of water is depicted. The freezing point of methanol is −143.5°F, and that of anhydrous ethanol is −173.9°F. Isopropanol has a freezing point of −126.0°F.

Thermal Stability: The alcohols are stable up to temperatures of 300°F or higher. The critical temperature of methanol is 384°F.

Chemical Stability: Alcohols tend to dehydrogenate at high temperatures. Certain metals, particularly copper, silver and their alloys, catalyze this reaction.

Surface Tension: A solution containing 50 per cent of methanol by weight has a surface tension of 33.9 dynes/cm; a similar isopropanol solution has a value of 24.4 dynes/cm. The surface tensions for pure methanol and isopropanol are 22.2 dynes/cm and 21.1 dynes/cm, respectively. All of the above data are for a temperature of 77°F.

Corrosion: The alcohols should not be used with lead or magnesium fittings. Aluminum is also attacked, but a protective layer of oxide is formed, preventing further corrosion.

Flammability:

Flash Points: Methanol: 62°F
Ethanol: 66°F
Isopropanol: 72°F

The anhydrous alcohols are extremely flammable and precautions against fire and explosion are necessary. Of course, aqueous solutions of alcohols present reduced fire hazards. When the water content is approximately 40 per cent or greater, there is no danger of fluid ignition.

Physiological Effects: These three materials are all toxic to some extent. Methanol, as a matter of fact, is labeled a poison. It is very important to insure that prolonged inhalation of the vapors and frequent contact with the skin are avoided.

Refractive Index: Methanol: 1.326
Isopropanol: 1.377

Color and Odor: These alcohols are colorless and have a characteristic odor.

Price Class: Methanol, Ethanol—B.
Isopropanol—C.

References

1. Union Carbide Chemical Corp., "Alcohols," 1958.
2. Commercial Solvents Corp., "Methanol" 1952.
3. Lucas, J. H., "Organic Chemistry," Second Edition, American Book Co., 1952.

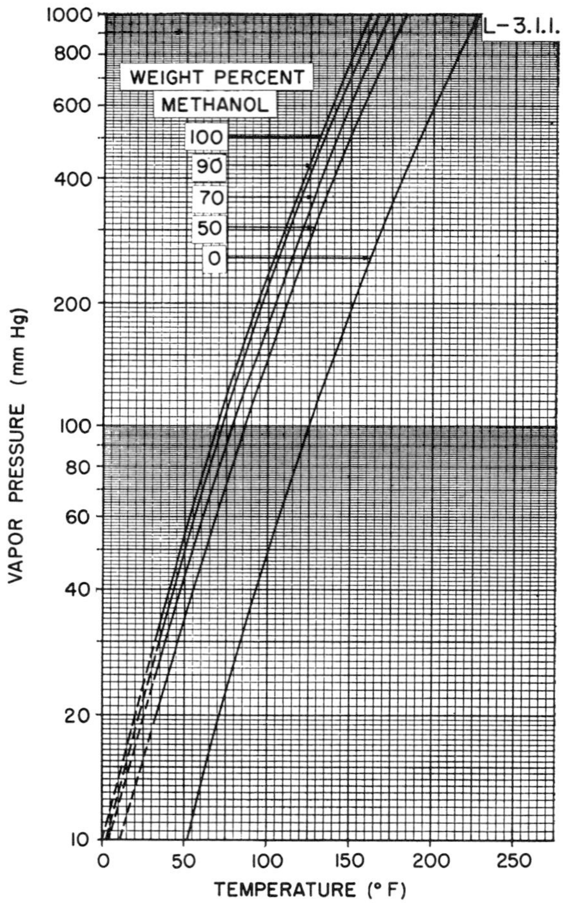

Vapor Pressure vs Temperature

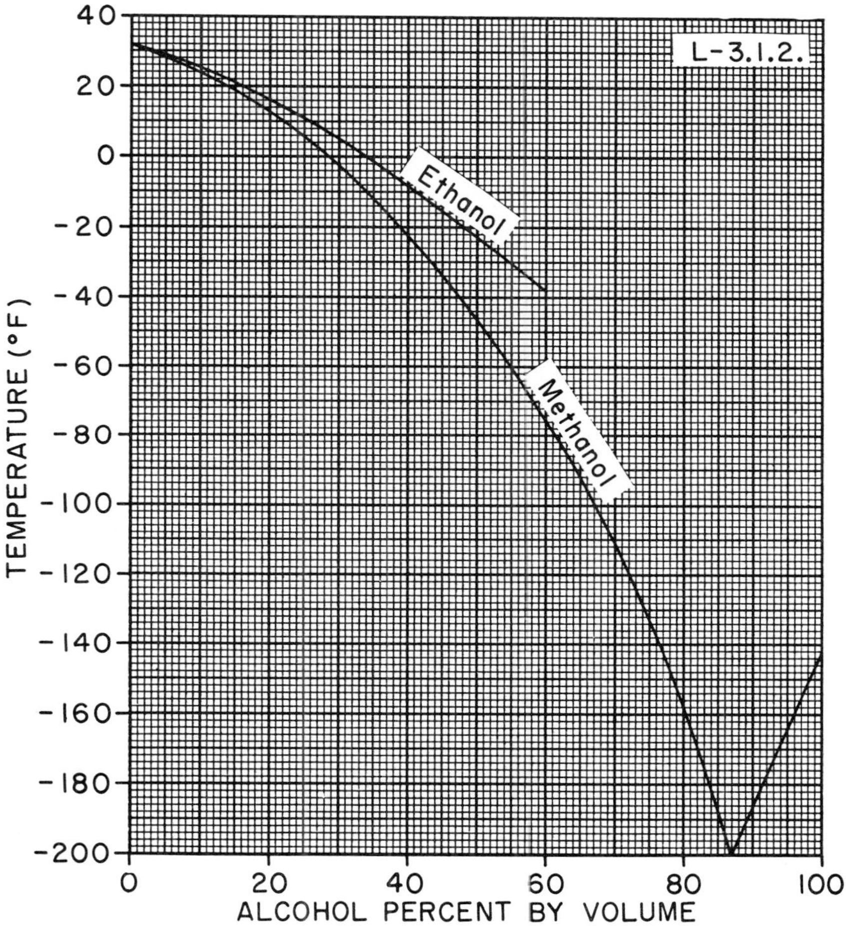

Freezing Point vs Per Cent Alcohol by Volume

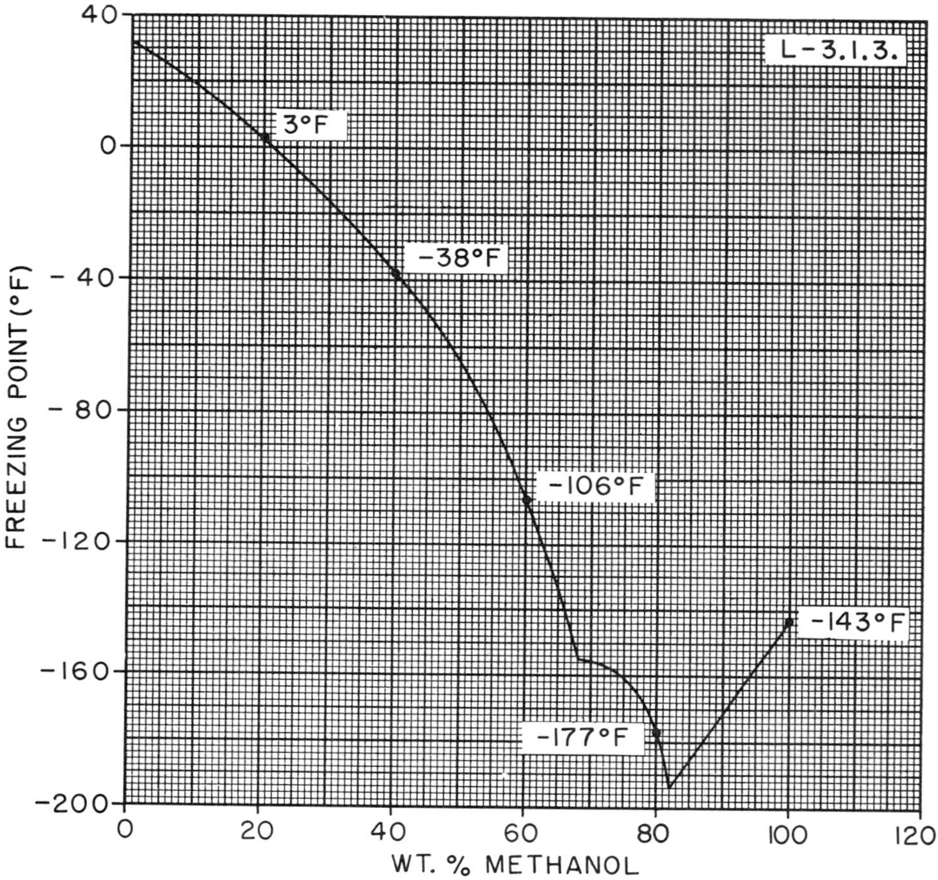

Freezing Point vs Weight Per Cent Methanol

L-4. ESTERS

Esters are neutral compounds formed along with water, by the reaction of acids and alcohols. Both organic and inorganic acids form esters. The reaction is reversible, and esters heated with water break down to yield the corresponding acid and alcohol. The temperature at which this hydrolysis reaction takes place varies widely, depending upon the molecular structure of the ester involved.

Since the number of acids and alcohols that may react to form esters is very great, an extremely large number of esters can be synthesized. Many esters occur naturally as fats and oils and are used as heat transfer agents in cooking. These and synthetic esters, particularly the phosphate and silicate esters, are of interest for heat transfer applications because of their wide fluid range, good thermal stability, and chemical neutrality.

L-4.1. ESTERS OF PHOSPHORIC ACID

The prospect of using esters, particularly those of phosphoric acid, as heat transfer media is a promising and intriguing one because of the high thermal stability of these compounds. These fluids are being utilized commercially as fire-resistant hydraulic fluids and synthetic lubricants.

Monsanto Chemical Company and Celanese Corporation of America have been quite active in developing phosphate esters for the aircraft industry. Two phosphate-based esters manufactured by Monsanto are Skydrol 7000, developed for propeller aircraft, and Skydrol 500A, used for higher altitude, low temperature operation of jets.

Celanese produces a series of synthetic lubricants for high performance jet aircraft from esters of the polyol, trimethylolpropane. The thermal stability of trimethylolpropane esters stems from a neopentyl structure—four carbon atoms each bonded to a fifth central carbon.

Phosphate ester and phosphate ester-based fluids such as Pydraul and the Cellulube Safety Series are excellent fire-resistant hydraulic fluids for equipment such as die-casting machines, conveying equipment handling molten metals, and other potentially dangerous hydraulic applications. The fact that these fluids are functional from −90° to 300°–350°F should spark further investigation as to how closely they satisfy the essential heat transfer criteria previously outlined.

References

1. Technical Product literature on fire-resistant hydraulic fluids, Monsanto Chemical Co., "Pydraul 60, Pydraul 150, Pydraul A-200, Pydraul F-9, and Pydraul 625."
2. Monsanto Chemical Co., "Pydraul AC Fire-Resistant Lubricant for Air Compressors."
3. Monsanto Chemical Co., "Skydrol 7000 and Skydrol 500A."
4. Monsanto Chemical Co., "Monsanto Functional Fluids for Specialized Engineering."
5. Union Carbide Chemicals Co., "Esters," 1959.
6. Celanese Corp. of America, "Celanese Synthetic Lubricants for Aircraft Gas Turbine Engines and Base Stocks," 1960.
7. Celanese Corp. of America, "Cellulube Fire-Resistant Synthetic Lubricants for Air Compressors," 1959.
8. Celanese Corp. of America, "Fire-Resistant Hydraulic Fluids and Synthetic Lubricants in Controlled Viscosities," 1958.
9. *Nuclear Science Abstr.*, 4420.

L-4.2. SILICATE ESTER

(Coolanol 45)

Coolanol 45 is the trade name of a fluid having a silicate ester base, manufactured by Monsanto Chemical Company. Gas and evaporative liquid cooling of electronic components are sometimes inadequate because of stringent demands for equipment reliability and the concomitant requirement of exact temperature control. Thus, liquid coolants such as Coolanol 45 were developed.

This fluid, because of its excellent dielectric strength and its capacity to dissipate heat rapidly and efficiently has found widespread application in cooling the complex electronic systems of missiles and aircraft. Coolanol 45 makes a one-fluid package feasible since, besides being a coolant, it can also function as a hydraulic fluid. Another important advantage stems from its extremely wide temperature range, −65° to 400°F, that makes Coolanol 45 suitable for high and low temperature applications. In summation, its dielectric, hydraulic, heat transfer, and lubricating properties are quite favorable.

Properties of Importance

Density: See Figure L-4.2.1.

Specific Heat: See Figure L-4.2.2.

Thermal Conductivity: See Figure L-4.2.3.

Viscosity: See Figure L-4.2.4.

Physical Stability: Coolanol 45 has a very low vapor pressure at relatively high temperatures. See Figure L-4.2.5. It has a pour point below −85°F. However, the minimum operating temperature is usually taken at −65°F, where the fluid has a viscosity of about 2600 centistokes.

Thermal Stability: The maximum recommended operating temperature for Coolanol 45 is 400°F.

Chemical Stability: Water contamination must be avoided because silicate esters will react in the presence of excessive amounts of water to form a precipitate or sludge. With the exception of silicones, Coolanol 45 is compatible with all common materials. However, Coolanol 45 is completely compatible with Dow-Corning's silicone elastomer, LS-53.

Surface Tension: Coolanol 45 can be readily retained in a system without any danger of leakage.

Corrosion: The corrosion and oxidation data presented in the following table shows that Coolanol 45 has only a slight effect on the common metals—aluminum, iron, copper, silver alloys, brass, cadmium-plated steels, and solders of brazing materials. The fluxes employed for joining metals may alter the fluid and these should be excluded from a system using Coolanol 45.

CORROSION AND OXIDATION TEST DATA ON COOLANOL 45

Metal	Weight Change of Metal (mg/cm^2) 250°F(1)	300°F	400°F(2)
Copper	−0.023	−0.07	−0.062
Iron	+0.023	+0.01	−0.065
Aluminum	+0.023	+0.01	+0.016
Magnesium	+0.023	+0.10	+0.016
Silver	—	—	+0.008
Fluid evaporation, %	1.8	4.0	—

1. Under conditions of MIL-H-5606 oxidation, corrosion tests.
2. Under conditions of MIL-H-8446 oxidation, corrosion tests.

Physiological Effects: Coolanol 45 is essentially nontoxic and presents no handling problem for maintenance personnel. The pure compound may be mildly irritating to eye tissues. This irritation may be easily removed by washing the eyes with clear water. Workers should not inhale the fumes during high temperature tests.

Electrical Properties: The favorable electrical properties of Coolanol 45 which make it serviceable under dielectric conditions, are shown below.

Dielectric strength		
20°C	27.0 kilovolts	
50°C	20.1 kilovolts	
85°C	18.0 kilovolts	
Leakage current	1 milliampere at 10,000 volts A.C, 60 cycles	
Specific resistivity(1)		
25°C	9×10^{10} ohm-cm	
50°C	2.65×10^{10} ohm-cm	
100°C	1.15×10^{10} ohm-cm	
Dielectric constant, Loss tangent at 25°C(2)		
Freq. cps	K	Tan
10^3	2.65	0.0336
10^4	2.65	0.0034
10^5	2.65	0.0021
10^6	2.65	0.0042
10^7	2.64	0.0041
10^8	2.63	0.0130
10^9	2.48	0.0482

1. IBM data.
2. MIT data.

Price Class: C.

References

1. Monsanto Chemical Co., "Coolanol 45, Coolant-Dielectric for Electronic Equipment," Jan. 1960.

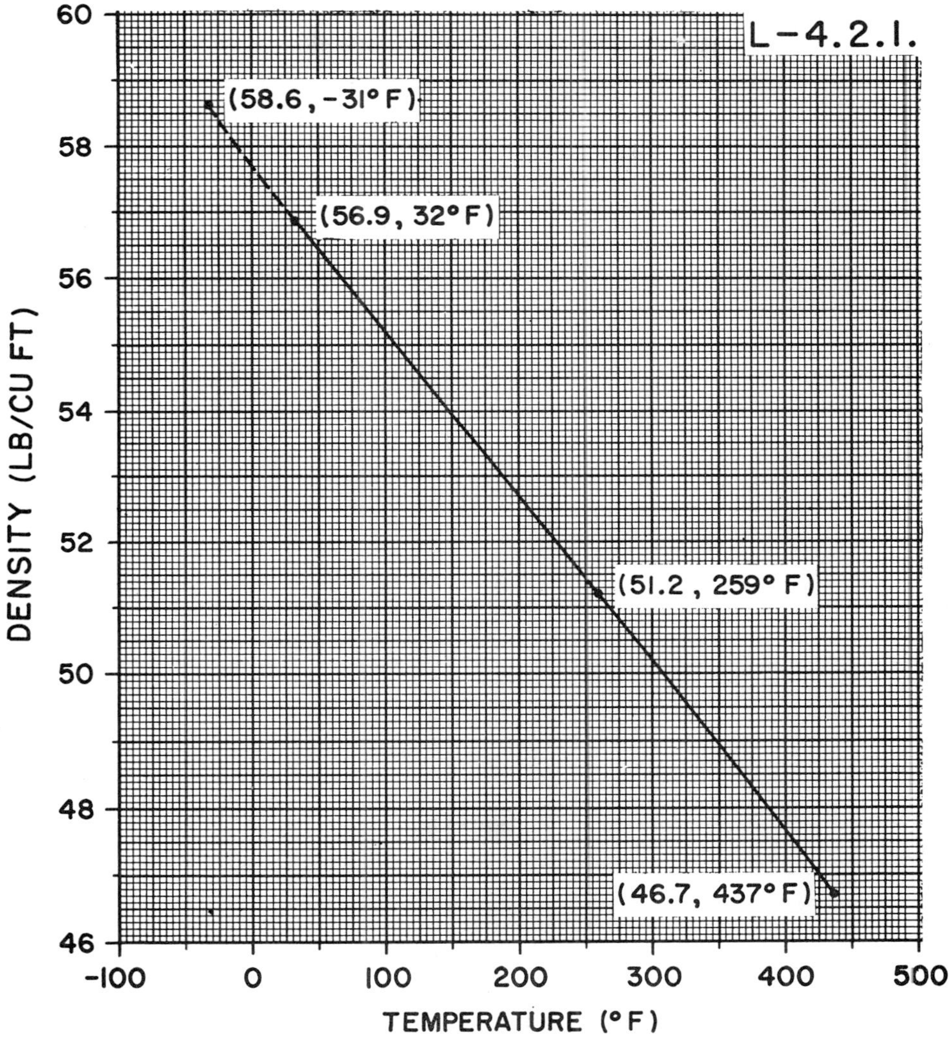

Density vs Temperature

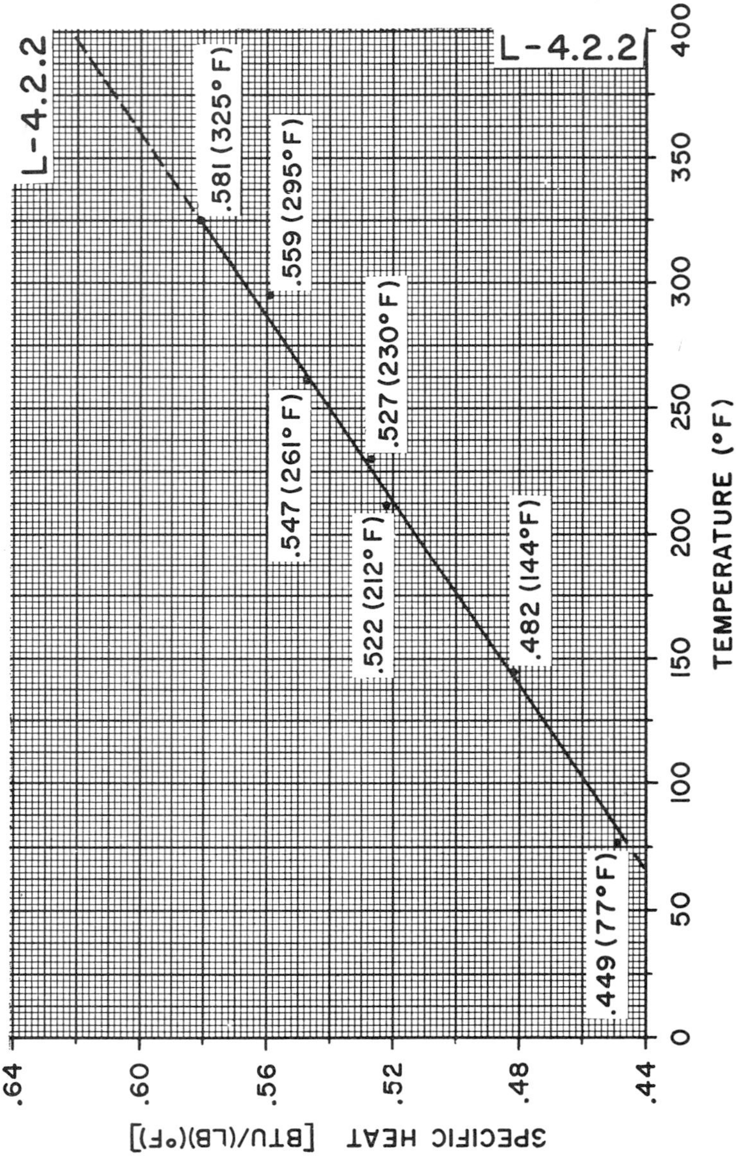

Specific Heat vs Temperature

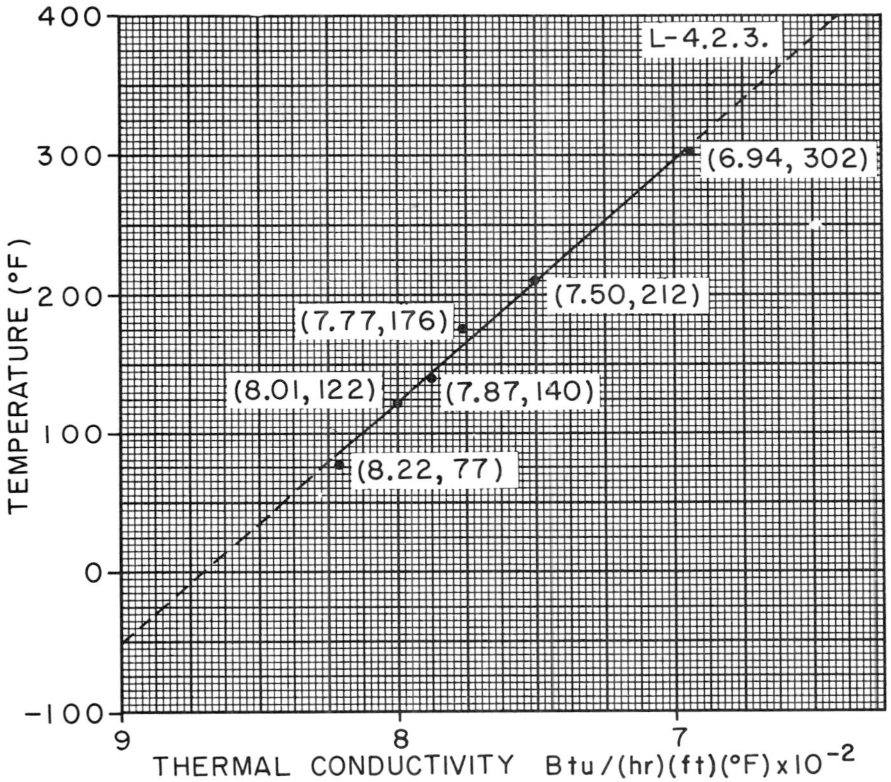

Thermal Conductivity vs Temperature

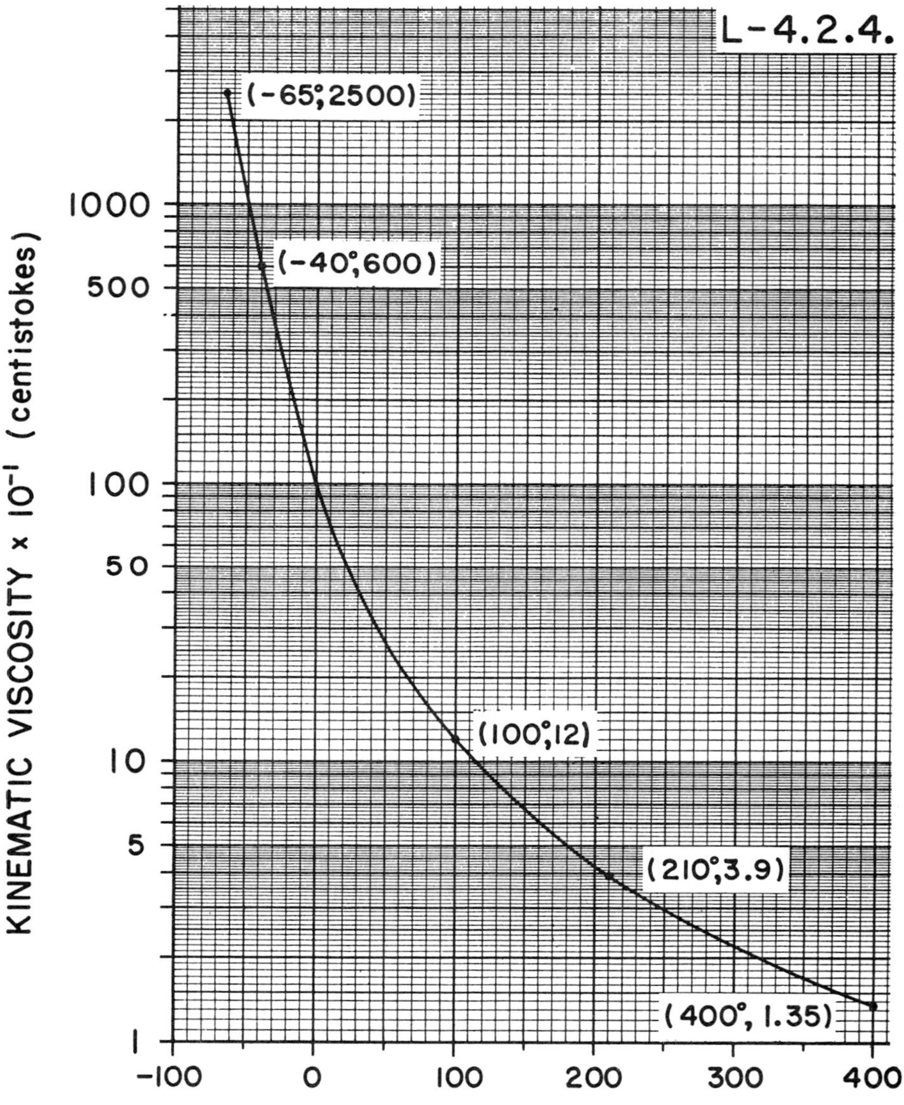

Kinematic Viscosity vs Temperature

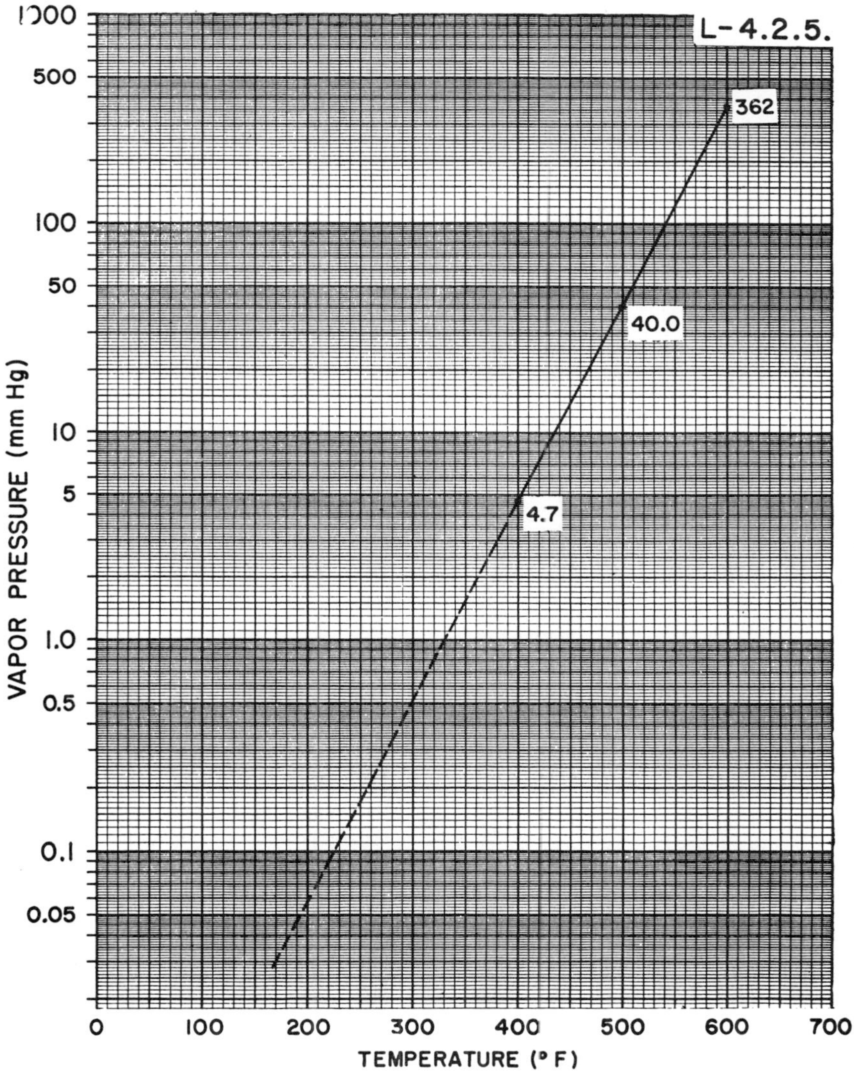

Vapor Pressure vs Temperature
Operating Range: −65° – 400°F

L-4.3. TRIARYL BORATE

Triaryl borate is an example of an organic salt, which in this case is an organometallic ester, used as a heat transfer medium. It is usually used at temperatures around 600°F but it is possible to use it at higher temperatures. At these higher temperatures the effective life of triaryl borate is shortened. The material can be regenerated fairly easily by vacuum distillation. Although it has been used as a liquid until now, it could probably be applied for heat transfer in the vapor phase.

Properties of Importance

Density: See Figure L-4.3.1.

Specific Heat: See Figure L-4.3.1.

Thermal Conductivity: The thermal conductivity is 0.068 Btu/(hr)(ft)(°F) at 201°F.

Viscosity: See Figure L-4.3.2.

Physical Stability: See Figure L-4.3.2.

Thermal Stability: The borate system should be heated in the absence of air. At temperatures close to and above 600°F degradation takes place after a time. In general, the higher the temperature used, the shorter the effective life.

Chemical Stability: Triaryl borate is readily hydrolyzed by water, liberating cresol. A means of drying the fluid must be introduced into the system. The hydrolysis reaction is reversible and the triaryl borate can be recovered by vacuum heating.

Corrosion: Triaryl borate is most commonly used in cast iron or steel piping systems.

Flammability: (Open Cup) 290°F.
Fire Point: 300°F.

Physiological Effects: Since triaryl borate liberates cresol when it comes in contact with moisture, it is a safety hazard. Any burns caused by it must be treated as cresol burns.

Reference

1. Hygrotherm Engineering, Ltd., London, England.

L-4.4. ANIMAL OIL

(Thermofluid)

Thermofluid is obtained from a fish oil, and is used as a heat carrier in the liquid state. It is claimed to be applicable in the temperature range of 14°- 662°F. This fluid was developed by Caliqua, S. A., of Paris, in order to make possible pressureless liquid heat transfer at temperatures between 400° and 650°F.

Properties of Importance

Density: See Figure L-4.4.1.

Specific Heat: See Figure L-4.4.2.

Thermal Conductivity: The thermal conductivity is a linear function of temperature. At 212°F the conductivity is 0.0806 Btu/(hr)(ft)(°F), and at 662°F is 0.0699 Btu/(hr)(ft)(°F).

Viscosity: See Figure L-4.4.2.

Physical Stability: See Figure L-4.4.3. Note that the pressure does not rise above atmospheric pressure within the recommended operating range. The pour point is 14°F.

Thermal Stability: Thermofluid remains stable at high temperatures. Its boiling point is in the vicinity of 700°F.

Chemical Stability: The material is very stable and there is no danger of explosion.

Corrosion: Thermofluid may be used with any of the common structural materials.

Physiological Effects: Thermofluid has no toxicity and no special precautions are necessary during handling.

Color: Colorless.

Price Class: B.

References

1. Fluitherma, Paris, "Fluitherma."
2. Caliqua Process and District Heating, "Thermofluid."

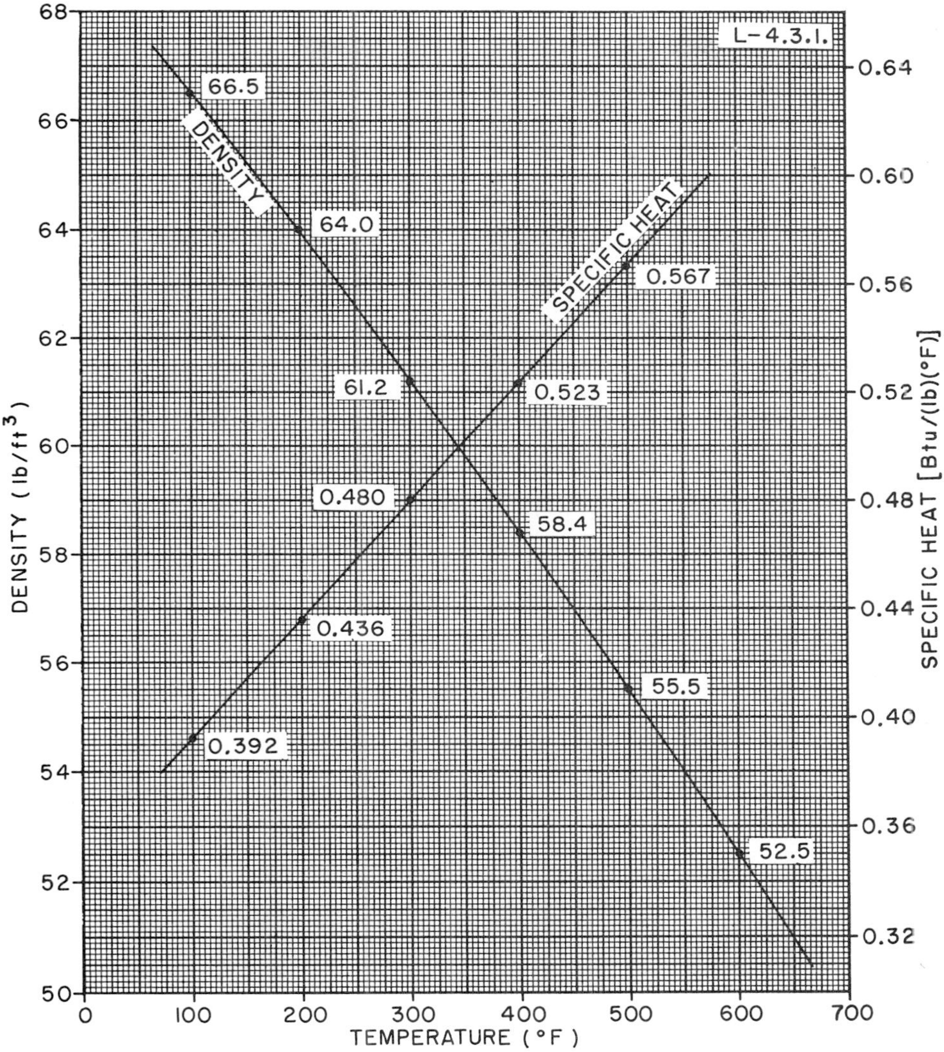

Density vs Temperature
Specific Heat vs Temperature

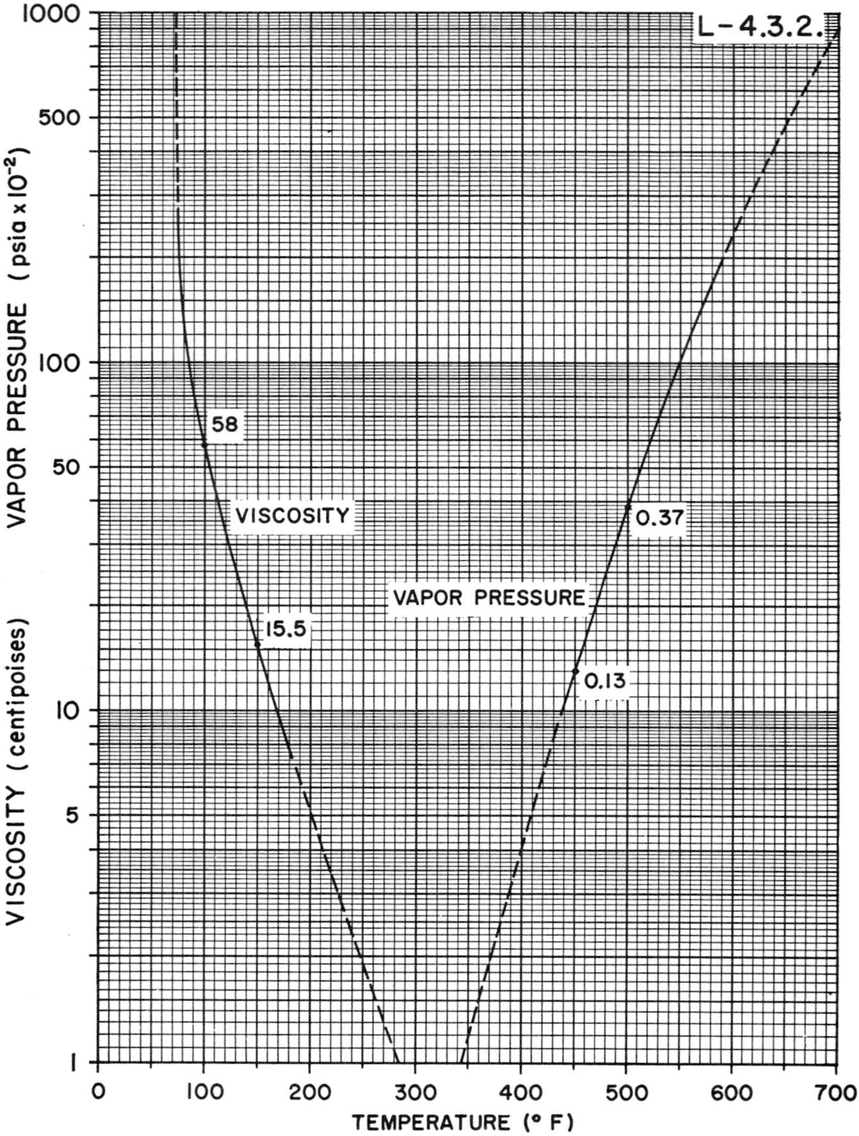

Vapor Pressure vs Temperature
Viscosity vs Temperature

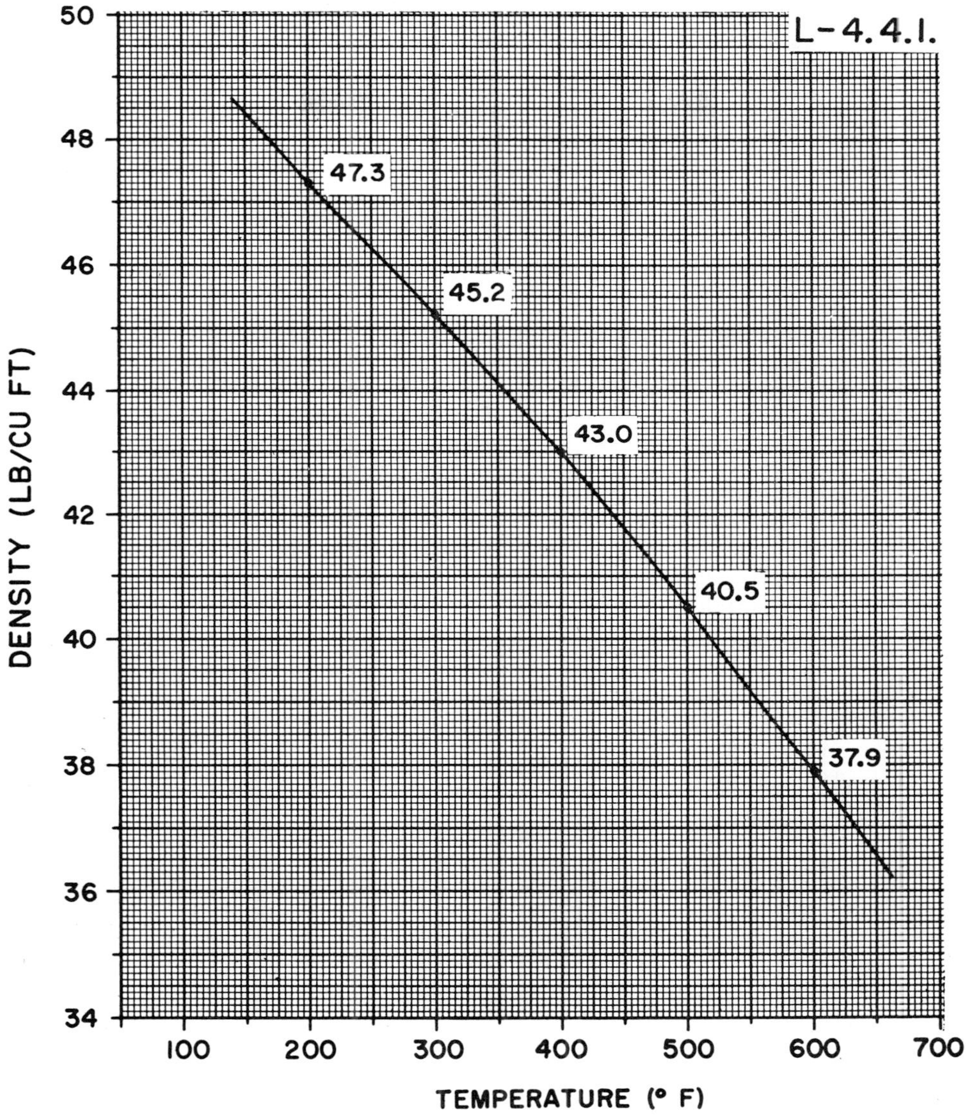

Density vs Temperature

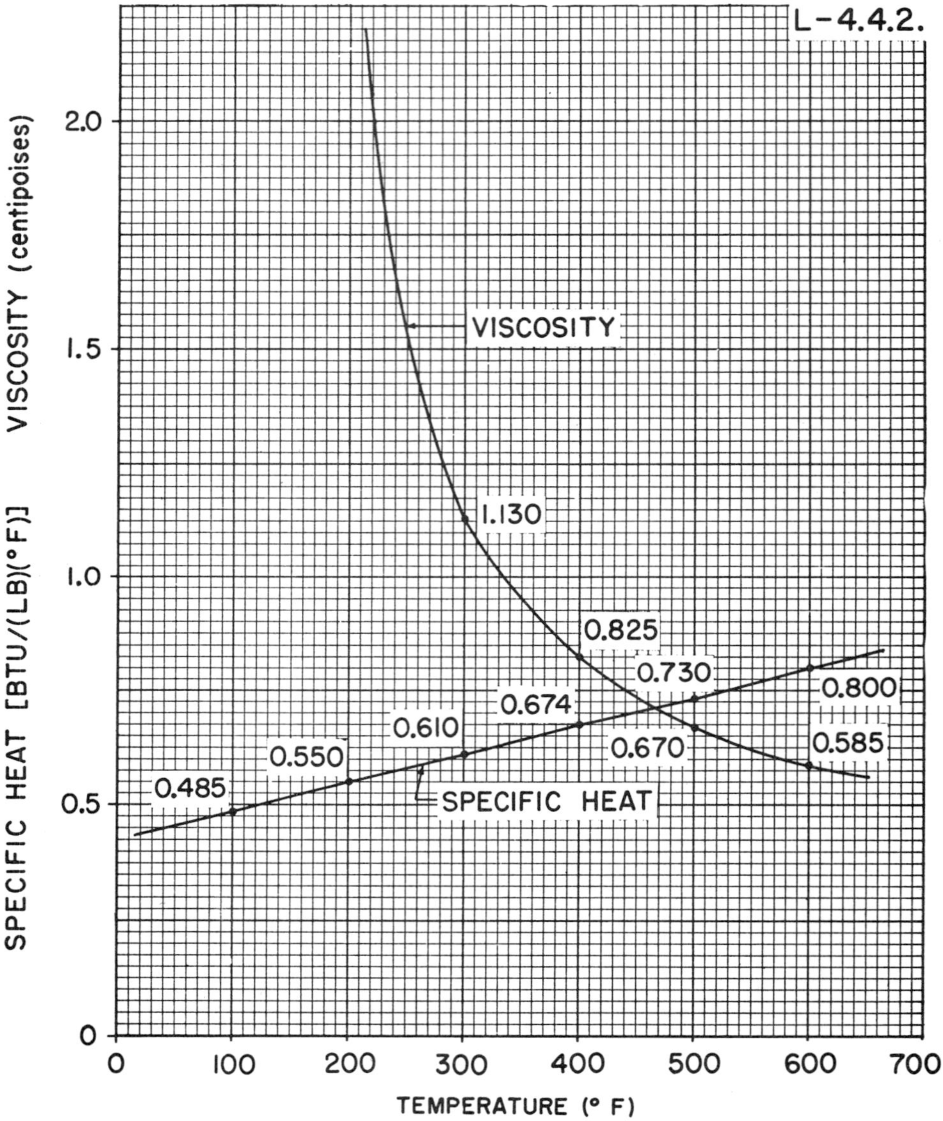

Viscosity and Specific Heat vs Temperature

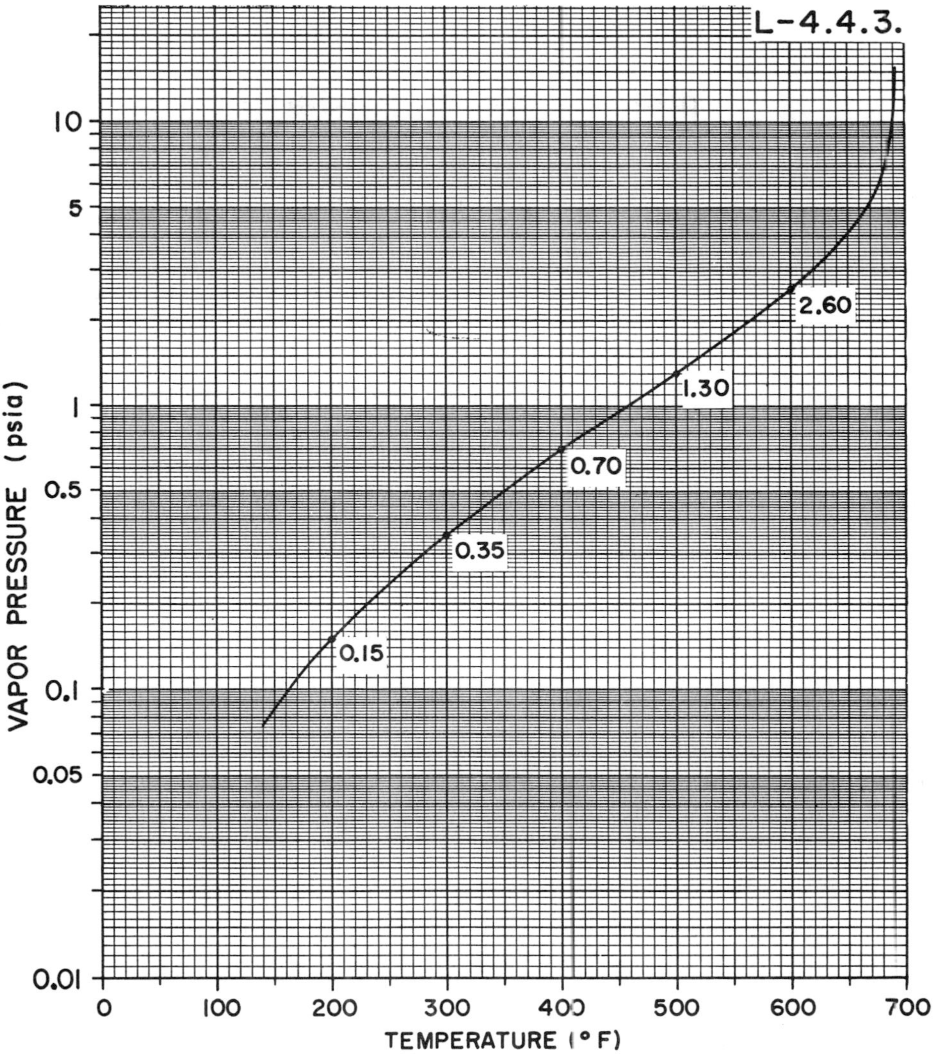

Vapor Pressure vs Temperature

L-5. ETHERS

Ethers may be regarded as organic oxides or as derivatives of water in which both hydrogen atoms are replaced by organic groups. Ethers are characterized by good chemical stability and wide fluid range. The pure aromatic (polyphenyl) ethers have excellent thermal stability, and claims have been made that they are stable up to 700°F. They have better lubricating properties than the polyphenyls, and any polyphenyl ether has better low temperature properties than its comparable (chain-length equivalent) polyphenyl.

A polyphenyl ether (diphenyl oxide) was one of the first synthetic materials used in this country as a high temperature heat transfer fluid, being the main ingredient in Dowtherm A.

L-5.1. ANISOLE (PHENYL METHYL ETHER)

Anisole, a high boiling aryl alkyl ether, has served as a heat carrier in the −30°–500°F range. The fluid may be applied in either the vapor or liquid phases.

For further discussion of Anisole, see page 70.

L-5.2. POLYPHENYL ETHERS

OS-124

The use of polyphenyl ethers for high temperature heat transfer applications is a relatively new idea and development work is still being conducted on this subject by many organizations. Very optimistic and favorable temperature ranges for this liquid are brought out. Research up to now indicates that the meta isomers have the best low temperature properties. However, for a particular application, one of the others might be more suitable. The lubricity of the polyphenyl ethers compares favorably with that of silicones, silicates, hydrocarbons, esters, and other commonly used high temperature lubricants. The data given below are for OS-124, a recently developed mixed isomeric five ring polyphenyl ether, whose properties may be considered typical of these compounds.

Properties of Importance

Density: See Figure L-5.2.1.

Specific Heat: See Figure L-5.2.1.

Thermal Conductivity: The conductivity of OS-124 has been measured only at 80°F, at which point it is 0.0775 Btu/(hr)(ft)(°F), and at 300°F, where the value is 0.0702 Btu/(hr)(ft)(°F). Extrapolations on these data lead to a value of 0.0642 at 500°F; 0.0613 at 600°F, and 0.0582 at 700°F.

Viscosity: The kinematic viscosity of OS-124 is shown in Figure L-5.2.2. In addition, kinematic viscosity of five-, six-, and seven-ringed meta polyphenyl ethers is shown in Figure L-5.2.3.

Physical Stability: See Figure L-5.2.2. The pour point of OS-124 is 40°F.

Thermal Stability: By the use of an isoteniscope, the temperature at which OS-124 is stable has been determined to be 847°F. A good measure of thermal stability is the useful life of a material at a high temperature. For example, the useful life of the all meta four-ringed ether is 80 hours at 900°F. The best presently used lubricants last only about 18 minutes. It has not been proved yet that these compounds could be usable at these extreme temperature conditions for industrial heat transfer applications.

Chemical Stability: The polyphenyl ethers are stable in water, acids, bases, and oxygen at temperatures considerably higher than any presently used lubricant.

Flammability:
Flash Point: 550°F.
Fire Point: 680°F.
Spontaneous Ignition Point: 1135°F.

Physiological Effects: OS-124 has a very low oral and respiratory toxicity. Skin absorption tests prove that no harm is done up to very large doses. These tests have shown also that the material is a nonirritant.

Electrical Properties: The dielectric constant of OS-124 at 68° F is 4.45; at 212° F it is 3.96.

Refractive Index: 1.6306 (N_D^{25})

Price Class: D.

References

1. Monsanto Chemical Co., "Preliminary Report on the Polyphenyl Ethers 700°-1000°F Fluids," May, 1960.
2. Monsanto Chemical Co., "OS-124 Preliminary Technical Data Sheet," February 1960.

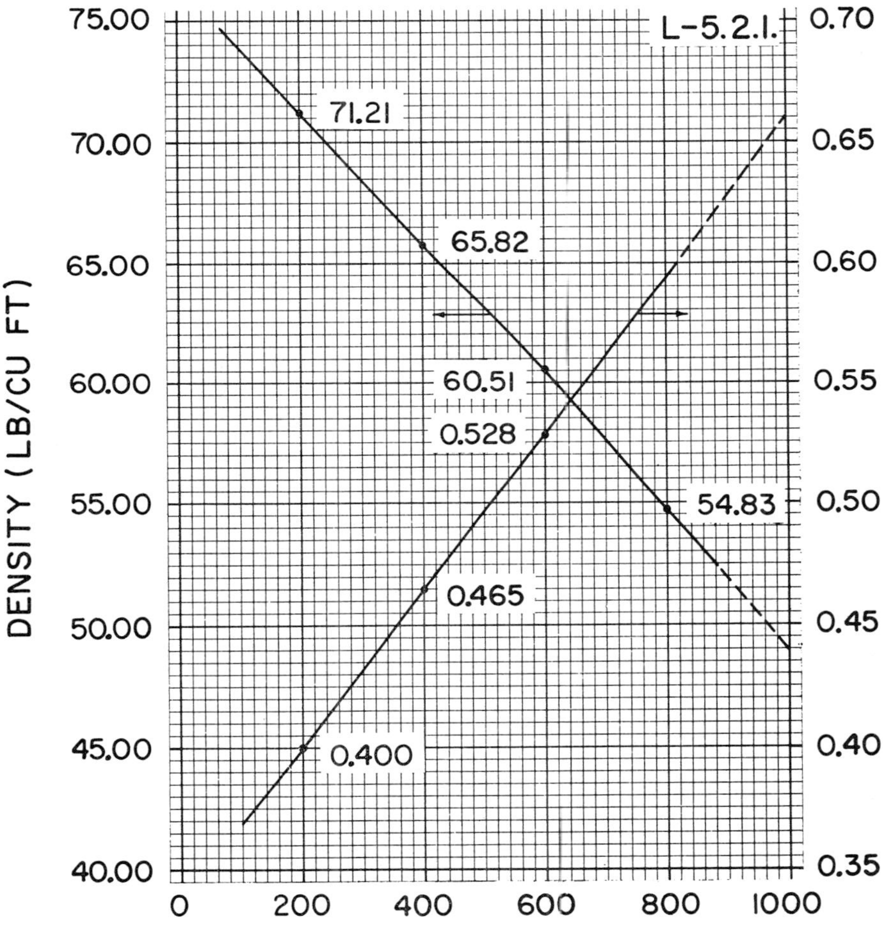

Density vs Temperature
Specific Heat vs Temperature

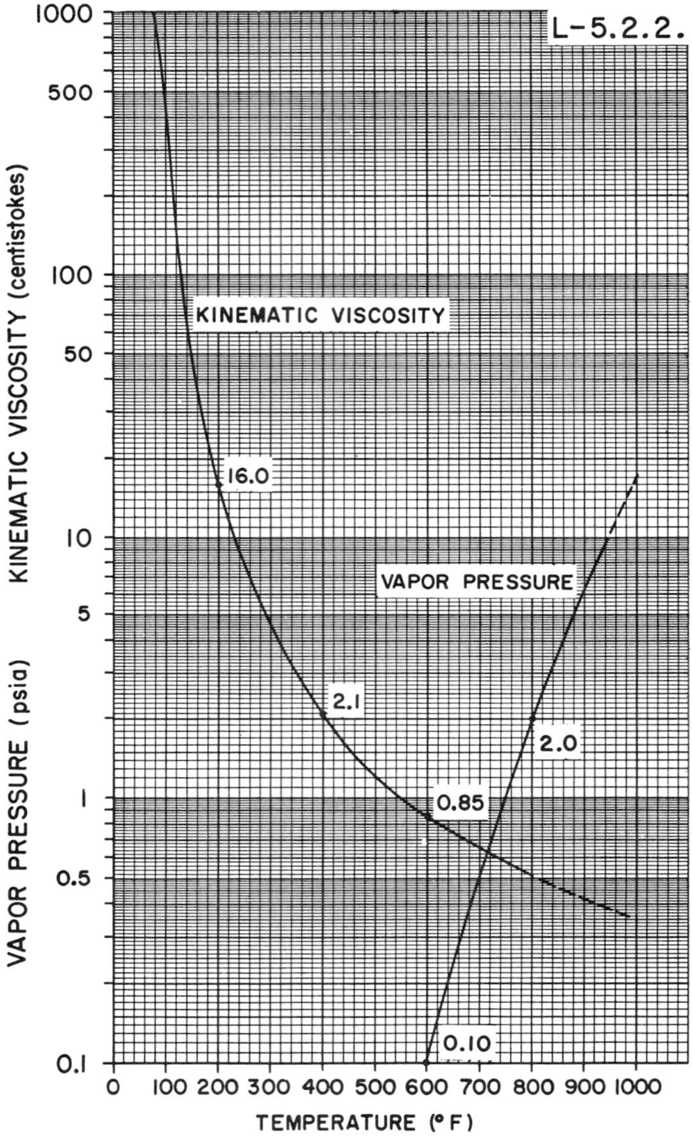

Kinematic Viscosity vs Temperature
Vapor Pressure vs Temperature

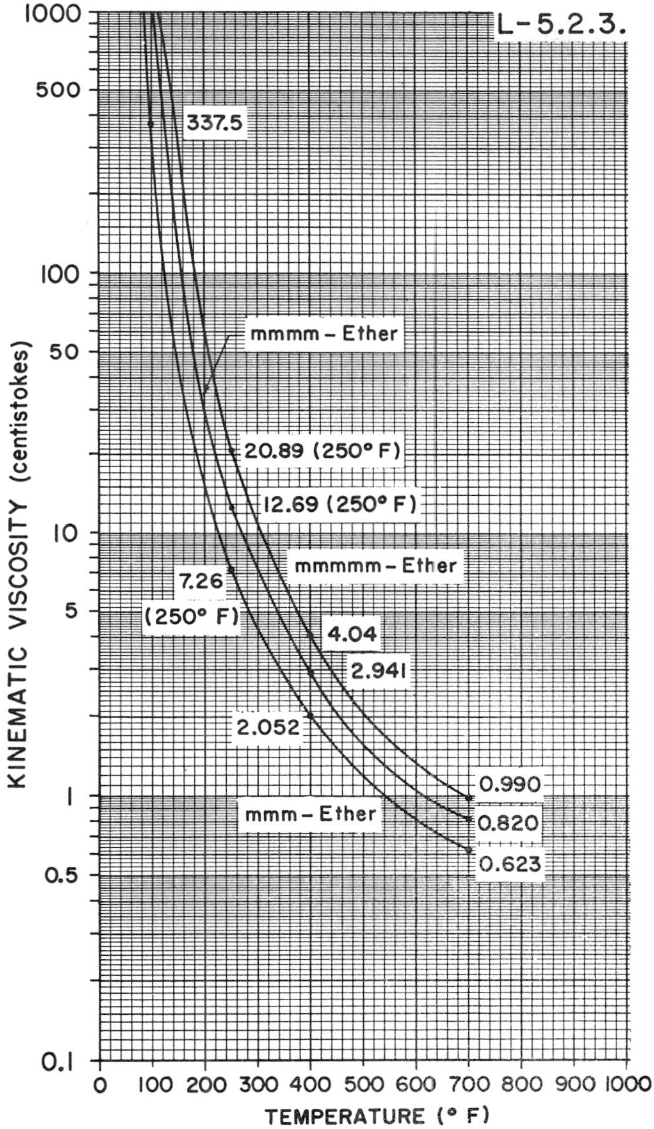

Kinematic Viscosity vs Temperature

L-6. GLYCOLS

The glycols are polyhydric alcohols and are sometimes referred to as diols. The lower glycols are used primarily in anti-freeze solutions. They are miscible with water, and the mixtures have evaporation points higher than water.

Certain polymeric glycols are effective heat transfer media at higher temperatures.

a. Glycol mixtures sometimes have more favorable properties than individual glycols. Other additives may be used to enhance the utilization of the mixture.

b. Ethylene and propylene glycols are excellent heat carriers over a wide range of temperatures between -50° to about 400°F.

c. Polyalkylene glycol derivatives are fine heat transfer media, and show some exceptionally favorable properties.

L-6.1. GLYCOL MIXTURES

(Dowtherm SR-1)

Dowtherm SR-1 is a mixture of glycols with a special inhibitor and some water added. It has been developed for snow and ice melting systems but it can also be used as a secondary coolant in air conditioning systems and cold storage trucks to protect against freezing in hot water heating systems which are periodically subjected to low temperatures. The inhibitor for Dowtherm SR-1 is designed to be most effective at the temperatures and conditions of snow melting installations.

Properties of Importance

Density: See Figure L-6.1.1.

Specific Heat: See Figure L-6.1.2.

Thermal Conductivity: See Figure L-6.1.3.

Viscosity: See Figure L-6.1.4.

Physical Stability: The freezing points of aqueous Dowtherm SR-1 solutions are shown in Figure L-6.1.5. The pour points, as determined by ASTM method D97-47, are given in the table below:

% Dowtherm SR-1	Pour Point (°F)
10	30
20	21
30	9
40	−11
50	−40
60	−94
70	−94
80	−94
90	−94
100	−31

Dowtherm SR-1 lowers the vapor pressure of water.

Surface Tension: A 50 per cent solution of Dowtherm SR-1, by weight, has a surface tension of approximately 55 dynes/cm; pure Dowtherm SR-1 has a surface tension of 46.7 dynes/cm, both at 77°F.

Thermal Stability: The equilibrium boiling point of Dowtherm SR-1 is 345°F. Dowtherm SR-1 may be operated at temperatures up to 250°F. If operated higher, special care should be taken in its handling.

Chemical Stability: After the fluid has been in use for some time, the inhibitor becomes exhausted and a fresh amount must be added.

Corrosion: Dowtherm SR-1 is less corrosive than ethylene glycol with all common structural metals.

Flammability:
Flash Point: 240°F.
Fire Point: 250°F.

Physiological Effects: The fluid has a low oral toxicity. It presents no particular handling problem. However, prolonged skin contact should be avoided.

Refractive Index: At 77°F, a 50 per cent solution by volume of Dowtherm SR-1 has a refractive index of 1.385; 100 per cent Dowtherm SR-1 has an index of 1.431.

Appearance: Dowtherm SR-1 is a red liquid.

Price Class: B.

References

1. Dow Chemical Co., "Dowtherm SR-1."
2. Dow Chemical Co., "Dowtherm SR-1 for Use in Thermal Snow and Ice Removal Systems."

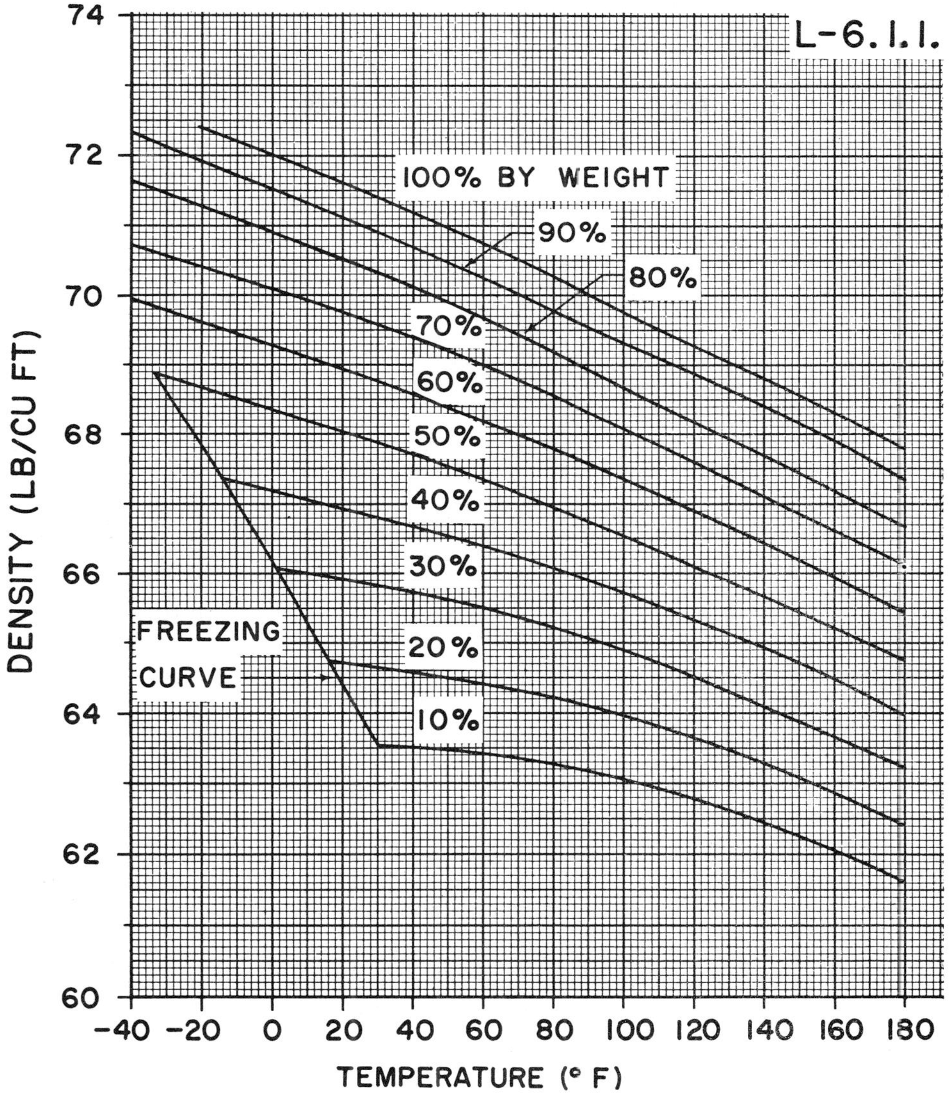

Density vs Temperature

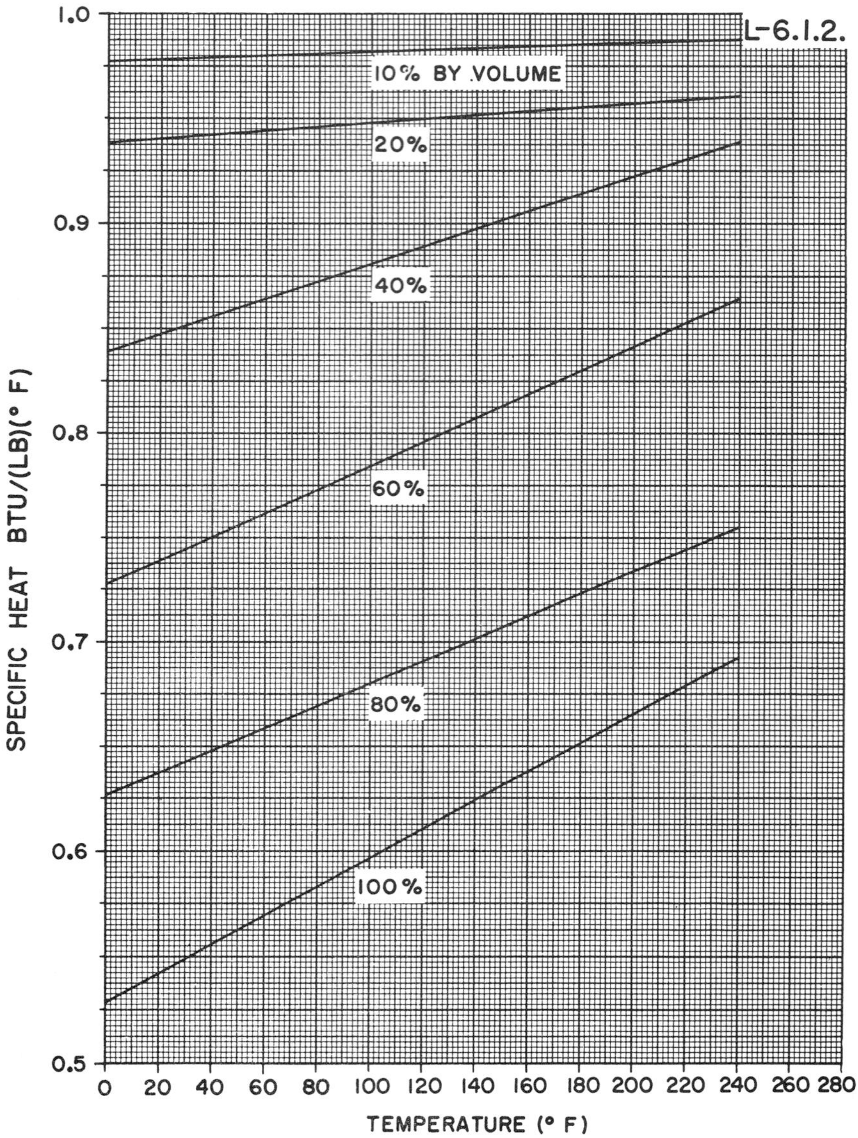

Specific Heat vs Temperature

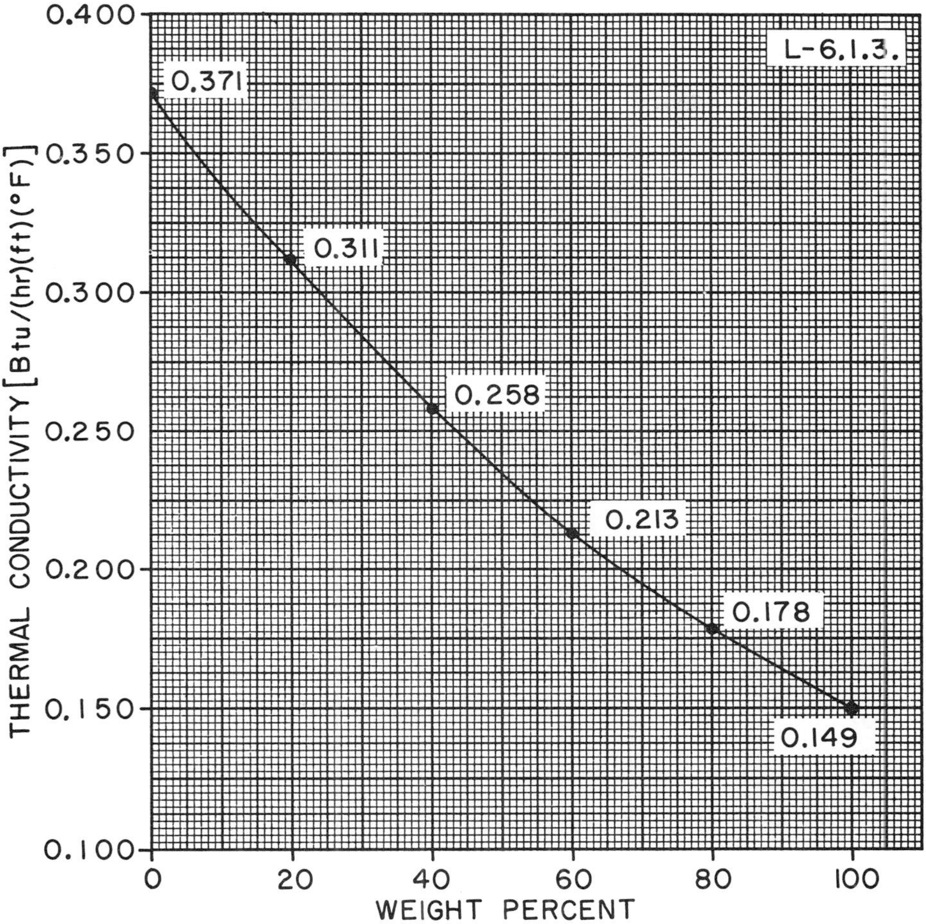

Thermal Conductivity vs Weight Per Cent (125 °F)

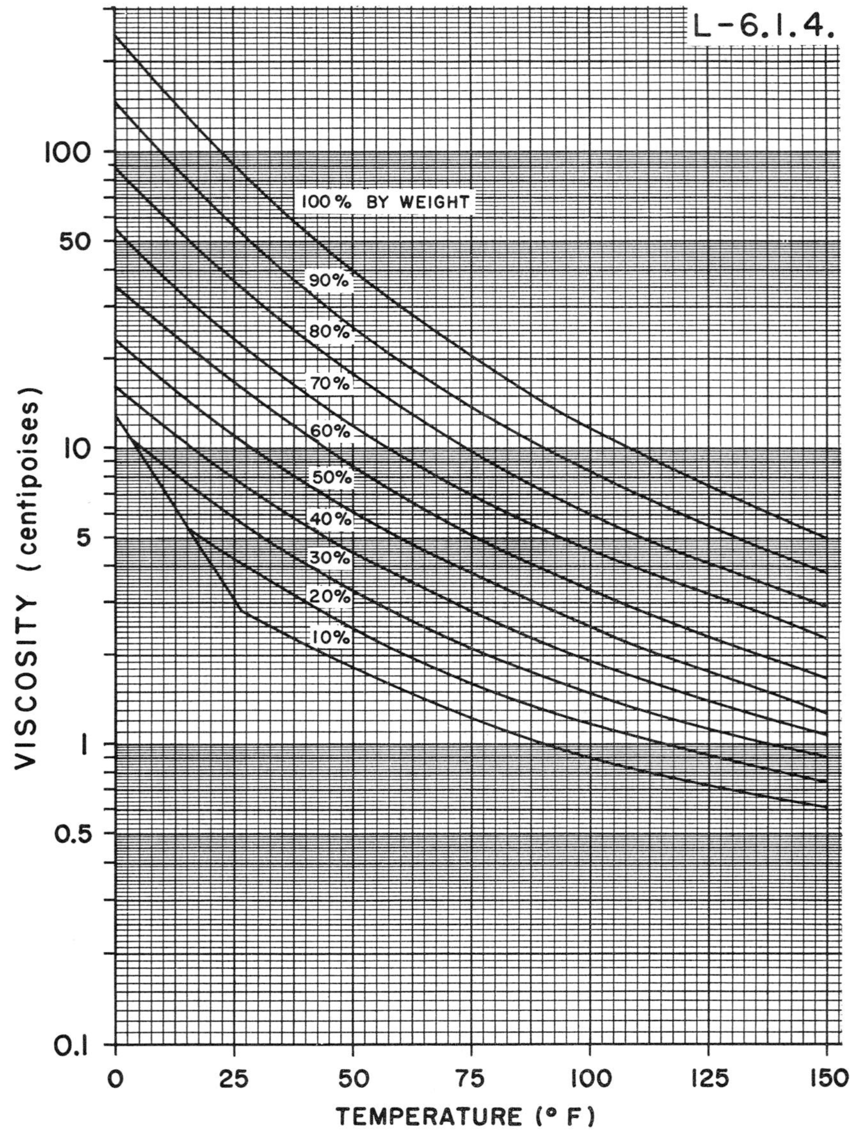

Viscosity vs Temperature

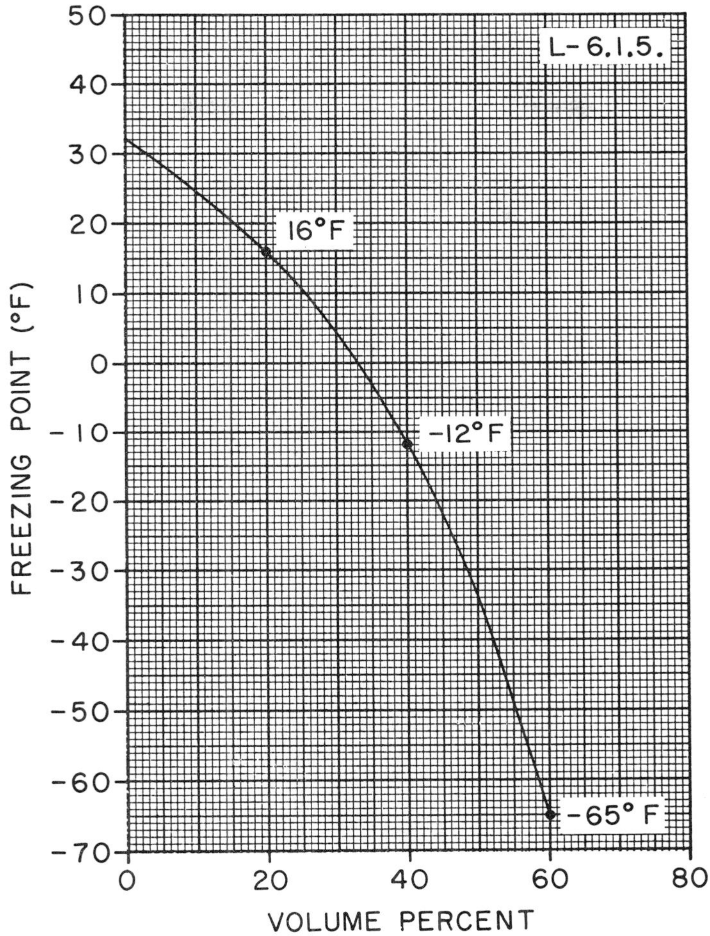

Freezing Point vs Volume Per Cent

L-6.2. GLYCOLS
(Ethylene and Propylene)

Glycols are used to improve one of the undesirable properties of water, i.e., its highest freezing point. Ethylene glycol, $HOCH_2CH_2OH$, and propylene glycol, $HOCH_2CH_2CH_2OH$, are two glycols commonly added to water in order to lower the freezing point. Most permanent type automobile anti-freezes have an ethylene glycol base. Propylene glycol, because of its low toxicity, is used for cooling applications in the food and pharmaceutical industries.

Properties of Importance

Specific Gravity: See Figure L-6.2.1.

Specific Heat: See Figures L-6.2.2 a, b, and c and L-6.2.3.

Thermal Conductivity: The thermal conductivity of an aqueous solution containing 50 per cent ethylene glycol by weight is 0.2419 Btu/(hr)(ft)(°F). For a 50 per cent solution of propylene glycol, the conductivity is 0.2177 Btu/(hr)(ft)(°F).

Viscosity: The addition of a glycol to water increases its viscosity. Pure ethylene glycol has a viscosity of 10.38 cp at 100°F and 61.7 cp at 30°F. The viscosity of pure propylene glycol is approximately 20 cp at 100°F and close to 250 cp at 32°F. As a comparison, the viscosity of water at 100° and 32°F is 0.68 cp and 1.79 cp, respectively. Viscosities of intermediate solutions fall between the values for the pure glycol and for water.

Vapor Pressure: See Figure L-6.2.4. The glycols lower the vapor pressure of water.

Freezing Point: See Figure L-6.2.5. There are no values for the freezing point of ethylene solutions with concentrations between 58 and 80 per cent ethylene glycol, because the nature of the solid phase in this range is doubtful and the results of various experiments differ.

Surface Tension: The surface tension of some glycol solutions is given in the following table in dynes/cm.

Wt. % Glycol (77°F)	Ethylene Glycol	Propylene Glycol
0	71.7	71.7
20	62.9	54.8
40	57.9	47.5
60	54.0	42.5
80	50.4	38.8
100	46.7	36.7

Corrosion: Ethylene glycol, by itself, or in aqueous solution, is quite corrosive to copper and steel, unless used with an effective inhibitor (and copper deactivator) and the residual alkalinity maintained at a safe level.

Thermal Stability: When the solution is heated above a temperature of about 220°F, the ethylene glycol seems to have a tendency to form a troublesome sludge unless a stabilizer is added.

Flammability: The flash points of 100 per cent ethylene glycol and propylene glycol are approximately 245° and 225°F, respectively. Ethylene glycol has a fire point of 250°F.

Physiological Effects: As mentioned above, the low toxicity of propylene glycol permits its use in the food industry.

Refractive Index: Ethylene glycol has a refractive index of 1.4316, and propylene glycol has a refractive index of 1.4326, both at 68°F.

Appearance: Both liquids are colorless.

Price Class: B.

References

1. Union Carbide Chemicals Co., "Glycols," 1955.
2. Cragoe, C. S., "Properties of Ethylene Glycol and Its Aqueous Solutions," Cooperative Fuel Research Committee of the Cooperative Research Council, 1943.
3. Factory Insurance Association, "Recommended Good Practice for Heat Transfer Mediums in Closed Systems," June 1954.

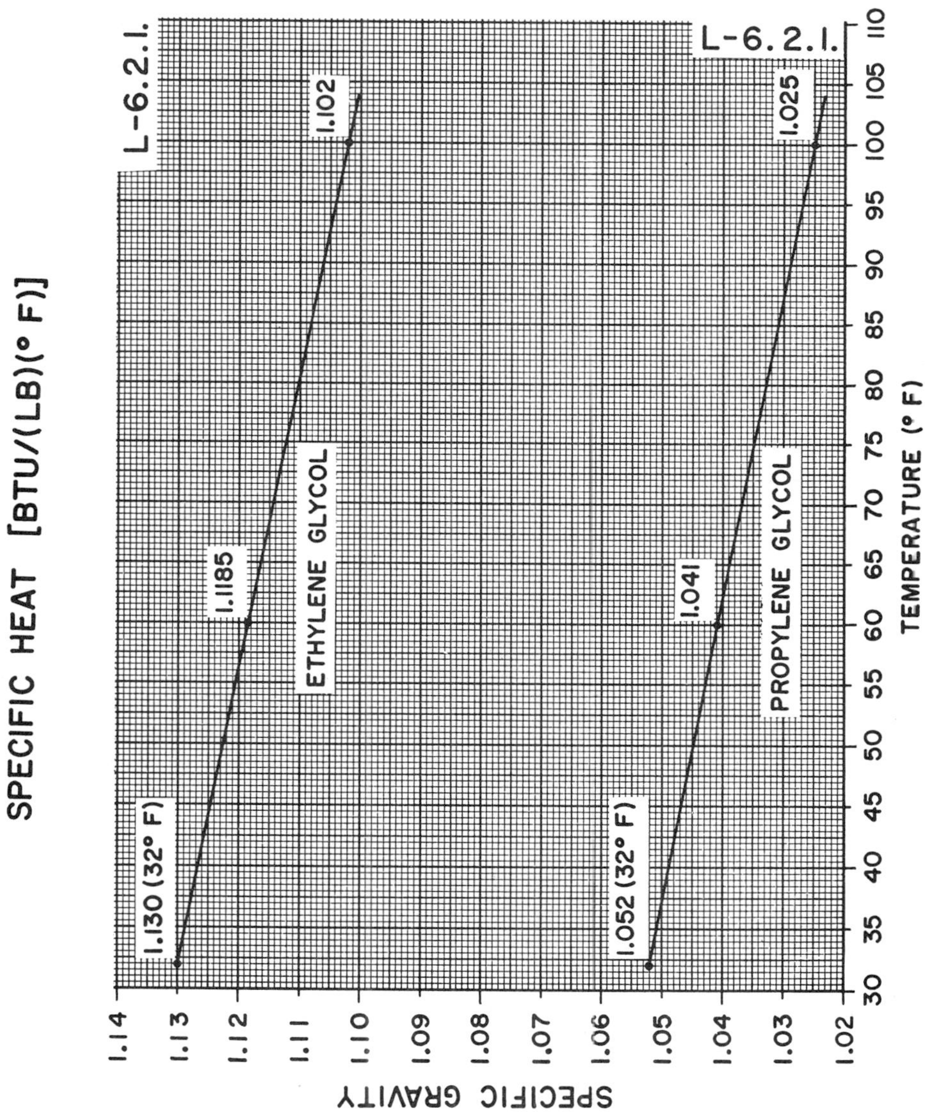

Specific Gravity vs Temperature

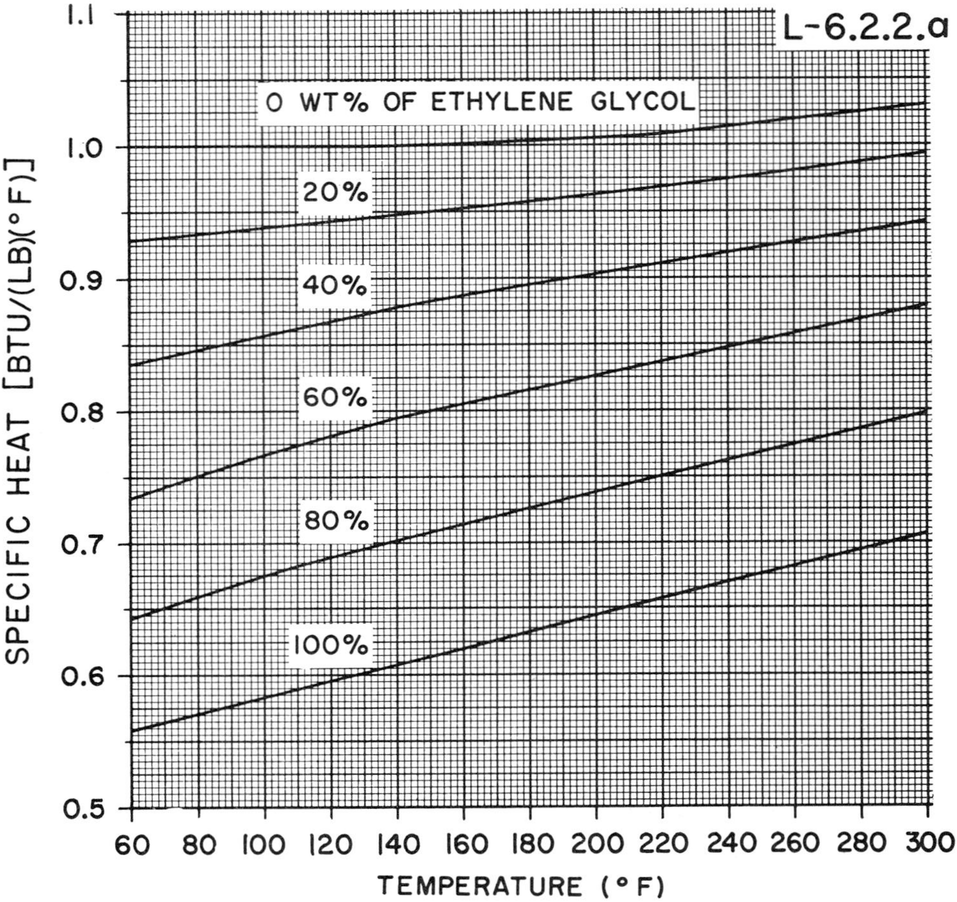

Specific Heat vs Temperature

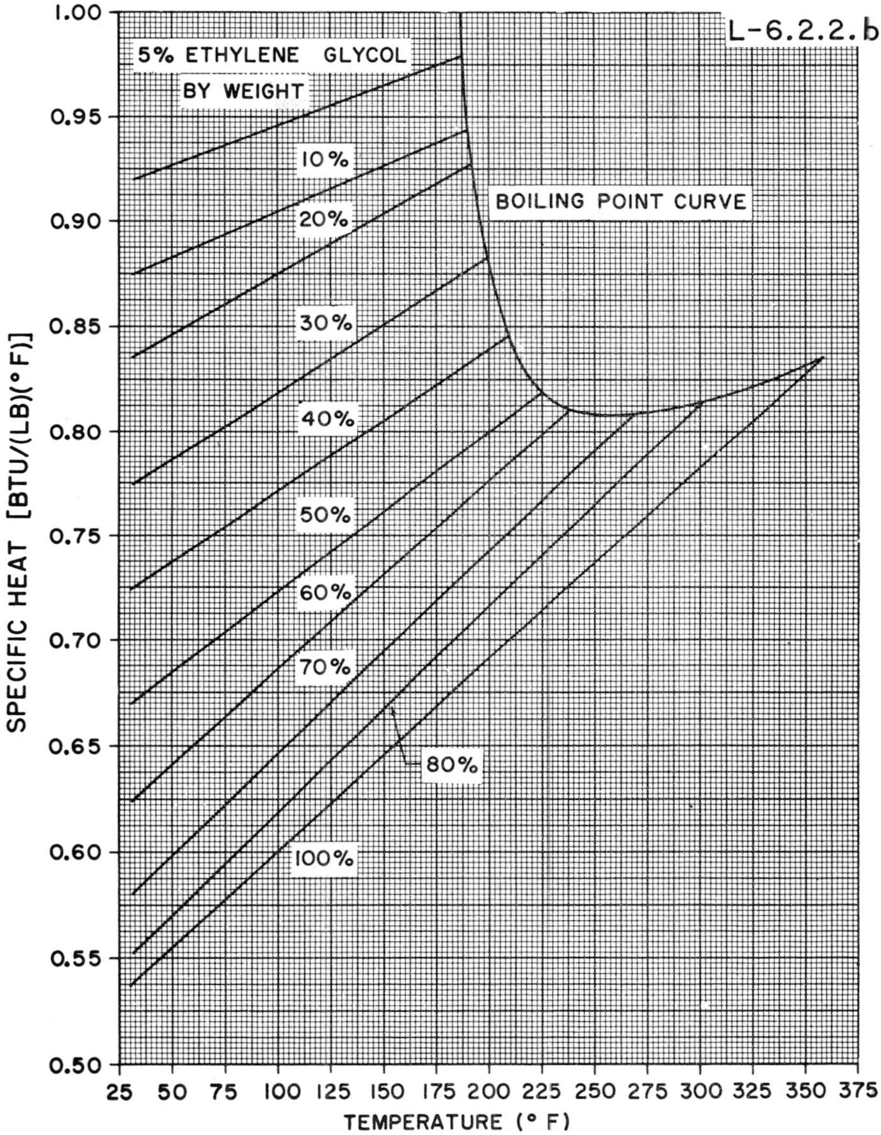

Specific Heat vs Temperature

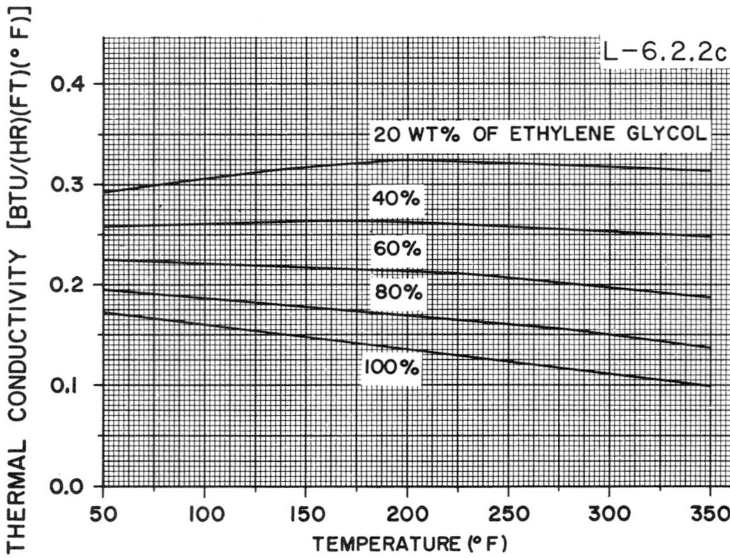

Thermal Conductivity vs Temperature

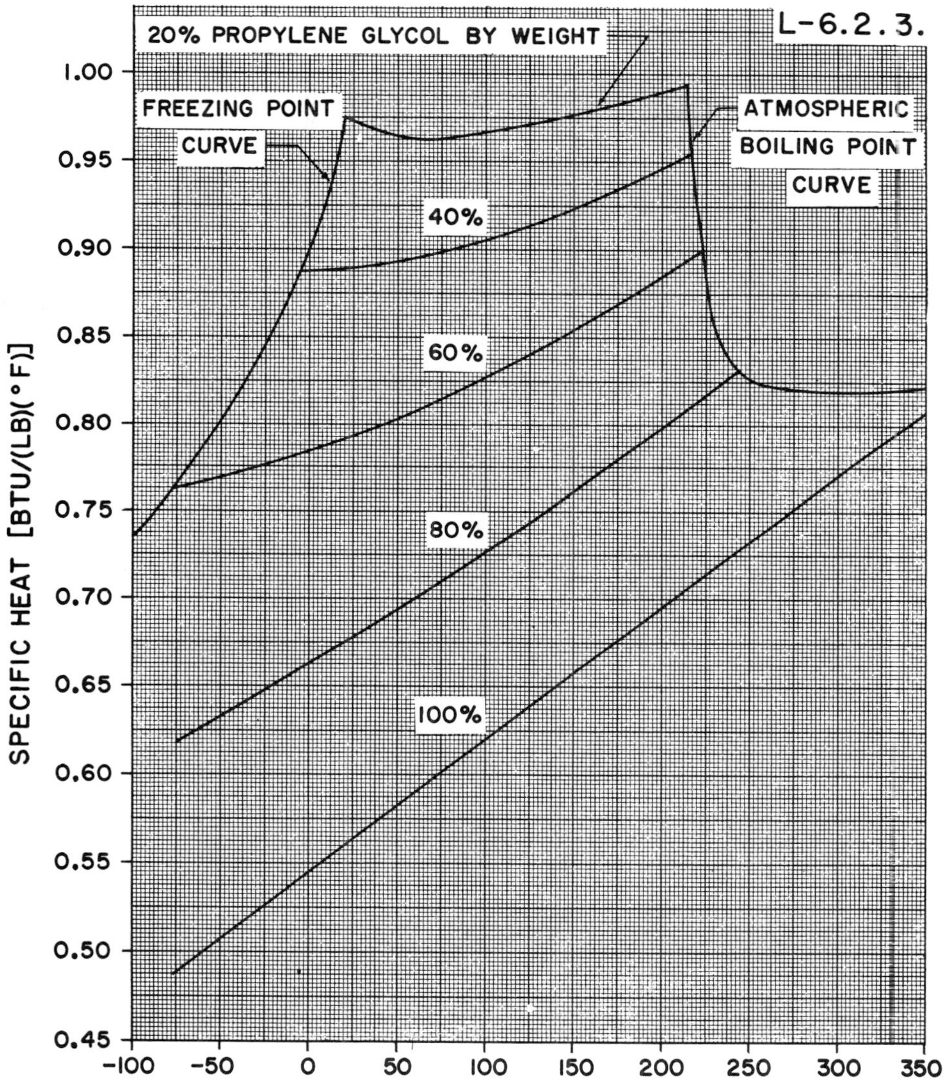

Specific Heat vs Temperature

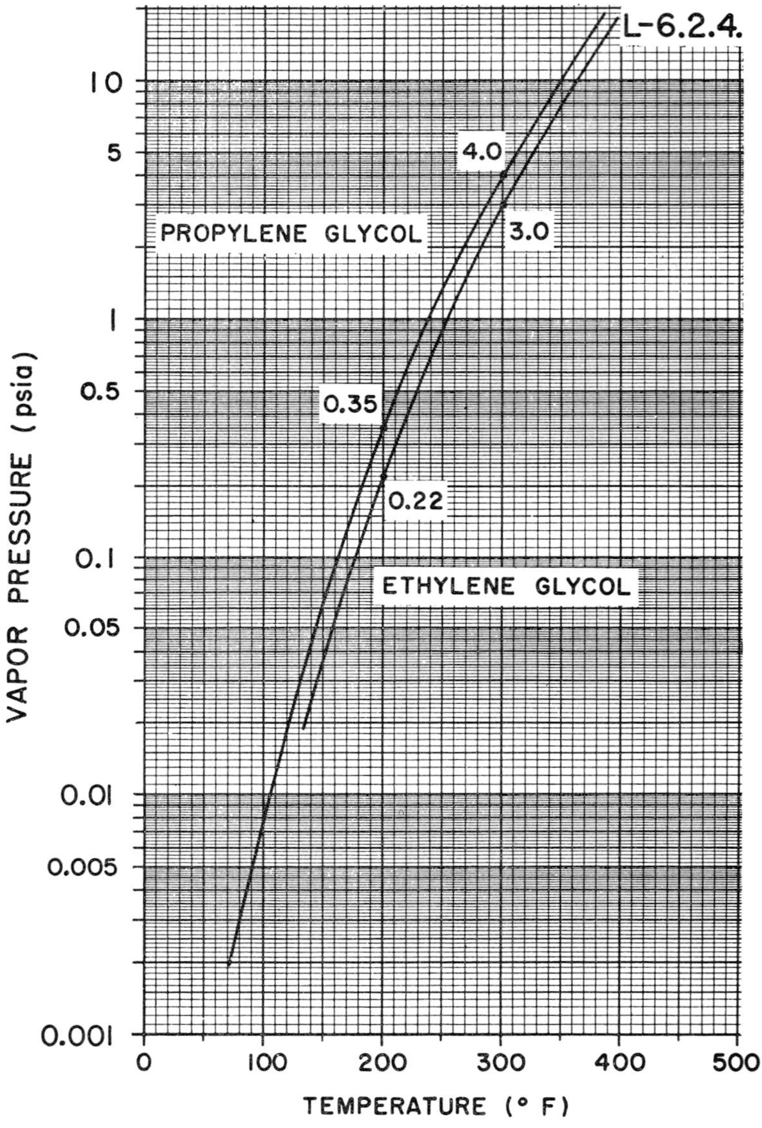

Vapor Pressure vs Temperature

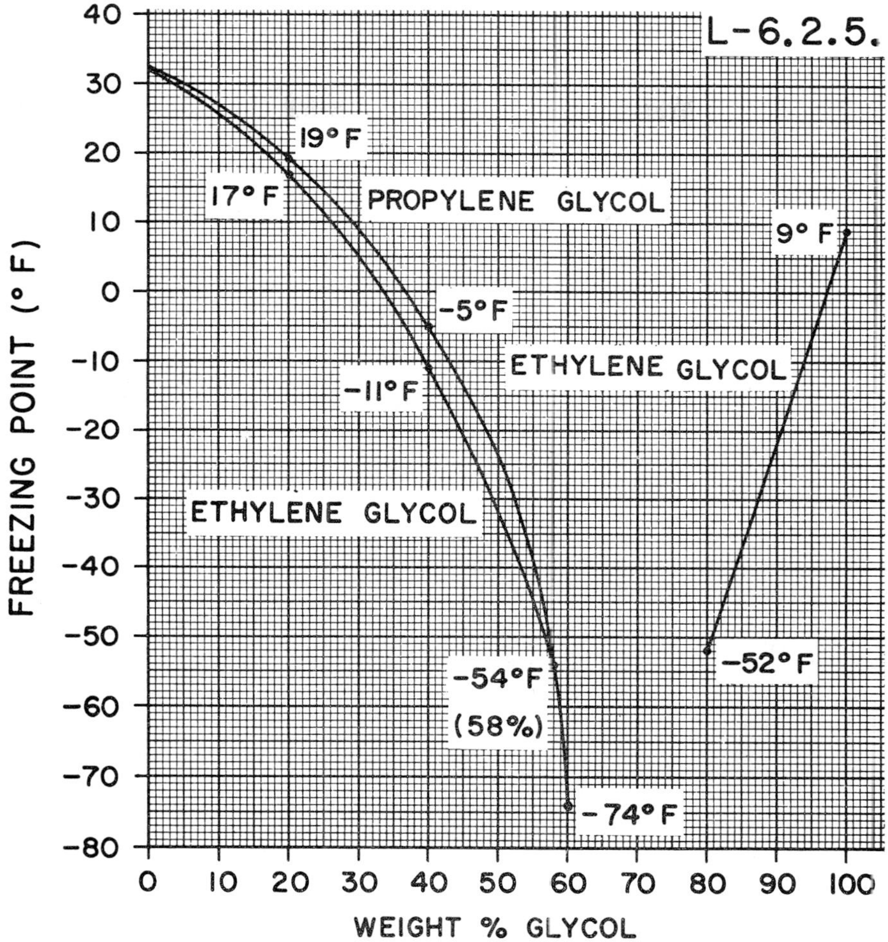

Freezing Point vs Weight Per Cent

L-6.3. POLYALKYLENE GLYCOL DERIVATIVES

(Ucon Fluids)

The Ucon fluids used for heat transfer are 50-HB-280-X and LB-300-X. The former is water soluble and the latter is water insoluble. These fluids are polyalkylene-glycol derivatives, and they are used at temperatures up to 500°F.

Properties of Importance

Specific Gravity: See Figure L-6.3.1.

Specific Heat: See Figure L-6.3.2.

Thermal Conductivity: See Figure L-6.3.1.

Viscosity: See Figure L-6.3.3.

Physical Stability: The pour point of Ucon 50-HB-280-X is −35°F. The pour point of LB-300-X is −40°F.

Thermal Stability: Ucon fluids have good thermal stability.

Chemical Stability: These fluids are very stable. However, even when they break down they tend to form soluble and volatile products instead of gummy sludges.

Surface Tension: At 68°F the surface tension of 50-HB-280-X is 35 to 40 dynes/cm; the surface tension of LB-300-X is 33 to 38 dynes/cm.

Corrosion: Ucon fluids can be used with any of the common structural metals.

Flammability:

Flash Point: The flash points of 50-HB-280-X and LB-300-X are 500° and 490°F, respectively.

Fire Point: LB-300-X has a fire point of 585°F and 50-HB-280-X has a fire point of 600°F.

Ignition Point: The ignition point of 50-HB-280-X is 752°F; the ignition point of LB-300-X is approximately 730°F.

Physiological Effects: These fluids have a low toxicity. When the systems are such that vapors are given off, adequate ventilation must be provided. The material is safe to handle.

Electrical Properties: At 77°F the dielectric constant of LB-300-X is 5.2 to 5.9. No data were reported on 50-HB-280-X. The resistivity of LB-300-X is 1030 to 8520 megohm-cm, the resistivity of 50-HB-280-X is 101 to 286 megohm-cm, both at 77°F.

Refractive Index: (n_d^{20}) The refractive index of LB-300-X is 1.452; that of 50-HB-280-X is not available.

Price Class: B.

References

1. Union Carbide and Carbon Corp., "Ucon Fluids and Lubricants," 1956.
2. Technical Information, Union Carbide and Carbon Corp., "Ucon Heat Transfer Fluids," February 1957.

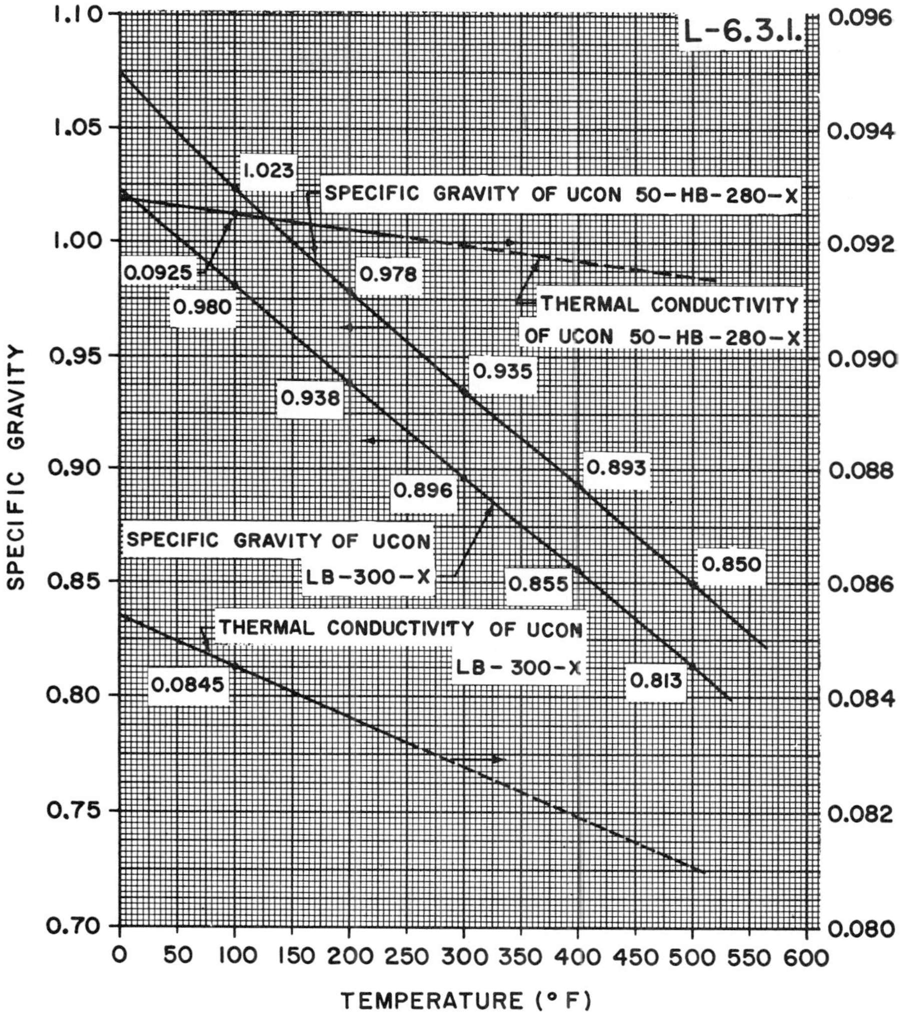

Thermal Conductivity vs Temperature
Specific Heat vs Temperature

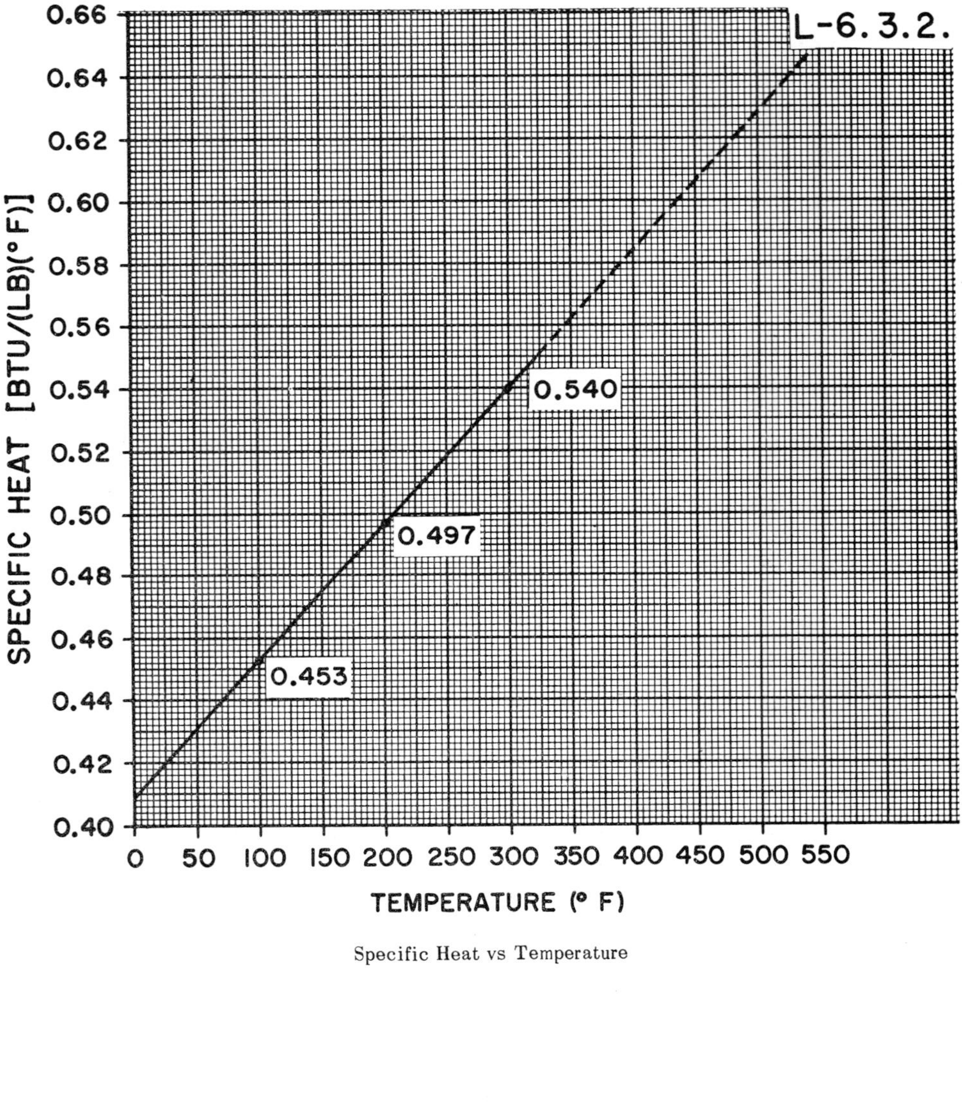

Specific Heat vs Temperature

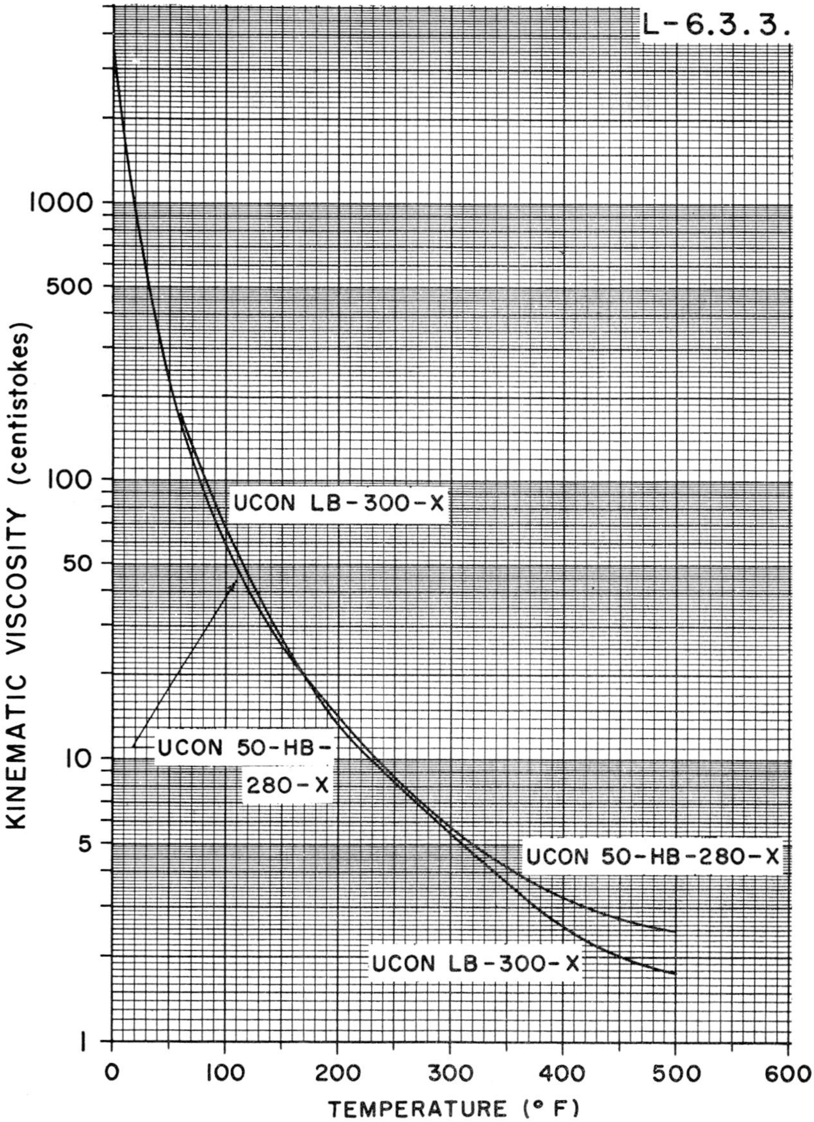

Kinematic Viscosity vs Temperature

L-7. HALOCARBONS

L-7.1. CHLORINATED BIPHENYLS

(Aroclors)

The Aroclor compounds are a series of chlorinated biphenyls and polyphenyls of increasing chlorine content, produced by Monsanto Chemical Company. These compounds range in form and appearance from Aroclor 1221, a colorless, mobile oil, to Aroclor 5460, a yellow, transparent resin. The last two numbers in each name represent the weight percent of chlorine in the compound. Many important physical characteristics such as viscosity, density, and pour point vary from lowest to highest molecular weight members of the series. Aroclor 1242, Aroclor 1248, and Aroclor 1254 have been employed as heat transfer media with great success, one reason being that they do not support combustion. The primary difference between the three lies in their pour points and their viscosities. Because of its low pour point (19.4°F) and its resistance to fire, Aroclor 1248 has been widely used at temperatures between 50° and 600°F. However, Aroclor 1242 and Aroclor 1254 may also be used in a number of special heat transfer applications. In addition to graphically presenting the important heat transfer properties of the Aroclor group, separate plots have been prepared for Aroclor 1248 (tetrachlorobiphenyl), the most eminent member of the family.

Properties of Importance

Density: See Figures L-7.1.1 for the density of Aroclor 1248, and Figure L-7.1.2 for the specific gravity of other Aroclor compounds.

Specific Heat: See Figures L-7.1.1 and L-7.1.3.

Enthalpy: See Figure L-7.1.4.

Thermal Conductivity:

180°F. – .0559 Btu/(hr)(ft)(°F)
80°F. – .0572 Btu/(hr)(ft)(°F)

Viscosity: See Figure L-7.1.5. The practical lower limit for pumping Aroclor 1248 is 50° to 55°F, when using centrifugal pumps.

Physical Stability

Vapor Pressure. See Figures L-7.1.6 and L-7.1.7. The practical lower limit for pumping Aroclor 1248 is 50°F.

Pour Point.

1242	1248	1254
2°F	19.4°F	50°F

Thermal Stability: The Aroclor compounds are all very stable within the recommended operating limit of 600°F. At higher temperatures, thermal decomposition takes place and higher polyphenyls are formed. Also, after a year or more of service slightly above 600°F, the fluid viscosity will increase and periodic change of the fluid is necessary. In a 454-day test in sealed glass under inert gas, Aroclor 1248 showed an increase in viscosity (at 77°F) from 155 cs to 221 cs and a low boiler content of 24 per cent.

Chemical Stability: Aroclor compounds are relatively inert liquids. They are resistant to the action of water, dilute alkalies, dilute acid solutions, air, and oxygen. They are readily soluble in most of the common organic solvents and drying oils.

Distillation Ranges: (ASTM D-20)

1242	1248	1254
617°-680°F	652°-725°F	690°-734°F

Flammability: The Aroclors do not support combustion, and their spontaneous ignition temperature is approximately 1200°F. Therefore, they can be considered fire-resistant. Flash points and fire points are tabulated below:

Flash Point: (*Cleveland Open Cup*)

1242	1248	1254
348°-356°F	379°-394°F	None

Fire Point: (*Cleveland Open Cup*)

1242	1248	1254
None	None	None

Physiological Effects: The use of Aroclor compounds in a typical closed heat exchange system does not present a problem. The level of toxicity of vapors of Aroclor substances at high temperatures is such that adequate provision must be made for ventilation where leakage occurs, and system vents should be removed from the normal operating area.

Although Aroclor compounds are not severe skin irritants, they should not be in repeated contact with the skin or for any length of time.

Electrical Properties: The Aroclors have high resistivities, high dielectric strength, and low power factors. The dielectric constant of Aroclor 1248 is 4.6 at 212 °F and 1000 cycles. These excellent properties make the Aroclor compounds outstanding fire-resistant transformer oils.

Refractive Index:

1242	1.627–1.629
1248	1.630–1.631
1254	1.639–1.641

Colors:

1242	Practically colorless mobile liquid.
1248	Yellow-tinted mobile liquid.
1254	Light-yellow viscous liquid.

Price Class: B.

A similar material is produced in Japan under the trade name of Kanechrol. However, this is a chlorinated diphenyl and chlorinated triphenyl.

Reference

1. Monsanto Chemical Co., "The Aroclors," Bulletin No. O-P-115.
2. Kanegafuchi Chemical Industry Co., Ltd., Tokyo, Japan.

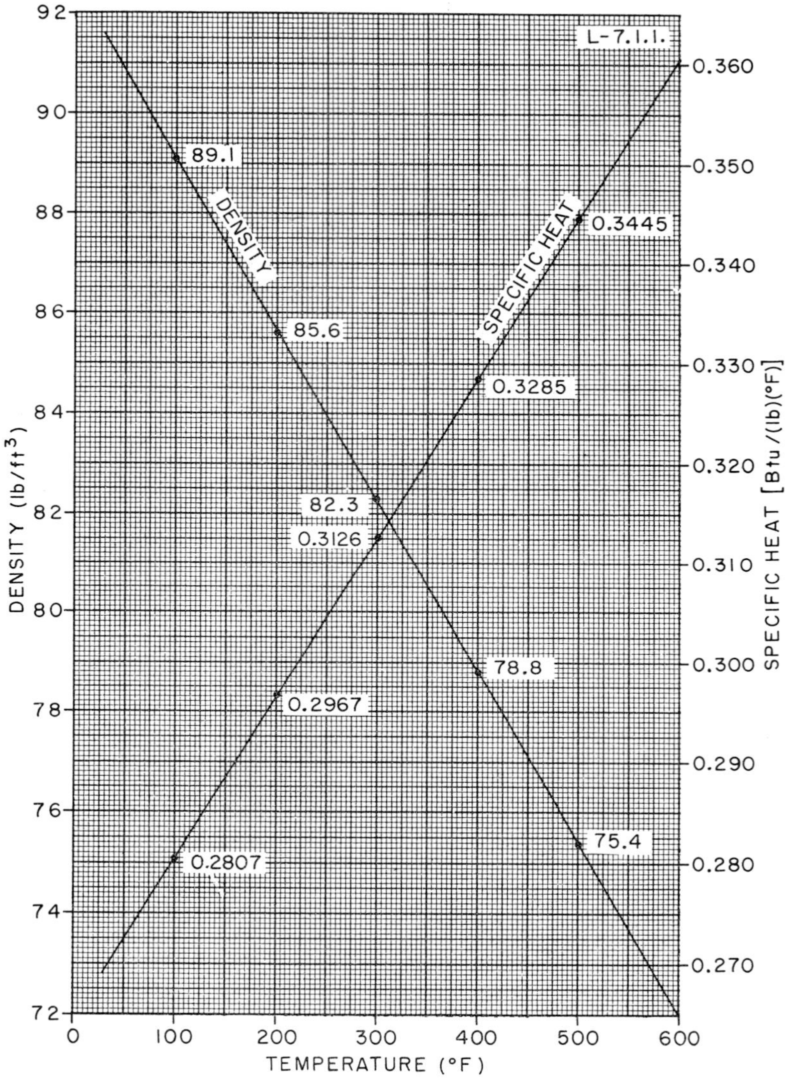

Specific Heat vs Temperature
Density vs Temperature

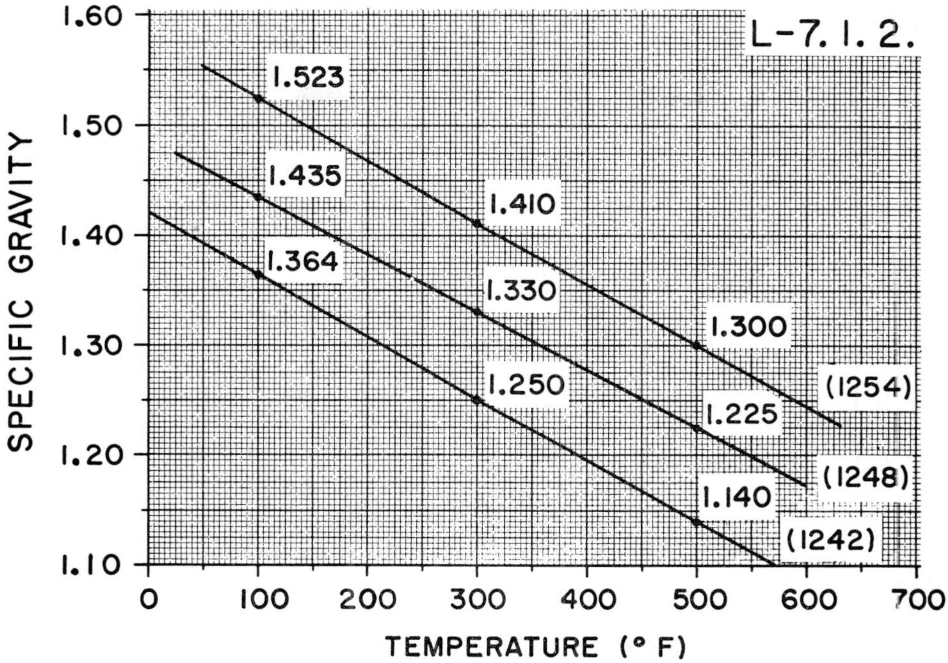

Specific Gravity vs Temperature

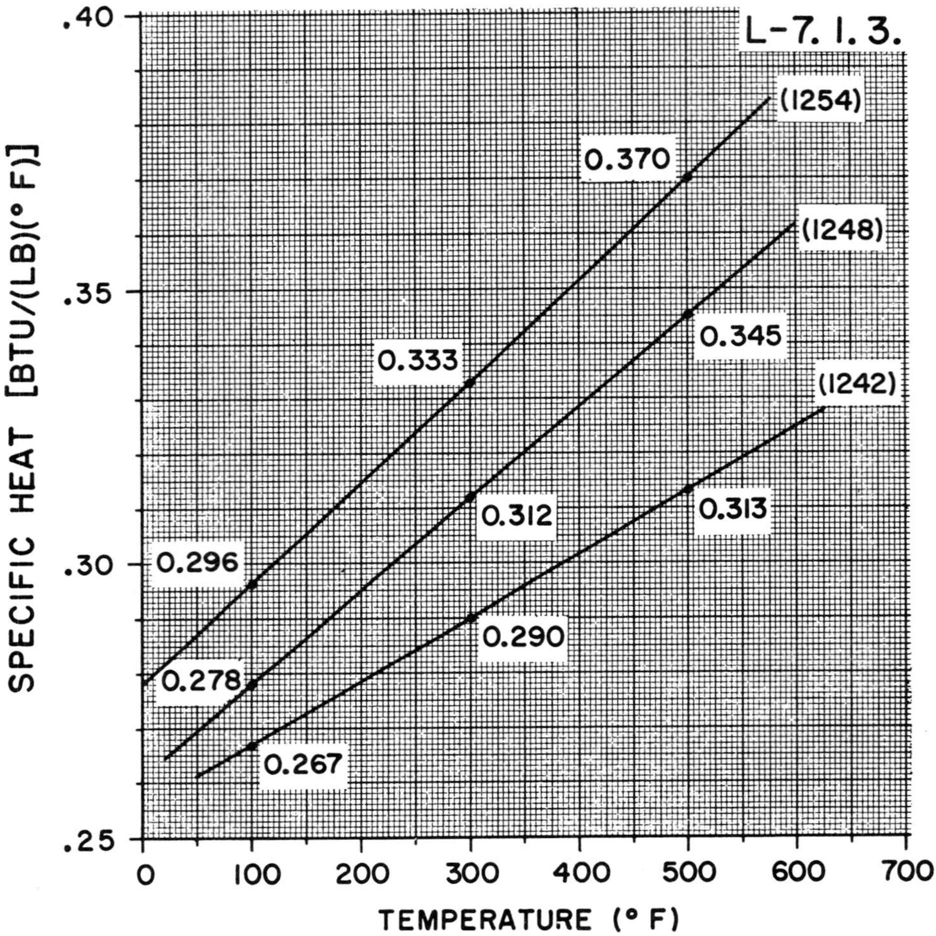

Specific Heat vs Temperature

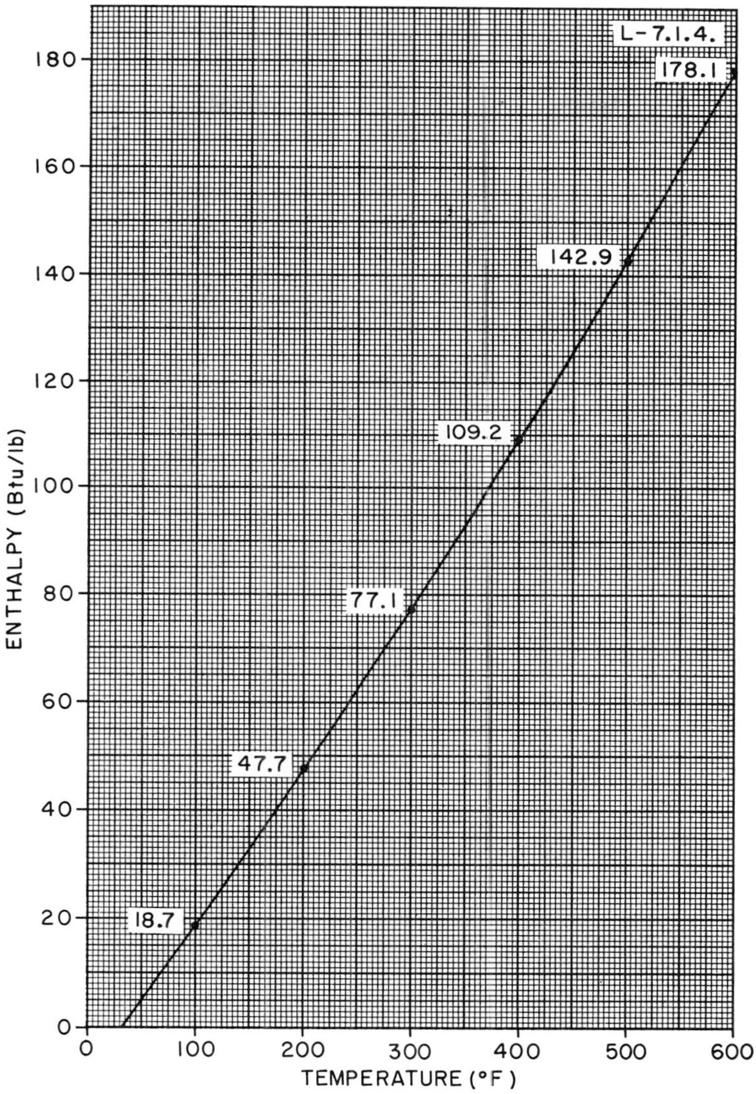

Enthalpy vs Temperature

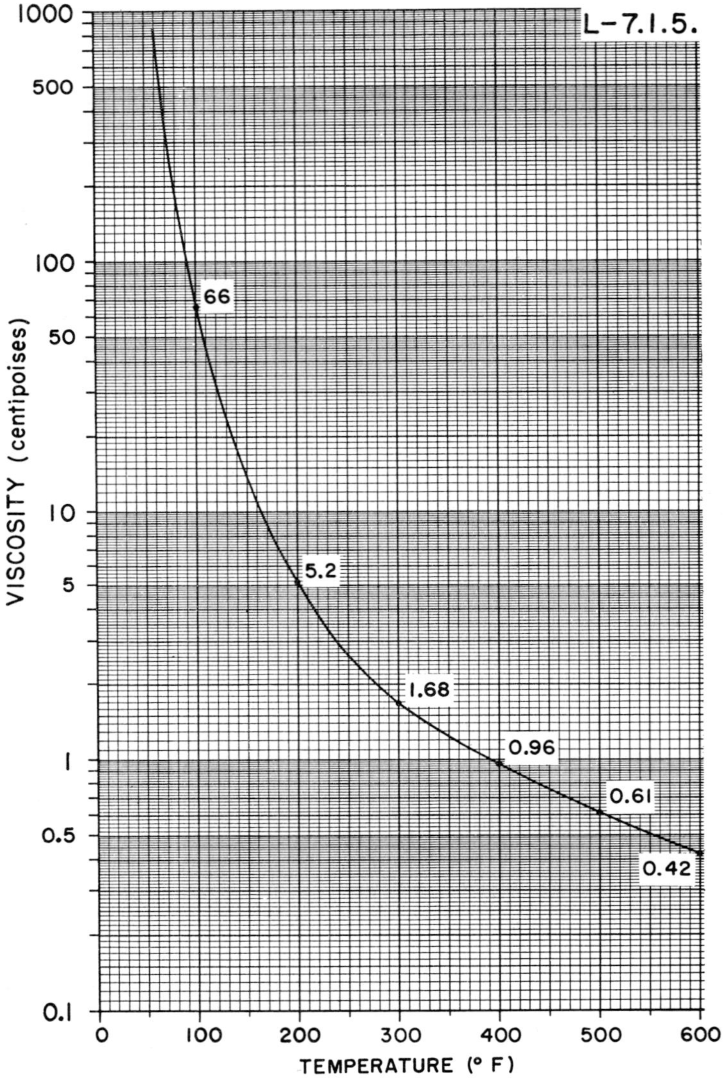

Viscosity vs Temperature

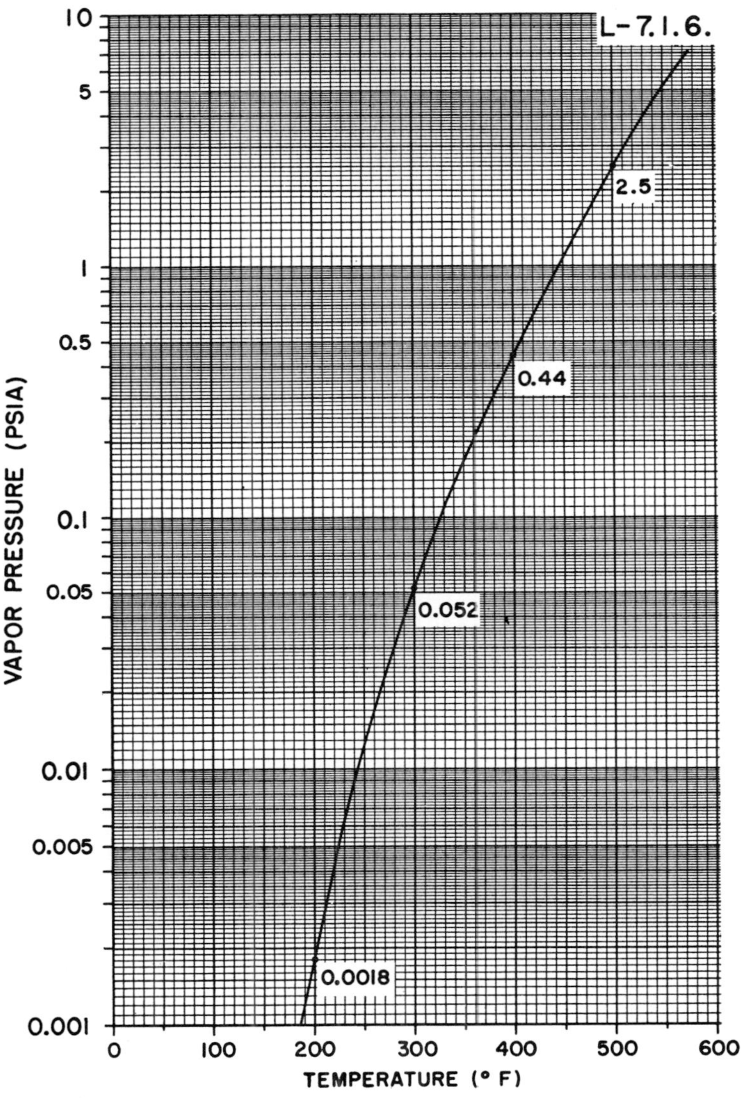

Vapor Pressure vs Temperature

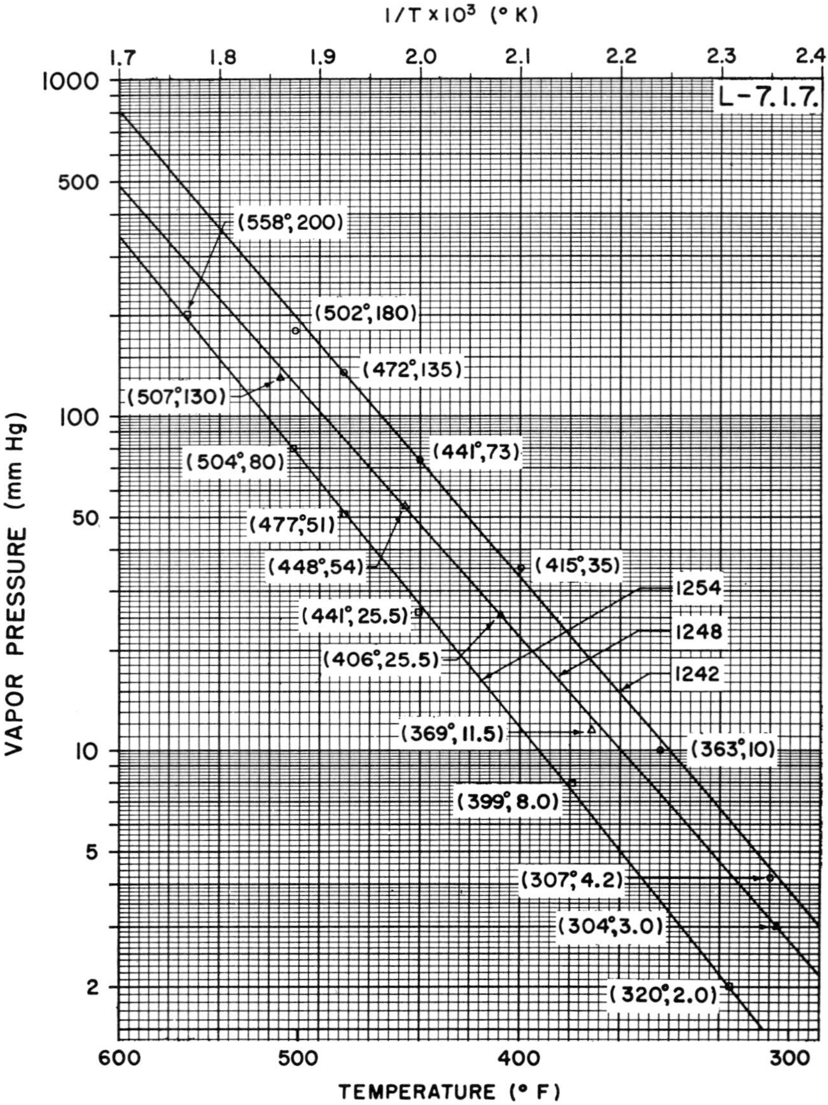
1/T x 10³ (° K)
1.7
1.8
1.9
2.0
2.1
2.2
2.3
2.4
L-7.1.7.
1000
500
100
50
10
5
2
(558°, 200)
(502°, 180)
(472°, 135)
(507°, 130)
(504°, 80)
(441°, 73)
(477°, 51)
(448°, 54)
(415°, 35)
(441°, 25.5)
(406°, 25.5)
1254
1248
1242
(369°, 11.5)
(363°, 10)
(399°, 8.0)
(307°, 4.2)
(304°, 3.0)
(320°, 2.0)
VAPOR PRESSURE (mm Hg)
600
500
400
300
TEMPERATURE (° F)

L-8. HYDROCARBONS

In this group are assembled liquids which contain compounds of carbon and hydrogen, in various chemical combinations. The simplest are the gaseous ones (previously discussed in Group G-4.), which are useful for heat transfer purposes as products of the combustion of gas, oil, and coal.

Liquid and solid hydrocarbons are more complicated. From the liquids found in nature, the mineral oils provide fine heat transfer media. Derivatives of the polyphenyl benzene ring are especially suited as heat carriers. They have received special attention due to their application for nuclear reactors.

The mineral oils and the polyphenyls are the major hydrocarbons used as liquid heat carriers.

Specially refined hydrocarbon oils or heat transfer oils find general use throughout the petroleum industry due to their low cost and ready availability. At elevated temperatures (450°-600°F) mineral oils undergo decomposition by two different mechanisms—cracking and oxidation.

In cracking, the carbon-hydrogen bonds are ruptured and the oils break down into heavier and lighter molecular weight fractions. Since heavy materials are insoluble, they deposit out on heat transfer surfaces, lowering thermal efficiency and blocking fluid flow. Lighter, volatile substances reduce the flash point of the oil by a substantial amount. Copper and its alloys catalyze thermal decomposition.

Rapid oxidation occurs when these oils are exposed to air at high temperatures. Coke-like oxy products are formed, blanketing and thermally insulating the walls of the system. Higher surface temperatures are required to preserve the original bulk oil temperatures; thus tube failure becomes likely. An oil at high temperature should not come into contact with air. Therefore, the expansion tanks should be kept cool in hot oil heat transfer systems.

The mineral oils can be divided into aliphatic and aromatic oils. Aliphatic oils have two major drawbacks which virtually eliminate them as candidate fluids: high viscosity and low thermal stability. For example, Pennsylvania cylinder oils possess Saybolt viscosities of the order of 200 seconds at 210°F. High viscosities cause trouble on starting pumps at low temperatures and substantially increase fluid resistance. These paraffinic oils are usually employed at temperatures as high as 450°F in open systems, while in closed systems, they can be applied in the 450°-600°F range.

Mineral oils of aromatic character, such as those found in coal tar, display greater thermal stability than ordinary aliphatic oils, because of a powerful ring structure. Lower viscosities further enhance the superiority of aromatic mineral oils over their aliphatic counterparts. Paraffinic oils enjoy a higher initial flash point, however, the production of volatile materials on cracking nullifies this advantage.

The upper temperature limit for aromatic oils has been the subject of considerable debate. Leading oil companies claim a maximum bulk oil temperature of 600°F under ideal operating conditions. However, independent tests indicate that 600°F is too optimistic a figure and a temperature range of 550°-575°F seems to be the absolute maximum. The exact maximum temperature depends on the system design and the type of operation desired.

L-8.1. MINERAL OIL

(Humble Therm 500)

Humble Therm 500 is a petroleum oil with excellent thermal stability and high oxidation resistance, specially processed for use as a heat transfer medium.

Properties of Importance

Specific Gravity: See Figure L-8.1.1.

Specific Heat: See Figure L-8.1.1.

Viscosity: See Figure L-8.1.2. A high viscosity index insures a continuity of film heat transfer, which is a valuable property of mineral oils.

Thermal Conductivity: See Figure L-8.1.1.

Physical Stability: The fluid remains pumpable down to 15°F.

Thermal Stability: In open systems exposed to the atmosphere, Humble Therm 500 can be applied at temperatures up to 400°F. The maximum temperature limit for operation in a closed system is 600°F.

Corrosion: The fluid has a low corrosivity. Copper and copper alloys are most susceptible to corrosion by Humble Therm 500, in fact, by all petroleum oils.

Flammability: Humble Therm 500 has a flash point of 425°F.

Physiological Effects: This fluid should be handled with the same precautions afforded any petroleum lubricating oil.

Appearance: The color is ASTM 1½; the odor level of Humble Therm 500 is low.

Price Class: A.

Reference

1. Correspondence with Humble Oil and Refining Co.

L-8.2. MINERAL OIL

(Mobiltherm Light and Mobiltherm 600)

Mobiltherm Light and Mobiltherm 600 are aromatic hydrocarbon mixtures used as liquid heat transfer materials. Mobiltherm Light is recommended for use in the temperature range of −15° to 460°F. Mobiltherm 600 is suggested for the 30°–600°F range.

Properties of Importance

Specific Gravity: See Figure L-8.2.1.

Specific Heat: See Figure L-8.2.2.

Thermal Conductivity: See Figure L-8.2.3.

Kinematic Viscosity: See Figure L-8.2.4.

Physical Stability: See Figure L-8.2.5. for the vapor pressure curve.

Pour Point: Mobiltherm Light: below −20°F
Mobiltherm 600: +20°F

Thermal Stability: Since Mobiltherm oils are composed almost entirely of aromatic hydrocarbons which are extremely resistant to cracking and thermal decomposition, they do not break down in long-time service in the temperature ranges recommended.

Chemical Stability: Copper catalyzes oxidation of the oils, therefore copper pipes and valves should be avoided.

Corrosion: Mobiltherm fluids should not be used with copper and copper alloys.

Flammability: The flash points of Mobiltherm Light and Mobiltherm 600 are 250° and 360°F, respectively.

Physiological Effects: The Mobiltherm oils are nontoxic.

Color: Both fluids are colorless.

Price Class: A.

References

1. Socony Mobil Oil Co., "Aromatic Heat Transfer Oils," 1957.
2. Socony Mobil Oil Co., "Heating with Mobiltherm," 1960.

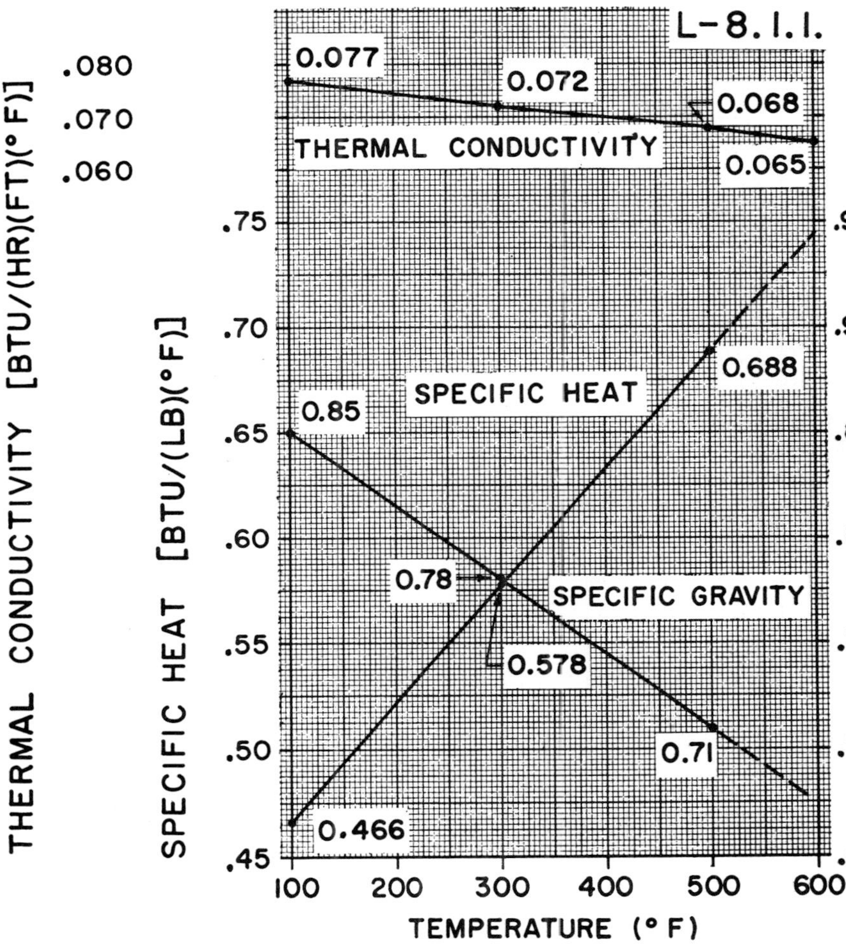

Thermal Conductivity, Specific Heat, and Density vs Temperature

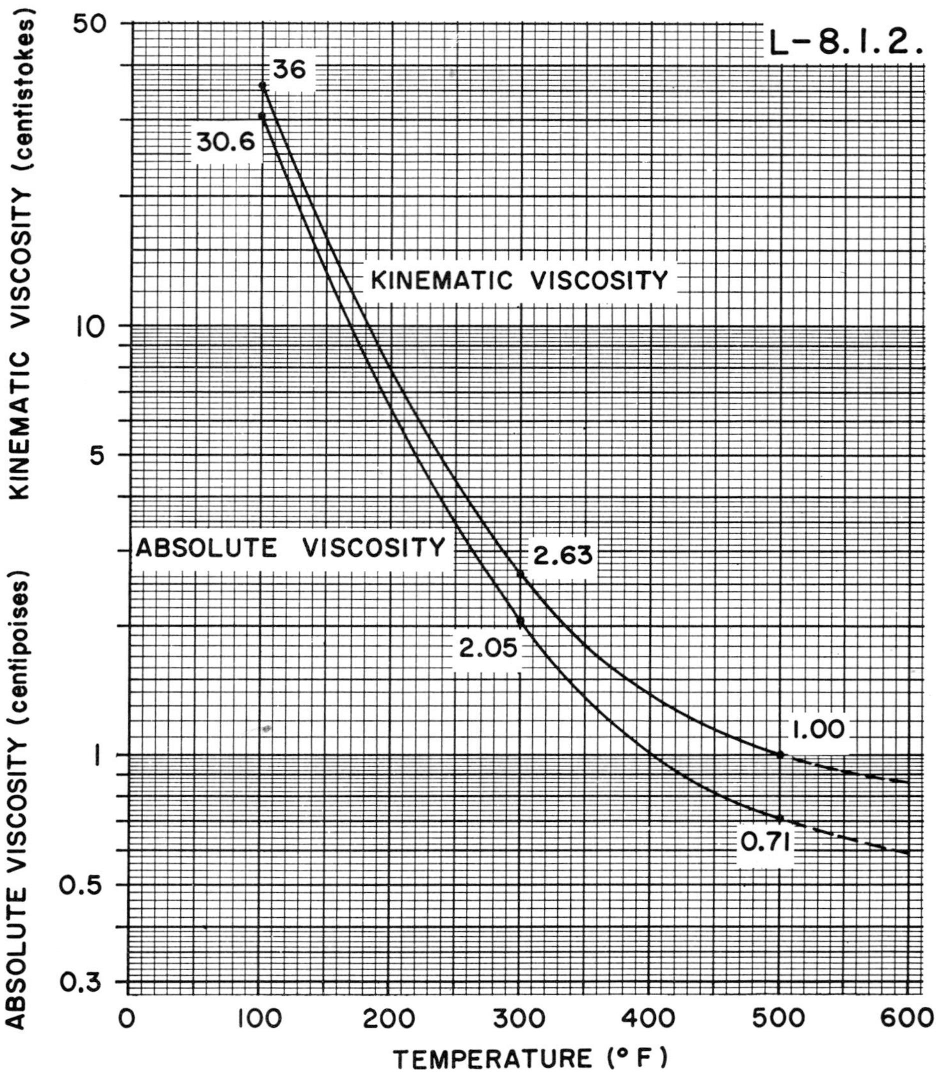

Kinematic Viscosity vs Temperature
Absolute Viscosity vs Temperature

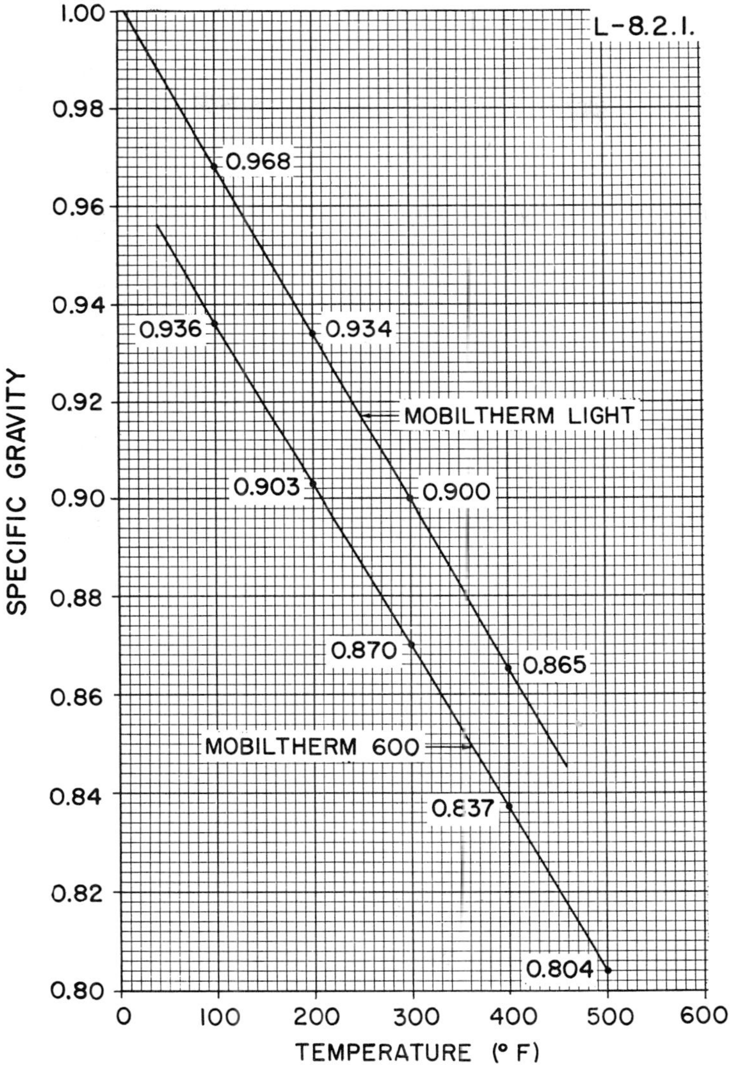

Specific Gravity vs Temperature

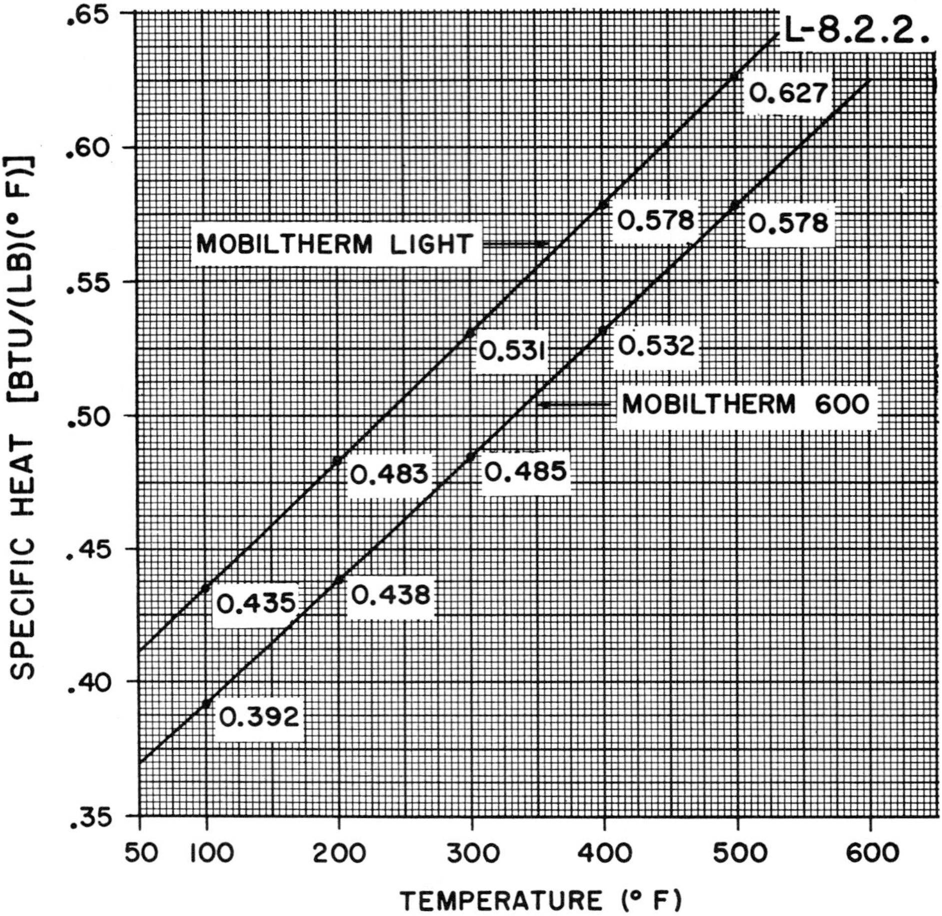

Specific Heat vs Temperature

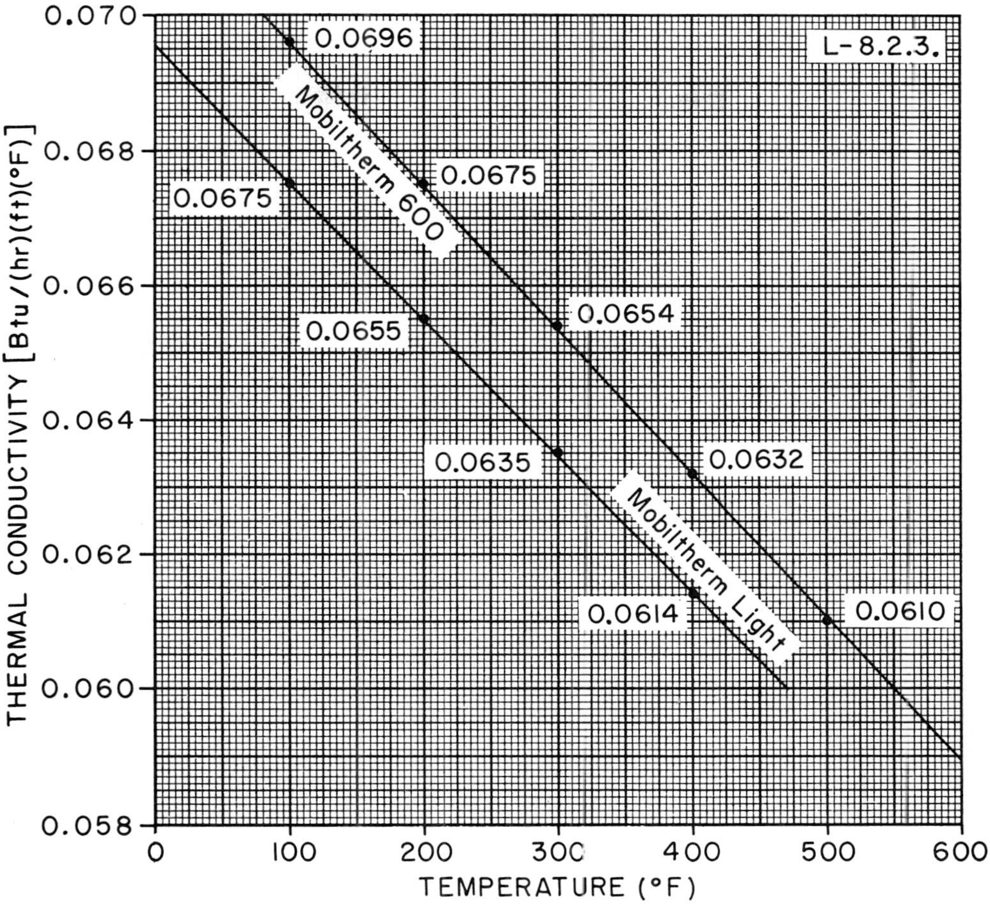

Thermal Conductivity vs Temperature

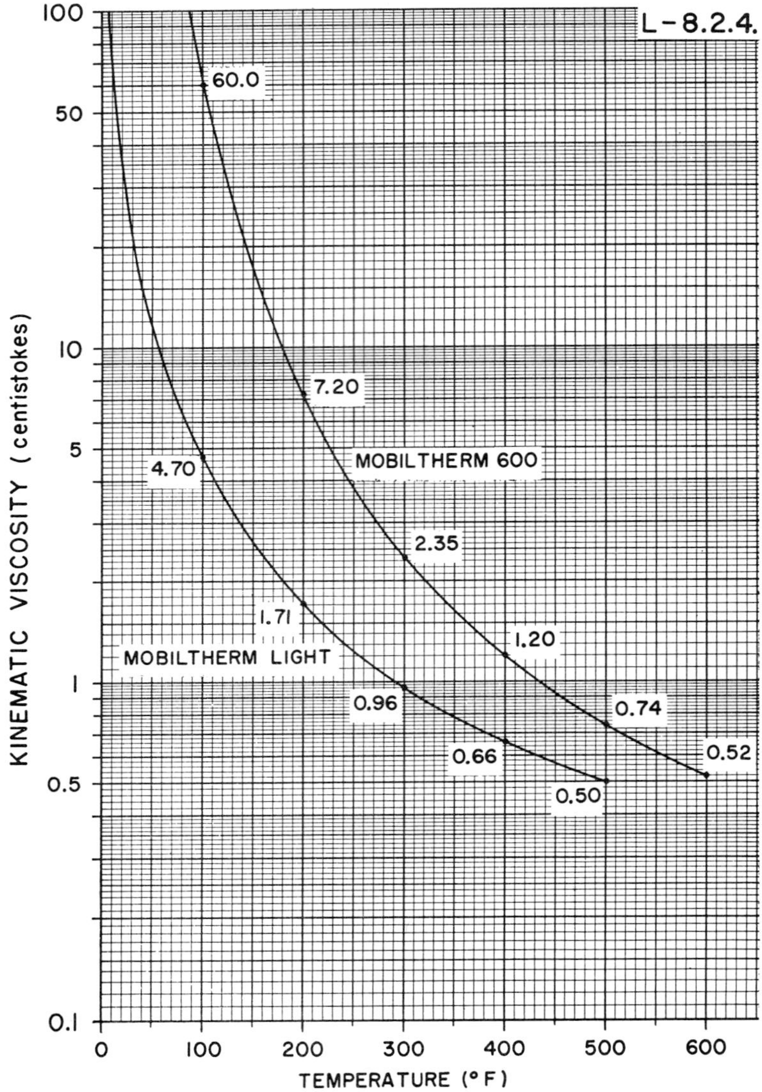

Kinematic Viscosity vs Temperature

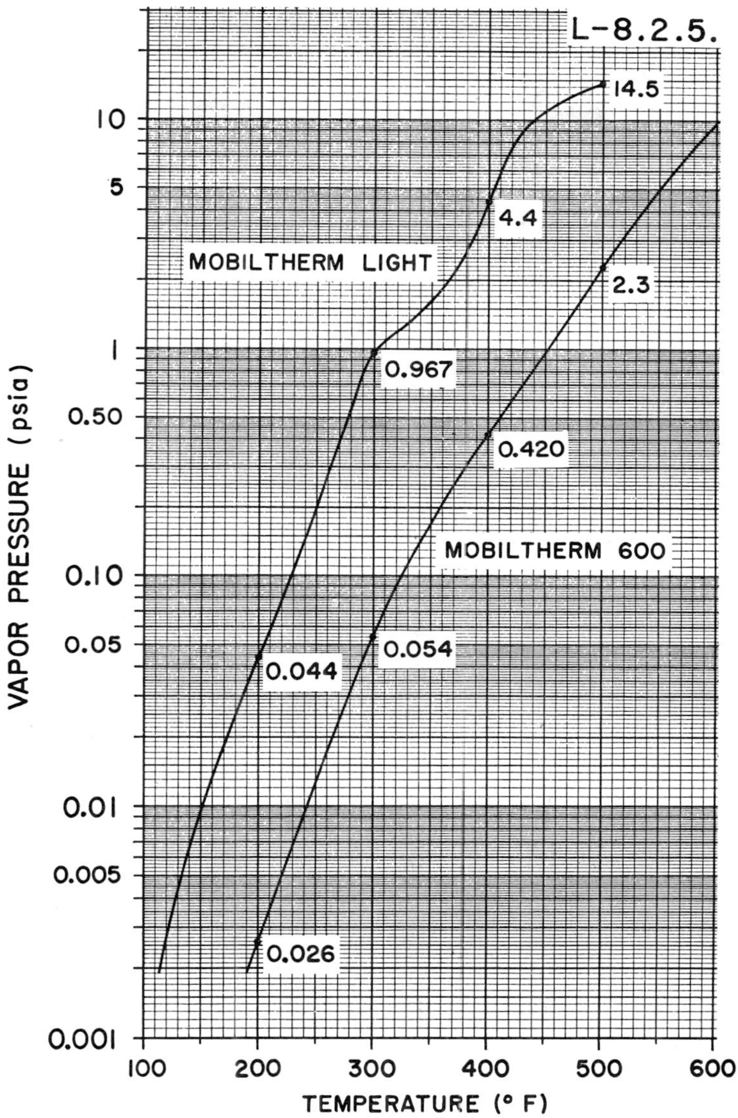

Vapor Pressure vs Temperature

L-8.3. MINERAL OIL

(Sun 21 Heat Transfer Oil)

Sun 21 Heat Transfer Oil is designed for use in closed circulating systems equipped with expansion tanks at operating temperatures up to 600 °F.

Properties of Importance

Specific Gravity: See Figure L-8.3.1.

Specific Heat: See Figure L-8.3.1.

Thermal Conductivity: See Figure L-8.3.1.

Viscosity: See Figure L-8.3.2.

Physical Stability: Sun 21 Oil has low vapor pressure at elevated operating temperatures and a maximum pour point of 0 °F.

Thermal Stability: Sun 21 is stable within the recommended operating range. Bomb tests have shown that after six hours at 700 °F, the pressure build-up caused by thermal breakdown to gas is only 4 psi.

Chemical Stability: Sun 21 Heat Transfer Oil will not form hard carbon and will not crack into light compounds under the conditions of recommended service.

Flammability: The flash point is 415 °F, minimum (Open Cup).

Price Class: A.

References

1. Sun Oil Co., "Sun 21 Heat-Transfer Oil," 1959.

L-8.4. POLYPHENYLS

Polyphenyls have extended the workable heat transfer range for liquid hydrocarbons up to temperatures approaching 900 °F. Only molten salts, liquid metals, and certain newly developed ethers seem to withstand these elevated temperatures. Polyphenyls have properties which have favored their use as coolants in nuclear reactors.

Properties of Importance

Density: See Figure L-8.4.1.

Specific Heat: See Figure L-8.4.2.

Enthalpy: The following table gives the liquid enthalpy of biphenyl at various temperatures:

t (°F)	Enthalpy (Btu/lb)
156.6	0.0
200	17.7
300	61.9
400	112.6
500	171.0
600	234.8
700	301.5
800	369.6
900	438.4
980	493.7

Latent Heat of Fusion: 53.1 Btu/lb
Latent Heat of Vaporization: 190.0 Btu/lb

Thermal Conductivity: As is typical of most organic liquids, the thermal conductivity of polyphenyls is relatively low:

Biphenyl	0.076 Btu/(hr)(ft)(°F)
p-Terphenyl	0.060 " " " "
m-Terphenyl	0.063 " " " "
p-Terphenyl	0.070 " " " "

Viscosity: See Figure L-8.4.3.

Physical Stability: See Figure L-8.4.4. Polyphenyls have low vapor pressures up to about 850 °F. They have relatively high melting points, as borne out by the following values:

Biphenyl	156 °F
o-Terphenyl	133 °F
m-Terphenyl	189 °F
p-Terphenyl	415 °F

Lower melting points can be obtained with a eutectic mixture of biphenyl and terphenyls.

Chemical Stability: No hazardous chemical reactions take place between uranium or water and polyphenyls.

Thermal Stability: Pyrolytic and radiolytic breakdowns follow similar patterns, resulting in similar decomposition products. The approximate bulk temperatures at which pyrolysis becomes significant are:

Biphenyl	815 °F
Terphenyl	795 °F

In nuclear applications, constant regeneration and partial replacement of the fluids allow the maintenance of higher temperatures. At these temperatures, polymerization to higher molecular weight polyphenyls slowly takes place. This increases the viscosity and reduces heat transfer. For example, when *m*-terphenyl is held at 842°F for 74 hours, 12 per cent polymer by weight is formed. Hydrogen gas is a by-product of the polymerization reaction.

Radiolytic Stability: Polyphenyls do not become very radioactive and do not have large neutron capture cross-sections. All of the polyphenyls have very nearly the same radiolytic decomposition rates with *p*-terphenyl having the best radiation stability. Nuclear radiation in a reactor tends to polymerize polyphenyls into longer chain terphenyl compounds, with evolution of hydrogen gas in the same way as in the pyrolysis reactions, as shown below.

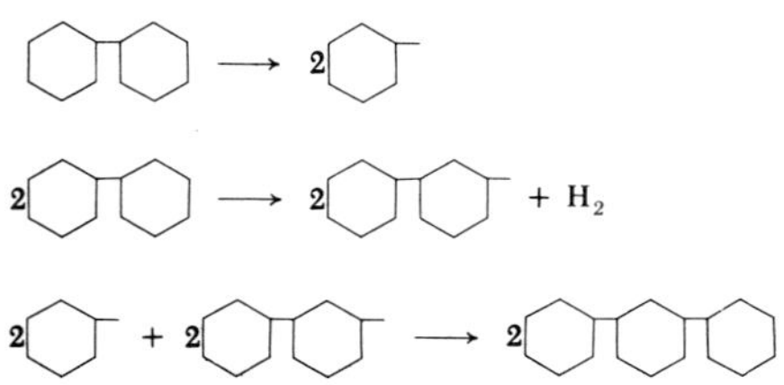

Surface Tension: 265°F – 39.5 dynes/cm

Corrosion: Polyphenyls do not attack common, low-cost construction materials.

Flammability:

	Flash Point	Fire Point
Biphenyl	235°F	253°F
o-Terphenyl	340°F	380°F
m-Terphenyl	405°F	445°F
p-Terphenyl	410°F	460°F

Physiological Effects: These compounds have a low order of toxicity. However, their vapor should not be inhaled in great quantities.

Electrical Properties:

	Dielectric Constant	Resistivity (ohm-cm)
o-Terphenyl	77°F (60 cps) 2.54	212°F 8200×10^9
m-Terphenyl	77°F (60 cps) 2.62	212°F 2600×10^9
p-Terphenyl		482°F 30×10^9

Refractive Index: Biphenyl (170.8) 1.58822

Appearance: Biphenyl is a pale yellow crystalline solid at room temperature.

Price Class: B.

References

1. Monsanto Chemical Co., "Monsanto Data Book."
2. Monsanto Chemical Co., "Radiation Resistant Fluids."
3. Monsanto Chemicals Ltd., England, "Diphenyl for Heat Transfer."
4. Ewing, C. T., Stone, J. P., Blachly, C. H., Walker, B. E., and Miller, R. R., "Naval Research Laboratory Report 4990," Nov. 13, 1957.
5. Colichman, E. L., Gercke, R. H., *Nucleonics*, **14**, (1956).
6. *Nucleonics*, **16**, 108 (1958).

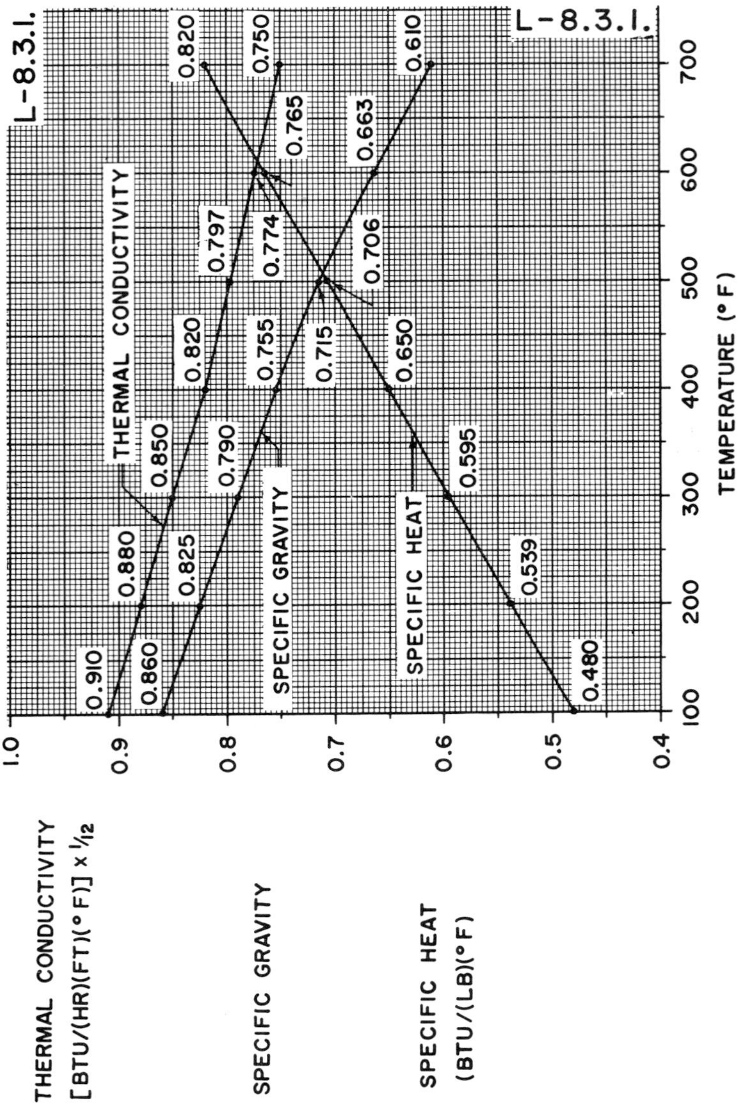
L-8.3.1.
L-8.3.1.
THERMAL CONDUCTIVITY
SPECIFIC GRAVITY
SPECIFIC HEAT
0.910
0.880
0.850
0.820
0.797
0.774
0.765
0.750
0.860
0.825
0.790
0.755
0.715
0.820
0.480
0.539
0.595
0.650
0.706
0.663
0.610
100
200
300
400
500
600
700
TEMPERATURE (° F)
1.0
0.9
0.8
0.7
0.6
0.5
0.4
THERMAL CONDUCTIVITY
[BTU/(HR)(FT)(° F)] x 1/12
SPECIFIC GRAVITY
SPECIFIC HEAT
(BTU/(LB)(° F)

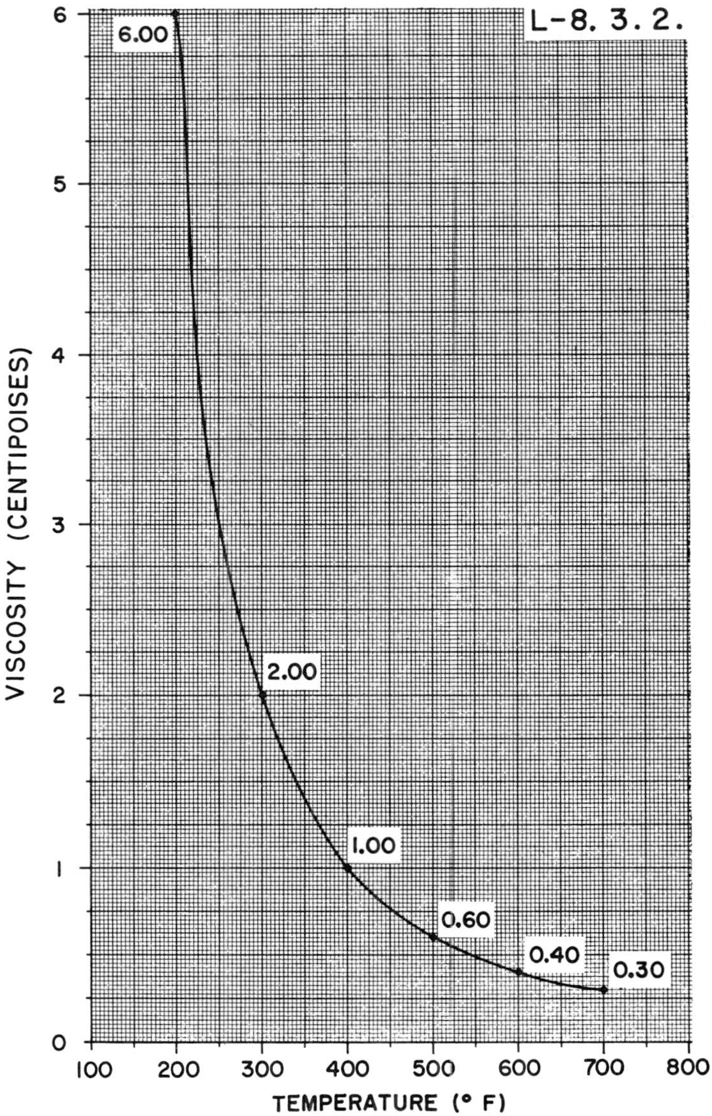

Viscosity vs Temperature

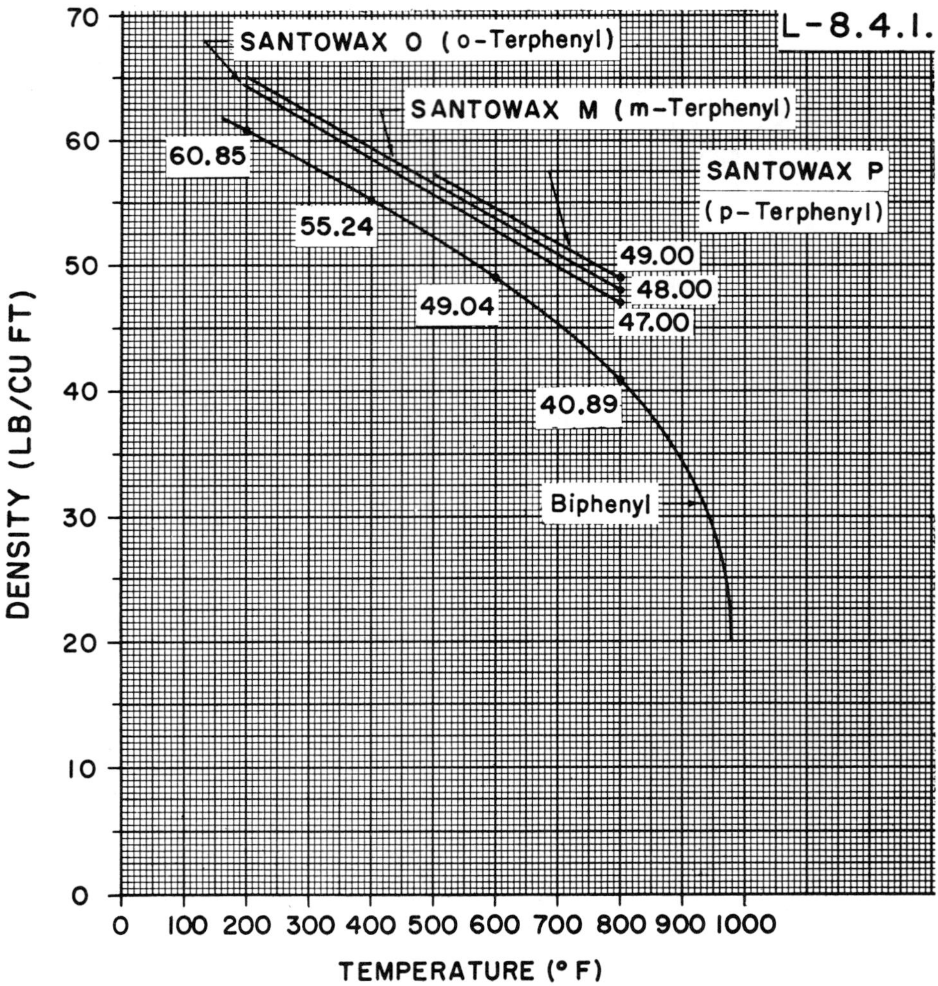

Density vs Temperature

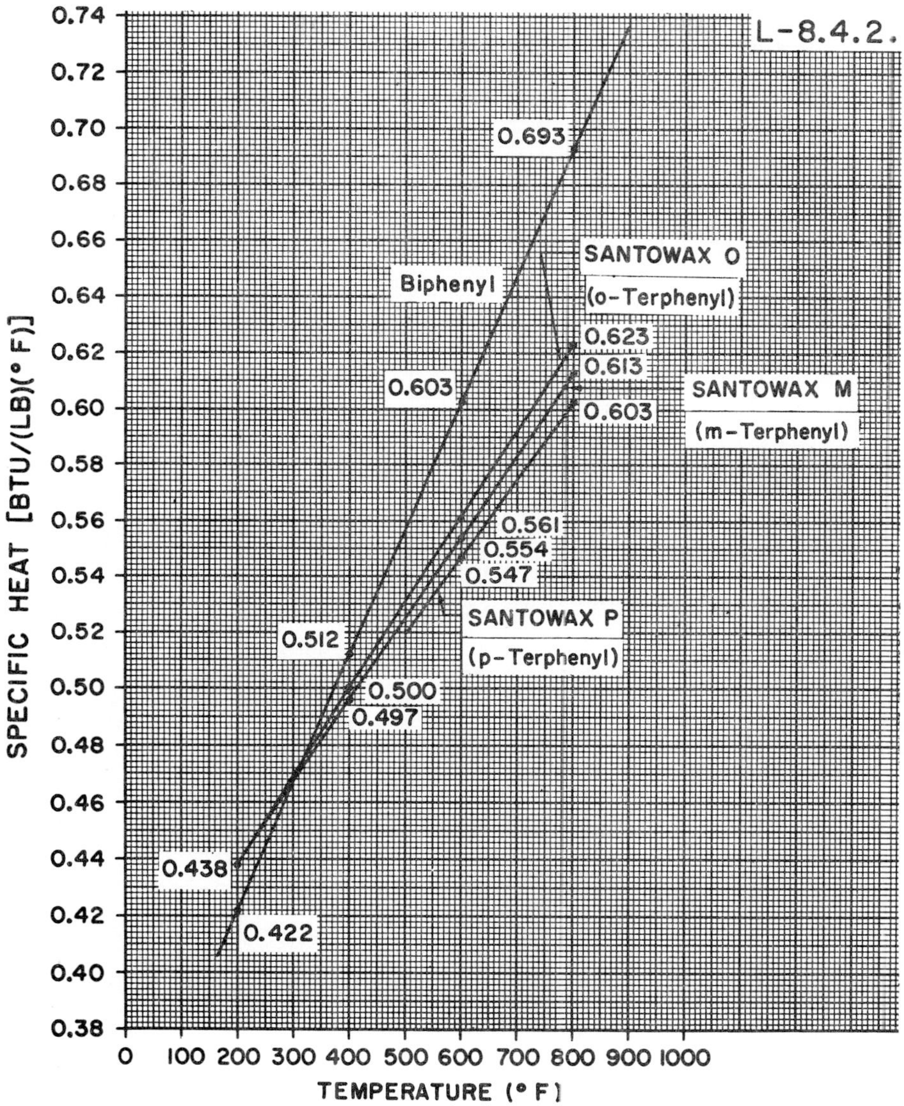

Specific Heat vs Temperature

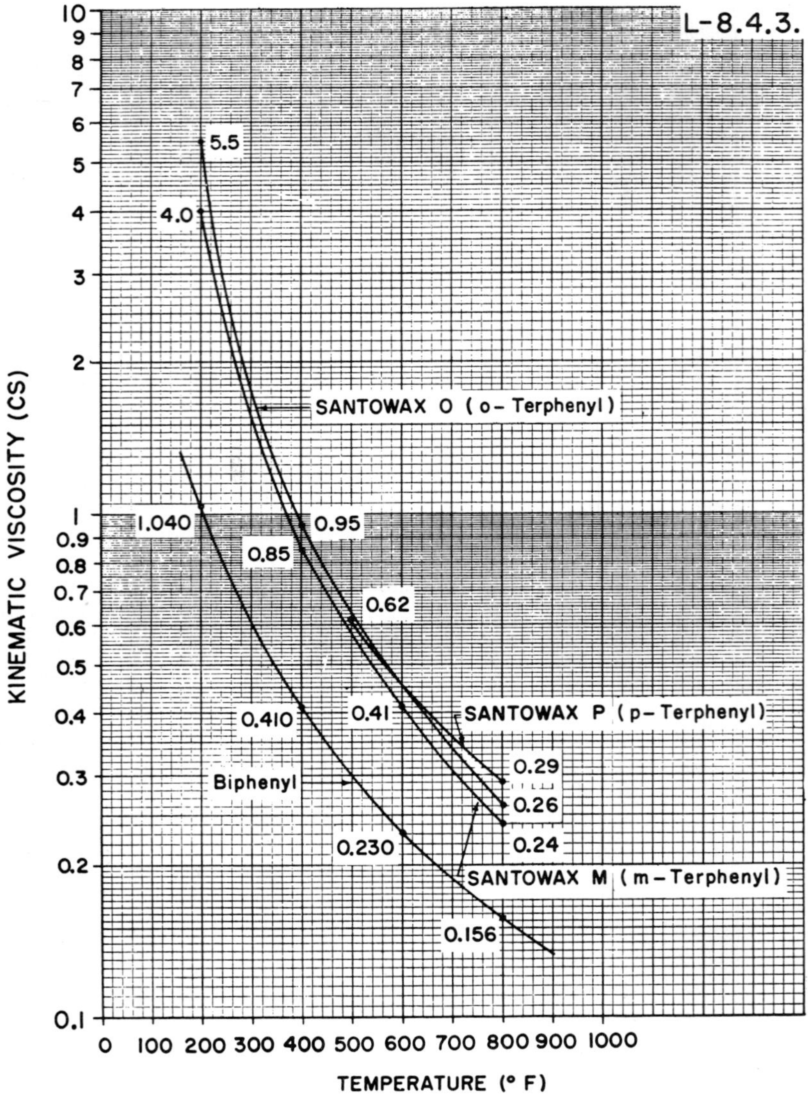

Kinematic Viscosity vs Temperature

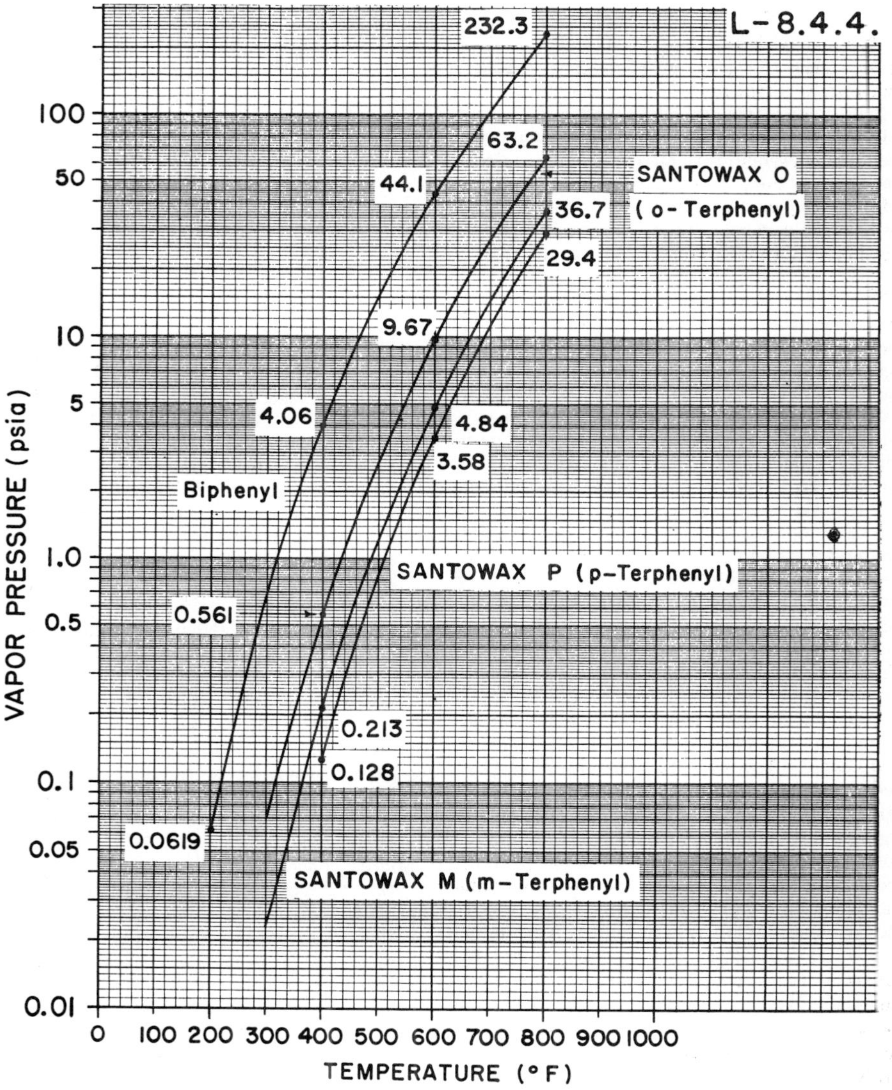

Vapor Pressure vs Temperature

L-9. LIQUID METALS

The use of liquid metals as heat carriers poses a number of unusual engineering problems. The design properties of liquid metals are excellent in every respect, while their construction and operating properties are less desirable and cause considerable difficulty in handling. However, since these two groups of properties are subject to possible improvement, much worthwhile effort has been devoted to exploiting the valuable heat transfer properties of liquid metals.

Heat Transfer

A major difference between liquid metals and more commonplace heat transfer media is their high thermal conductivity and consequently, low Prandtl numbers. In general, liquid metals have Prandtl numbers of the order of 0.001 to 0.1, which are about 1/1000 to 1/10 of those of most ordinary liquids. However, the high Reynolds numbers still permit good heat transference. The thermal conductivity of a liquid metal is composed of two parts: an electronic portion consisting of the motion of free electrons, and a nonmetallic or molecular portion. In a liquid metal, heat is conducted not only by molecular momentum and eddy diffusion as in gases, organic liquids, and water, but also by electron migration, thus explaining their high thermal conductivities.

From theoretical considerations, it can be shown that the Prandtl number is the ratio of momentum diffusivity to thermal diffusivity. Reduced Prandtl moduli indicate that, by molecular motion, liquid metals are able to transmit large quantities of heat with small temperature driving forces. Figure L-9. shows the temperature profiles of fluids, possessing high, medium and low Prandtl numbers.

Even in the turbulent core region, conduction by electron diffusivity predominates over heat transfer by macroscopic motion of eddies. From this it naturally follows that empirical correlations for turbulent heat transfer, valid for Prandtl numbers from 0.7 to 200, fail to describe accurately the behavior of low Prandtl number fluids such as liquid metals. Small scale deposits on walls, surface wetting of a wall by the metal, and solubility effects due to the presence of entrained gases, such as oxygen,

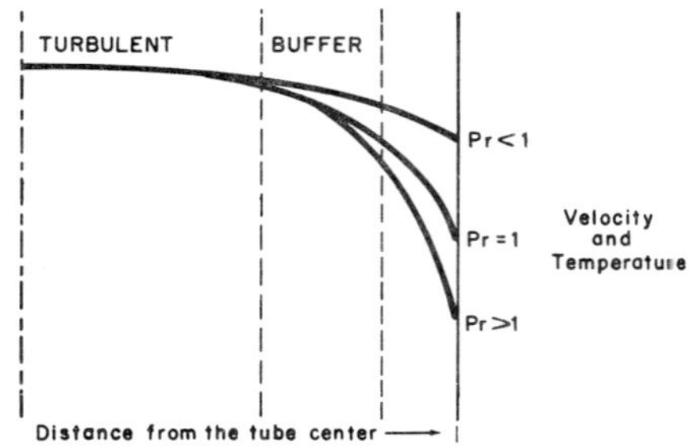

Figure L-9

in the liquid, further complicate determinations of liquid metal heat transfer coefficients. Considerable research on this subject has been carried out in the United States, and new design formulas have been developed for heat transfer by natural circulation, forced convection, boiling, and condensation. At any rate, even the most pessimistic investigators have confirmed the extraordinarily high heat transfer coefficients of liquid metals.

Handling

The handling of these materials presents a number of formidable obstacles. The question of corrosion is a critical one. In heat transfer loops of a nuclear reactor, dynamic corrosion takes place, i.e., material is dissolved in the hot zone of the system and is then transported by the flowing liquid metal coolant to colder sections, where it is deposited. In work of this nature, the term 'material transport' is usually employed in preference to corrosion. System failures generally result from blockages caused by this random erosion and diffusion of material, rather than by the usual structural instability of weakened materials. In designing liquid metal systems, high temperature sections are completely welded throughout, while valves and flanges are confined to cooler areas. In early liquid metal heat transfer applications, cost was secondary to time requirements, so most units were fabricated out of the safest available material: austenitic stainless steels. However, cheaper ferritic alloy steels with higher thermal conductivity and lower coefficients of thermal expansion are now being considered for use in nuclear power plants. Material transport in nuclear power systems can also lead to atomic migration from the intensely radioactive reactor core to unshielded parts of the system. Large increases in radioactivity may take place on pumps, heat exchangers and pipe surfaces, thus hampering the accessibility to the system and delaying maintenance. Liquid metals generally become highly radioactive after sustained usage.

When utilizing alkali metals, a major fire and safety problem exists because of their violent reaction with oxygen, water, halocarbons, and many common construction materials. Inert gas blankets prevent alkalis from coming into contact with the atmosphere. This problem and many others are thoroughly discussed in a series of three excellent "*Liquid Metals Handbooks*" published by the Atomic Energy Commission and the Department of the Navy.

A virtue of handling liquid metals is that their high electrical conductivity makes feasible the use of electromagnetic pumps. These pumps utilize the pressure created by passing a current through a molten metal at right angles to a magnetic field. None of the various pumps developed have any moving parts, eliminating the problems of sealing and lubrication encountered in ordinary pumps. A permanent magnetic flow meter is another piece of equipment made convenient by the high electrical conductivity of liquid metals.

Other Factors

Other advantages of liquid metals are their broad temperature range, high thermal stability and radiant stability. The distance between their melting and boiling points, with the exception of mercury, is about 1000° to 2000°F. In addition, the low vapor pressures of liquid metals recommend them for high temperature work, removing the necessity for pressurized equipment. In summation, liquid metals are particularly well suited for high heat flux, or low temperature difference processes, such as those prevailing in the atomic energy field. One disadvantage of liquid metals is that their specific heat is lower than most organic liquids.

For the purposes of this Handbook, only the low melting liquid metals are considered. These include: bismuth, lead, mercury, sodium, potassium and lithium. In addition, alloys such as Na-K and Pb-Bi, which have attained considerable prominence in liquid metal technology, will be discussed.

References

1. Lyon, R. N., Editor, "Liquid Metals Handbook," 1st Edition, 1950.
2. Lyon, R. N., Editor, "Liquid Metals Handbook," 2nd Edition, 1952.
3. Jackson, B. B., Editor, "Liquid Metals Handbook," Na-K Supplement, 3rd Edition, 1955.
4. *Am. Inst. Chem. Engrs.*, "Liquid Metals Technology"-Part I, C.E.P. Symposium Series, No. 20, 53, 1957.

L-9.1. BISMUTH

Properties of Importance

Density: See Figure L-9.1.

Specific Heat: See Figure L-9.2.

Latent Heat: Heat of Fusion: 21.6 Btu/lb. Heat of Vaporization: 367.74 Btu/lb.

Thermal Conductivity: See Figure L-9.3. (It has about the lowest thermal conductivity of any metal except mercury.)

Viscosity: See Figure L-9.4.

Physical Stability: See Figure L-9.5. for the vapor pressure curve.

Melting Point	Boiling Point
520°F	2691°F

Thermal Stability: Bismuth has excellent thermal stability.

Chemical Stability: Bismuth undergoes superficial oxidation in wet air at room temperatures, but protective film readily forms, preventing further oxidation. At temperatures above its melting point, an oxide film forms, and at higher temperatures bismuth burns to its trioxide. Bismuth combines directly with sulfur and the halogens, and decomposes steam at red heat. Insoluble permanent gases such as hydrogen, nitrogen, argon, and helium are used to protect bismuth from the atmosphere.

Nuclear Properties: Absorption cross-section: 0.21 (Neutron absorption cross-section of an element for a neutron velocity of 2200 m/sec).

Scattering cross-section: 9 (Neutron scattering cross-section for thermal neutrons).

Surface Tension:

(dynes/cm)	t(°F)
376	572
363	932

Corrosion: Table L-9 summarizes the results of various quantitative measurements of the extent of corrosion resulting from contact with various structural materials. It is important to reiterate that these tests are only as meaningful as the purity of the liquid metal and test material involved. Inhibitors for mass-transfer attacks at high temperatures have not been fully investigated.

Flammability: None.

Physiological Effects: Workmen should be protected from the fumes of molten or burning bismuth. There are no recorded cases of bismuth poisoning, and although its metallic fumes are uncomfortable, they are not toxic.

Electrical Resistivity:

t(°F)	Resistance (micro-ohms)
572	128.9
1382	153.53

Price Class: B per lb.

L-9.2. MERCURY

Mercury is a rather unusual heat carrier because, unlike other metals, it is a liquid at ambient conditions. Of all liquid metals, it is the most fusible and the most volatile, melting at −37.97°F., and boiling at 675°F. Due to its low temperature range, the technology of liquid mercury has been thoroughly investigated. The earliest applications of liquid mercury date back to 1913, when General Electric first proposed a two-fluid power generation system. Later, in the 1920's, mercury vapor was used by the Sun Oil Company for distillation of crude oil and for mercury-vapor cracking units. Besides its application as a pure metal, its lead and cadmium amalgams have also been employed. Amalgams are expensive; they raise boiling and freezing points of mercury, and also increase corrosion.

Properties of Importance

Density: Mercury is the densest of the liquid metals. See Figure L-9.1.

Specific Heat: See Figure L-9.2.

Latent Heat: Fusion: 5.04 Btu/lb. Vaporization: 125.5 Btu/lb.

Thermal Conductivity: See Figure L-9.3.

Viscosity: See Figure L-9.4.

Physical Stability: See Figure L-9.6.

Thermal Stability: Mercury has excellent thermal stability.

Chemical Stability: Mercury is a noble metal. It is prepared from its compounds by heating, and from its aqueous salt solutions by reducing agents. At about 650°F mercury is oxidized to mercuric oxide, which breaks down at higher temperatures. Gases have low solubilities in mercury.

Surface Tension: Typical values are given below:

t(°F)	Surface Tension (dynes/cm)
68	465
392	436
669	394

Mercury holding is a rather serious problem, which caused the decline of two-phase power production, and makes mercury surfaces extremely difficult to clean, raising plant costs. Many investigations have been conducted to study the reduction of holding.

Corrosion: The corrosion resistance of common engineering materials to Hg, is shown in Table L-9. It has been found that addition of controlled amounts of metals such as Ti and Mg inhibits mercury corrosion. For example, addition of titanium postpones attacks on mild steel from 400° to 650°C; while magnesium eliminates corrosion-causing oxygen from the system. Magnesium and titanium cause wetting of mild steel, which provides nucleation centers for boiling, thus increasing heat transfer coefficients of the liquid metal. Valve and gasket designs are not as critical as with other liquid metals.

Flammability: Mercury is nonflammable.

Electrical Resistivity: Typical values are given below.

t(°F)	Resistance (micro-ohms)
122	98.4
392	114.2
662	135.5

Physiological Effects: Mercury is exceedingly toxic. The maximum allowable concentration of mercury in air for a daily exposure of eight hours or less is 0.1 mg/m^3. Mercury vapor in air is commonly detected by ultra-violet absorption of vapor.

Color: Silver-gray.

Price Class: B per lb.

L-9.3. LEAD

Properties of Importance

Density: See Figure L-9.1.

Specific Heat: See Figure L-9.2.

Latent Heat: Fusion: 1060 Btu/lb. Vaporization: 368.3 Btu/lb.

Thermal Conductivity: See Figure L-9.3.

Viscosity: See Figure L-9.4.

Physical Stability: See Figure L-9.5. Melting Point: 621°F. Boiling Point: 3159°F.

Thermal Stability: Excellent.

Chemical Stability: Lead is one of the most permanent metals at ambient temperatures. Lead does not react with dry air at ordinary temperatures, but at elevated temperatures such as 600°F, a protective oxide coating forms. At bright red heat, rapid air oxidation takes place and the resulting oxide is molten. At white heat, water vapor oxidizes lead. In presence of moisture, CO_2 reacts with lead to form a carbonate. Gases such as H_2, N_2, CO, A and He are generally employed as protective atmospheres.

Nuclear Properties: Absorption cross-section 0.02. Scattering cross-section 11.

Surface Tension:

t(°F)	(dynes/cm)
662	442
932	431

Corrosion: Table L-9 shows the resistance of materials to liquid lead. The presence of impurities such as oxygen and antimony, strongly enhances and accelerates the destruction of steel, iron and other structural metals. At present, quantitative data for the effects of these impurities are unavailable.

Flammability: None.

Physiological Effects: Lead can be exceedingly toxic. Lead poisoning usually occurs as a result of inhaling or swallowing lead fumes or very fine lead dusts. Unfortunately, lead poisoning is cumulative and following circulation in the blood stream, it is often stored in the mineral structure of the bones. Adequate ventilation and cleanliness, and periodic medical examinations of personnel are required in all lead applications. The maximum allowable lead concentration is 0.15 mg/m^3.

Electrical Resistivity:

t(°F)	Resistance (micro-ohms)
627	94.6
1832	125.7

Price Class: A per lb.

L-9.4. LEAD—BISMUTH EUTECTIC

This eutectic mixture of 55.5 wt % Bi and 44.5 wt % Pb has a low melting point, high specific heat, and a high heat transfer coefficient.

Properties of Importance

Density: See Figure L-9.1.

Specific Heat: See Figure L-9.2.

Thermal Conductivity: See Figure L-9.3.

Viscosity: See Figure L-9.4.

Physical Stability: Vapor Pressure: The vapor pressure values fall between those of its pure components.

Melting Point 257°F
Boiling Point 3038°F

Thermal Stability: This eutectic mixture has the excellent thermal stability of its pure constituents.

Chemical Stability: Estimates of the extent of oxidation of this alloy have been made only from properties of its pure components. No detailed data have been published, yet it appears that this eutectic mixture should be insoluble in N_2, H_2, A or He, and any one of these gases may be used as a protective atmosphere.

Nuclear Properties: Absorption cross-section A = 0.17. Scattering cross-section S = 9.9.

Surface Tension:

t(°F)	(dynes/cm)
1472	367
1832	356

Corrosion: Table L-9 shows the resistance of common engineering materials to Pb-Bi attack. High chromium steels appear to be satisfactory for handling the liquid alloy.

Flammability: None.

Physiological Effects: All precautions outlined in the section for lead must be adhered to for any high lead alloy.

Electrical Resistivity: Electrical resistivity falls in the range 113–128 micro-ohms resistance, between temperatures of 392° and 932°F.

Price Class: B per lb.

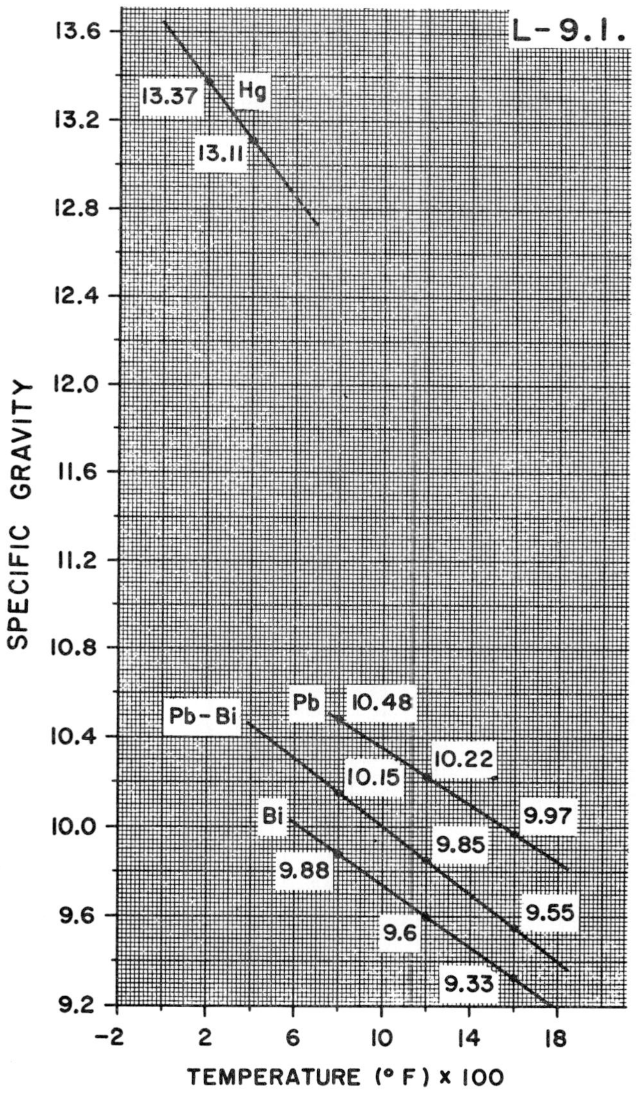

Specific Gravity vs Temperature

Table L.9

CONDENSED SUMMARY OF RESISTANCE OF MATERIALS TO LIQUID METALS AT 300°C, 600°C, AND 800°C

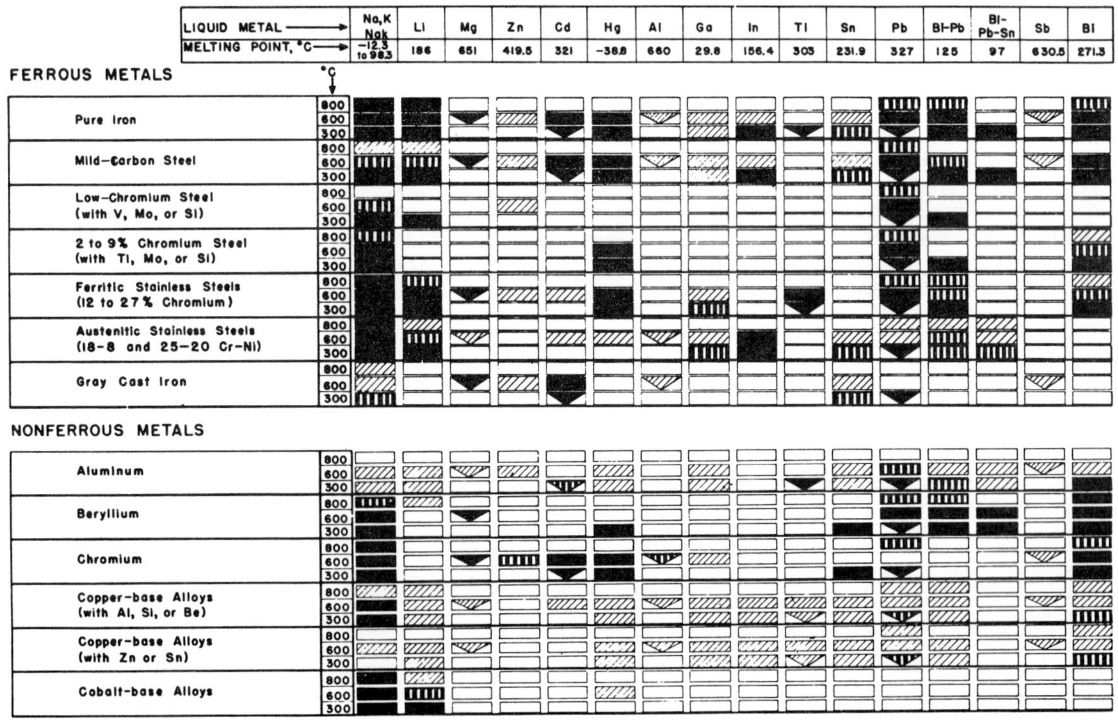

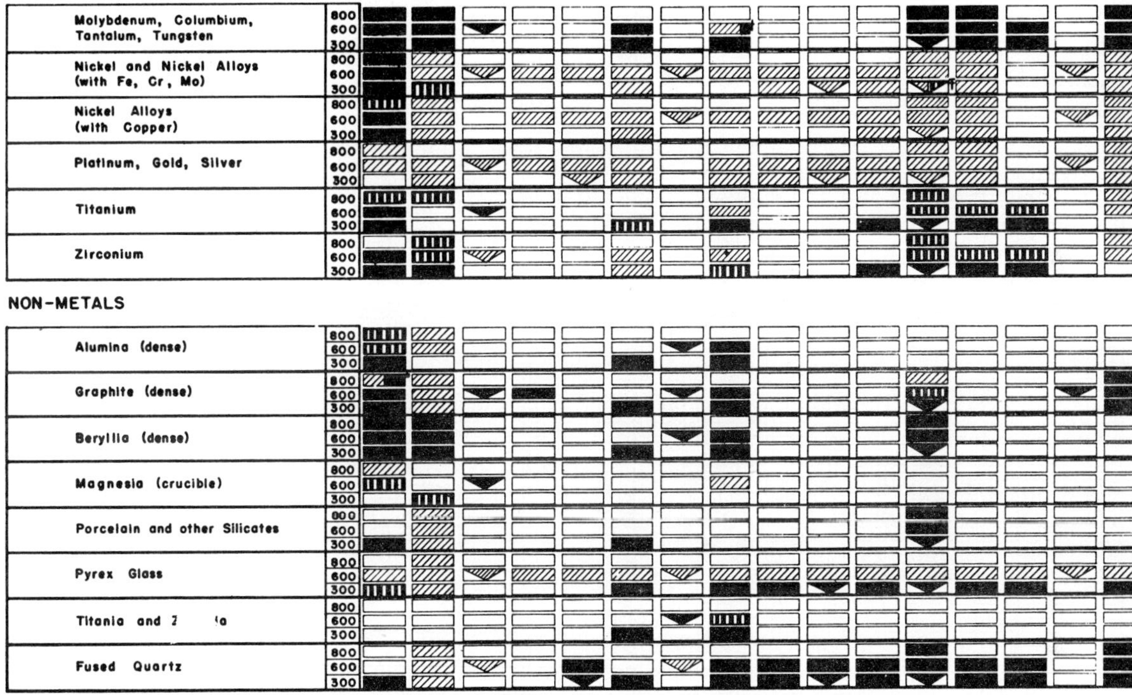

Resistance Ratings (These ratings refer to liquid-metal resistance only— not to temperature-dependent mechanical strength or metallurgical stability. See text for further discussion.): [black]— Good resistance; [vertical lines]—limited resistance; [hatched]— poor resistance; [blank]— unknown resistance.

[triangle]—Liquid freezes above this temperature, shading in triangle indicates degree of resistance to liquid at the melting point.

• —Indicates: Poor with potassium, good with sodium; † — Tungsten, good; ‡— Inconel, limited.

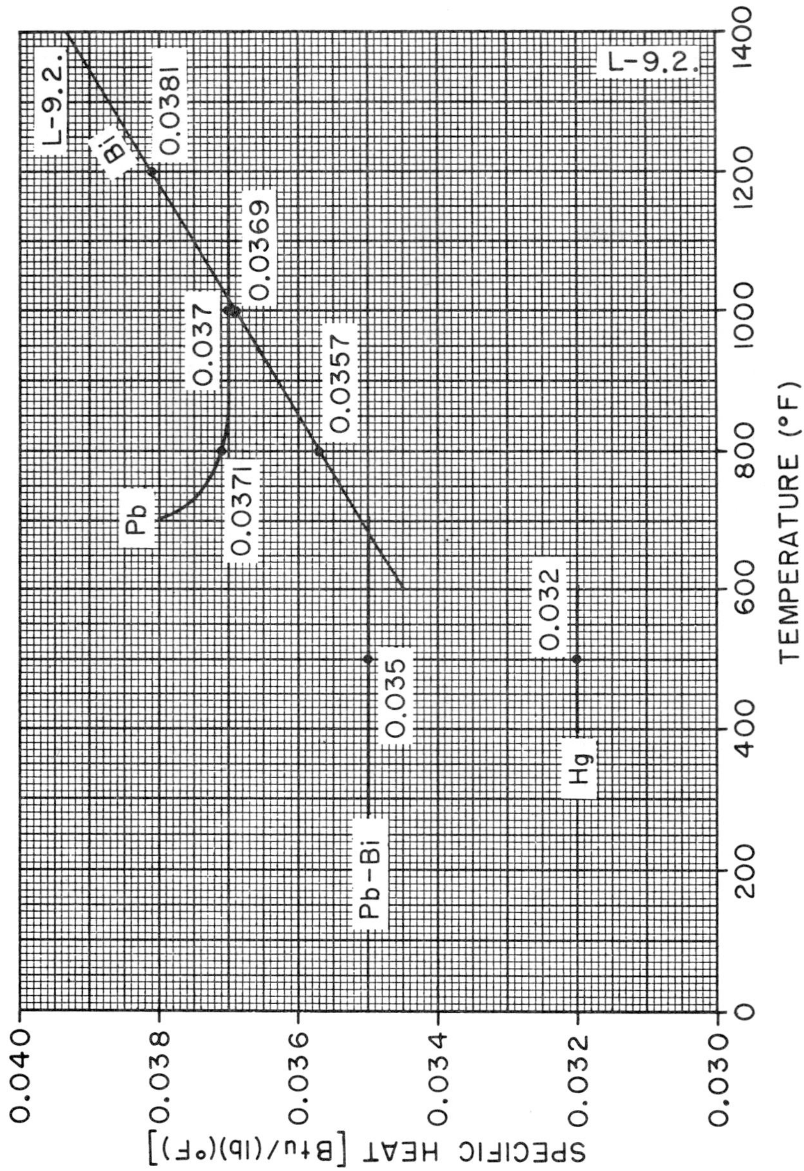

Specific Heat [Btu/(lb)(° f)]

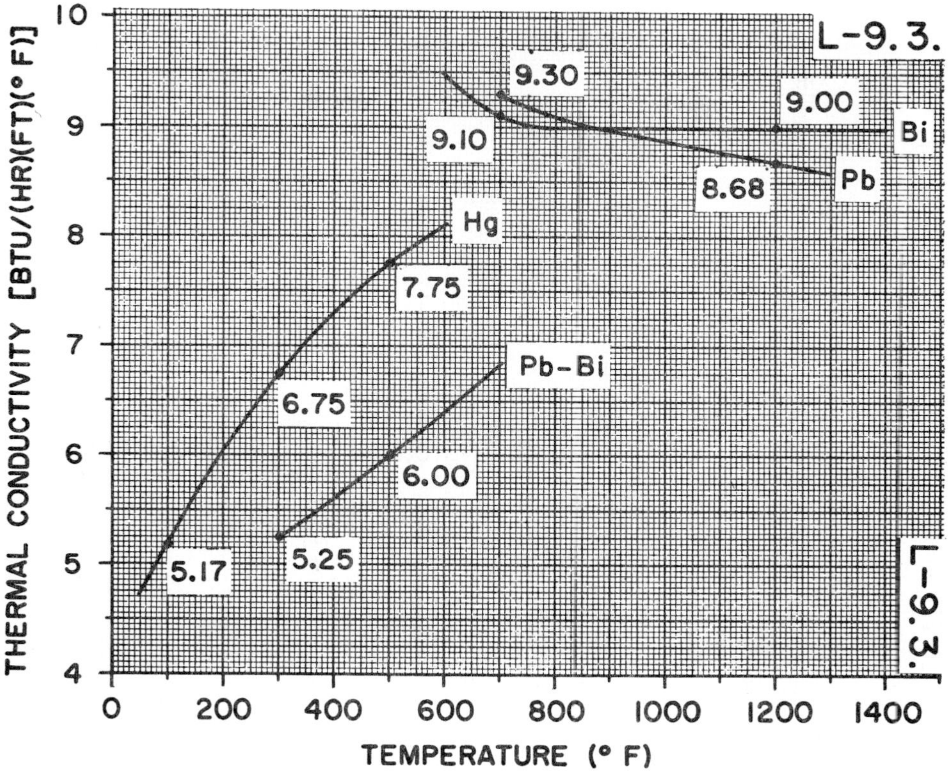

Thermal Conductivity vs Temperature

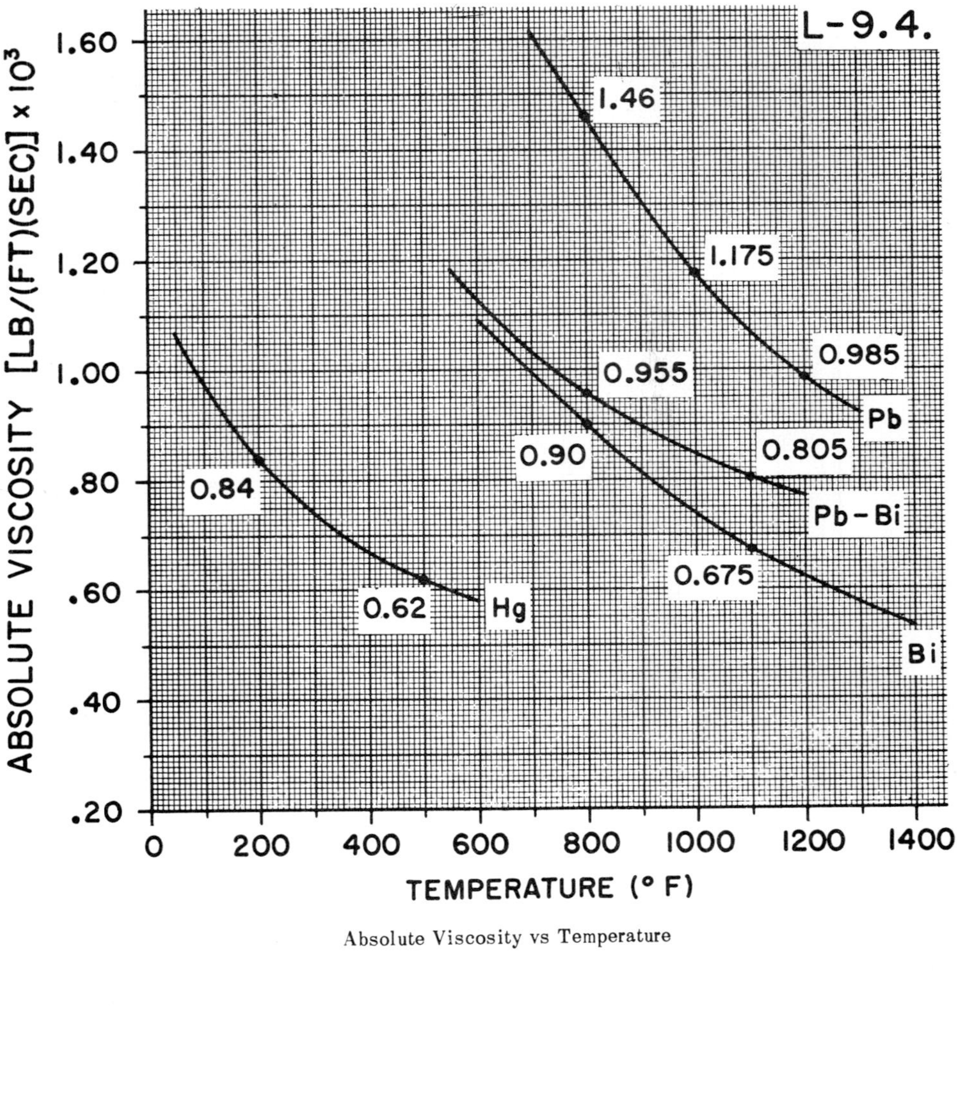

Absolute Viscosity vs Temperature

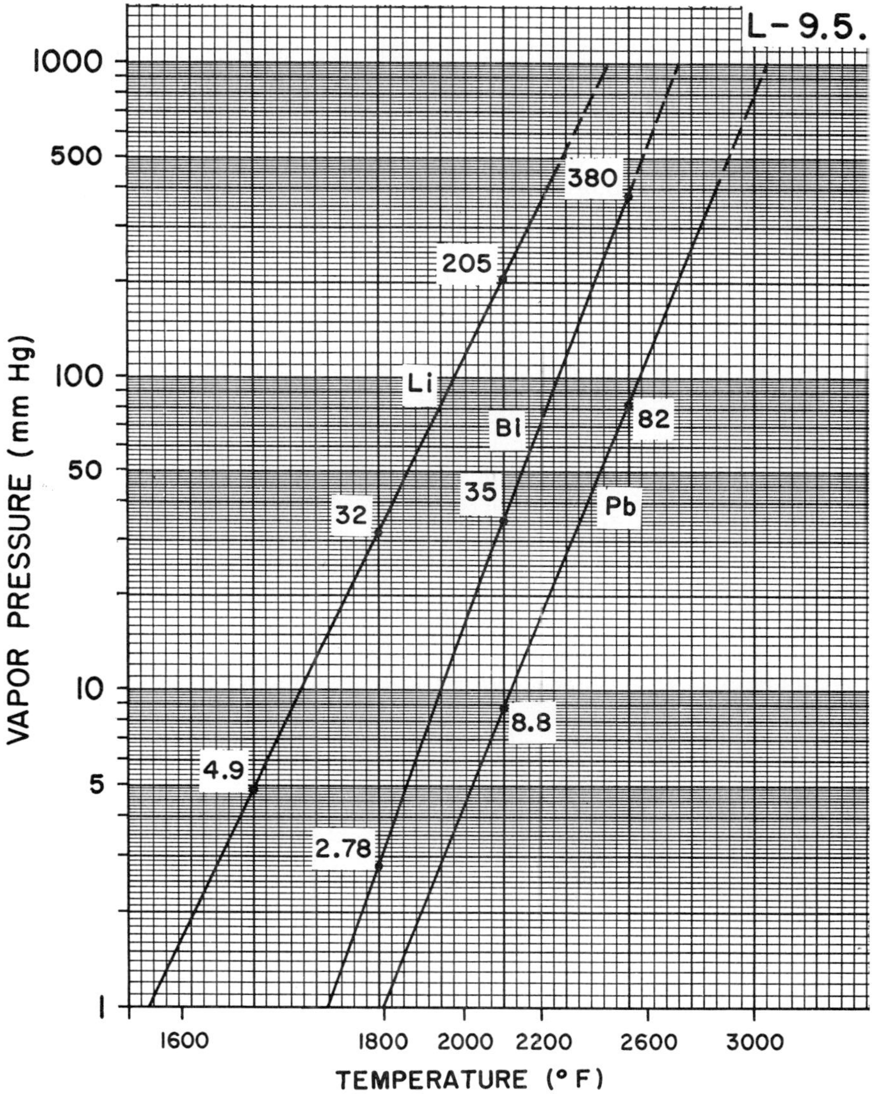

Vapor Pressure vs Temperature

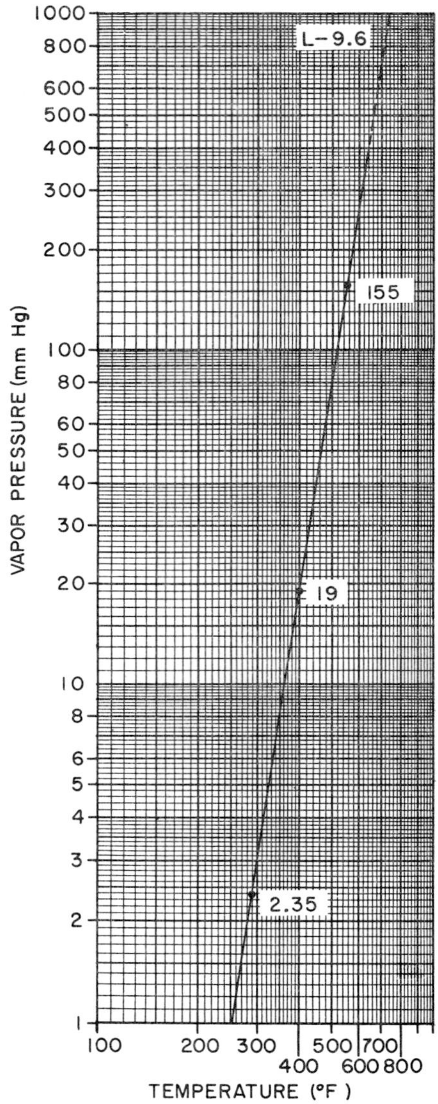

Vapor Pressure vs Temperature

L-9.5. GALLIUM

The use of gallium for heat transmission is a theoretically intriguing one because of its low melting point and wide liquid range. Nevertheless, the metal is found in such small concentrations that it is obtained only as a by-product of the refining of other metals.

Properties of Importance

Density: See Figure L-9.7.

Specific Heat: The specific heat of gallium is constant at 0.082 Btu/(lb)(°F).

Latent Heat of Fusion: 34.4 Btu/lb.

Latent Heat of Vaporization: 1825.2 Btu/lb.

Thermal Conductivity: At its melting point gallium has a thermal conductivity of 19.5 Btu/(hr)(ft)(°F).

Viscosity: See Figure L-9.8.

Physical Stability: Gallium has the second widest temperature range of a liquid metal; surpassed only by tin. It melts at 85.9°F and boils at 3601°F.

Chemical Stability: The chemical properties of gallium are quite similar to those of aluminum. For example, on contact with an acid or base, it exhibits the amphoteric character of aluminum. Gallium undergoes superficial oxidation only when subjected to red hot heat.

Thermal Stability: Gallium has excellent thermal stability.

Radiolytic Stability:
Absorption cross-section: 2.2
Scattering cross-section: 4

Corrosion: Gallium is one of the most aggressive liquid metals in its effect on structural metals. Metals such as tungsten or tantalum can contain it at high temperatures.

Surface Tension: (86°–104°F) 735 dynes/cm

Electrical Properties: The following values were obtained.

t(°F)	Resistance (micro-ohm)
85.55	25.9
86.54	27.2
114.98	28.4

Price Class: D per lb.

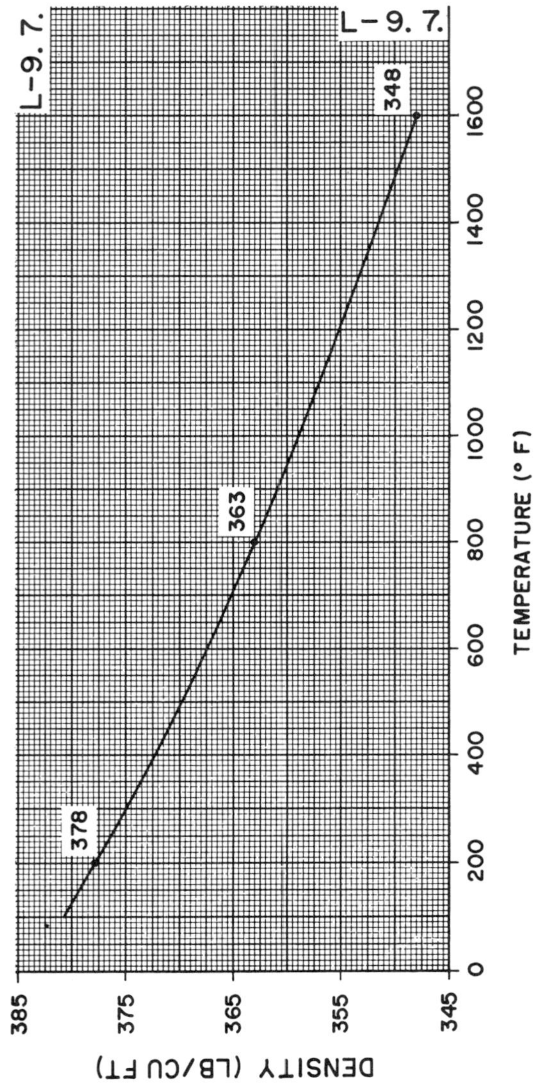

Density vs Temperature

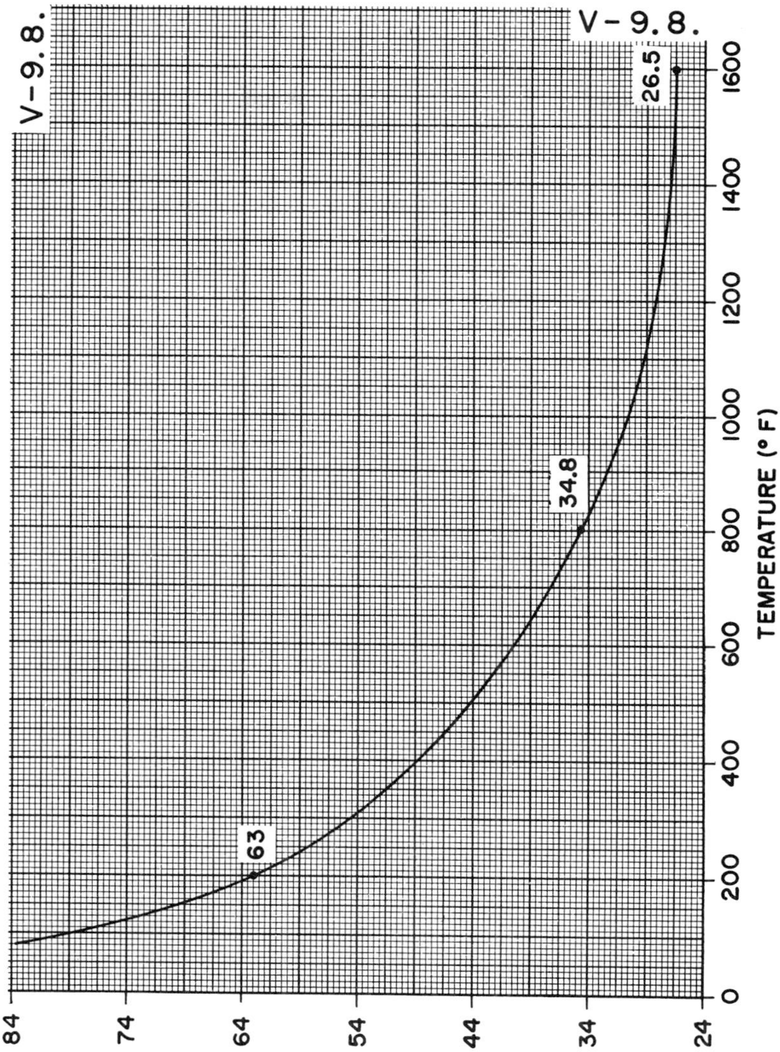

Viscosity vs Temperature

THE ALKALI METALS

(Lithium, Sodium and Sodium-Potassium Alloys)

Lithium is the least reactive member of the alkali metals group. It also has the lowest molecular weight of any metal used for heat transfer.

Because of their high heat coefficients and low pumping power requirements, sodium and sodium potassium alloys are extremely desirable heat transfer fluids. However, the extraordinarily high reactivity of these materials, combined with dangers of thermal stress, radioactivity and corrosion, has created many unusual design problems. The sodium-NaK supplement of the "Liquid Metals Handbook" is an excellent guide for projected applications of sodium and sodium-potassium alloys.

L-9.6. LITHIUM

Lithium, the rarest of the alkali metals, has several desirable heat transfer properties; namely, low density, low melting point, high specific heat, and high boiling point. However, the problem of concentrating lithium from the ore to the metal has kept it the most expensive of the alkali metals. Lithium metal is currently produced by the electrolysis of lithium chloride.

Properties of Importance

Density: Lithium is the least dense of the normal metals. See Figure L-9.9.

Specific Heat: The specific heat of lithium is about 1 $\frac{\text{Btu}}{(\text{lb})(^\circ\text{F})}$ at all temperatures.

Latent Heat:

Fusion: 284.4 Btu/lb
Vaporization: 8424 Btu/lb

Thermal Conductivity: 22 Btu/(hr)(°F)(ft) at 400°F

Viscosity: Values

t (°F)	(lb)/(ft)(sec) × 10^3
400	0.40
600	0.34
800	0.37
1200	0.29
1800	0.28

Physical Stability: See Figure L-9.5. Lithium has an exceedingly wide range of application: 158°–2403°F (normal boiling point).

Thermal Stability: Lithium has excellent thermal stability.

Chemical Stability: Lithium is the least reactive and least representative member of the alkali family. In contrast to sodium and potassium, it does not react with dry O_2 below 212°F, and undergoes only a slow reaction with cold water because of the presence of insoluble $LI(OH)_2$. However, at elevated temperatures, lithium combines with all known molecular gases. Inert gases, free of O_2, H_2O and N_2, can be employed as protective atmospheres.

Surface Tension: Between 662° and 932°F, the surface tension of Li varies between 442 and 431 dynes/cm.

Corrosion: Tests on lithium of questionable purity, indicate that it is more violent in its corrosive attacks on common materials than either Na or NaK alloys. The results of these tests are summarized in Table L-9. Main harmful impurities are O_2 and N_2. Before it is used, new and more sensitive analytical methods for detecting impurities must be developed.

Thoroughly clean mild steel is recommended for handling molten lithium. However, welded seams and joints suffer penetration at high temperatures. Steel can be used up to about 930°F, whereupon it is decarburized. Lithium attacks stainless steel and dissolves nickel at high temperatures.

Flammability: Lithium presents a definite fire hazard. On contact with air, lithium, near its melting point, may ignite and burn with an intense white flame. Fires should be smothered with graphite powder, as burning lithium reacts with sand.

Physiological Effects: Adequate protection from leaking and spattering of liquid lithium should be provided. On contacting moist skin, molten lithium burns with an intense flame. A protective shield is generally used when handling lithium.

Electrical Resistivity: The following values have been obtained:

t(°F)	Resistance (micro-ohms)
446	45.25

Nuclear Properties: Li_7 has a substantially lower neutron absorption cross-section than natural lithium, or Li_6. Thus, isotope separation must be perfected prior to nuclear application.

Absorption cross-section A-65
Scattering cross-section S-1.5

Price Class: C per lb.

L-9.7. SODIUM

Properties of Importance

Density: See Figure L-9.9. and L-9.10. Among the liquid metals, only lithium and potassium have lower densities.

Specific Heat: See Figure L-9.11.

Enthalpy: Latent Heat of Fusion: 49 Btu/lb
Latent Heat of Vaporization: 1809 Btu/lb

Thermal Conductivity: See Figure L-9.12.

Viscosity: See Figures L-9.13. and L-9.14. Liquid sodium has about the same fluidity as water.

Physical Stability: Sodium has an unusually wide liquid range. At 208°F, the solid is transformed into a silver white liquid, while at 1618°F, a partially dimeric blue vapor boils off at atmospheric pressure. See Figure L-9.15. for P-T characteristics.

Thermal Stability: As a basic element, sodium does not decompose with rising temperature.

Chemical Stability: The chief characteristic of sodium is its pronounced reactivity with nearly all liquids and gases. When exposed to the atmosphere, sodium burns fiercely and produces dense fumes of Na_2O. Most hydrogen containing gases, with the exception of hydrocarbons, react with the displacement of hydrogen by the sodium and the formation of a sodium salt. In the presence of large surfaces, sodium reacts explosively and instantaneously with water. The liquid metal undergoes a vigorous reaction with inorganic and organic halides. Sodium also forms alloys with many common metals (Ba, Bi, Cd, Ca, Cu, Fe, K, Hg, etc.).

Sodium is inert towards most hydrocarbons at temperatures below their cracking temperatures. The metal is protected from atmospheric exposure with inert gas blankets of N_2, A and He. Helium is preferred for nuclear reactors, since it is the only one of the three gases which does not become radioactive.

Radiolytic Stability: Absorption cross-section A = 0.45. Scattering cross-section S = 4.

Surface Tension: The following values have been obtained:

t (°F)	Surface Tension (dynes/cm)
212	190.4
572	170.6
842	156.0

Corrosion: Table L-9. gives the resistance of ordinary construction materials to liquid sodium. Recent results of static and dynamic corrosion tests on sodium heat transfer systems have shown that carbon and low alloy steels display good resistance to attack up to 1000°F, and can thus supplant more expensive stainless steel alloys.

Research has revealed that the presence of oxygen impurities above 100 ppm will markedly accelerate corrosion. In addition, precipitation of insoluble Na_2O in low temperature areas of the system may lead to plugging, followed by increasing temperatures and burnout. I.L. Gray, R. L. Neal, and B. G. Voorhees,[1] describe the development of a plugging indicator, for determining the oxygen saturation temperature of a liquid sodium system and a sodium cold trap for removal of oxides until a safe concentration is reached. Carbon and calcium are impurities which influence corrosion to a far lesser extent.

Flammability: Because of its high degree of reactivity, sodium constitutes a safety hazard. The heat and hydrogen released by the reaction with water, generally lead to an explosion. Above 237°F, sodium is combustible and will ignite spontaneously with air. Any leak will produce a fire.

Physiological Effects: Alkali metals react with the skin and eyes, causing thermal and alkali burns. Oxide smoke or hydroxide mist is very irritating and corrosive to the throat and lungs.

t (°F)	Resistivity of Sodium (microhms-cm)
32	4.477
104	5.314
176	6.279
208	6.749

Appearance: As a solid, sodium is a soft, ductile, silver white metal. In the liquid phase, it is also silver white, while as a vapor, it is blue.

Price Class: A per lb.

Reference

1. Gray, I. L., Nearl, R. L., and Voorhees, B. G., "Control of Oxygen in Sodium Heat Transfer Systems", Liquid Metals Technology-Part I, *Chem. Eng. Progr.* **53**, No. 20, 11, 1957.

L-9.8. NaK ALLOYS

Sodium-potassium alloys are used for heat transfer, instead of the pure elements, because these mixtures have lower melting points. The two most popular alloys are composed of:

56 wt% Na
44 wt% K

22 wt% Na
78 wt% K

The latter is approximately the eutectic composition. The phase diagram in Figure L-9.16. illustrates how melting points of NaK alloys rise from a minimum value at the eutectic point, and gradually approach those of the pure constituents, with a slight discontinuity prevailing at the Na_2K peritectic. Only for the eutectic composition have most of the important physical properties been ascertained.

It is only reasonable to expect that physical and chemical properties of alloys formed from two similar members of the same periodic group should obey ideal mixing predictions, such as Raoult's law for vapor pressures. At a given temperature, alloys have greater chemical activity than the elements, because of the presence of a liquid state. In comparison to the pure metals, NaK alloys have lower electrical and thermal conductivities, while vapor pressure, density, specific heat, enthalpy, and viscosity fall in between values of the two components.

Properties of Importance

Density: See Figures L-9.9. and L-9.10.

Specific Heat: See Figure L-9.11.

Thermal Conductivity: See Figure L-9.12.

Viscosity: See Figures L-9.13. and L-9.14.

Physical Stability: Data taken at the Naval Research Laboratory indicate that the alloy obeys Raoult's law for ideal mixtures. See Figure L-9.15.

Thermal Stability: Excellent.

Chemical Stability: The chemical properties of NaK alloys are almost identical to those of sodium; however, the alloys are more reactive. Increased chemical activity can be ascribed to the presence of potassium, which has a greater reactivity than sodium. Upon contact with the atmosphere, potassium is oxidized to the superoxide, KO_2, and explosions have been caused by reaction of the superoxide with hydrocarbon gases used as cleaning media. Experiments at 400°F have revealed that sodium monoxide, rather than potassium superoxide, precipitates from NaK alloys. Other reactions of potassium not duplicated by sodium are an attack on silicon and formation of an explosive carbonyl on contact with carbon monoxide.

Surface Tension:

WT % K	t (°F)	Surface Tension (dynes/cm)
44	m.p. to 480°F	110–100
78	m.p. to 480°F	120–110

Corrosion: Table L.9. summarizes the effects of NaK alloy on structural materials. The same problems in constructing a sodium system must also be dealt with, when handling the liquid alloy.

Flammability: The chief danger peculiar to NaK systems is the explosive reaction between KO_2 and hydrocarbons used for equipment cleaning. Even though NaO_2 is preferentially produced by oxidation, KO_2 may appear on surfaces and interfaces. The use of hydrocarbons in NaK systems should be discouraged.

Physiological Effects: Same precautions as required for a sodium system.

Electrical Resistivity:

t (°F)	Electrical Resistivity (microhms-cm)	
	44 wt % K	78 wt % K
212	41.61	145,63
572	54.33	58.58
932	69.37	73.48
1472	99.68	104.51

Price Class: B per lb.

Reference

1. "Liquid Metals Handbook," A.E.C., June 1952.

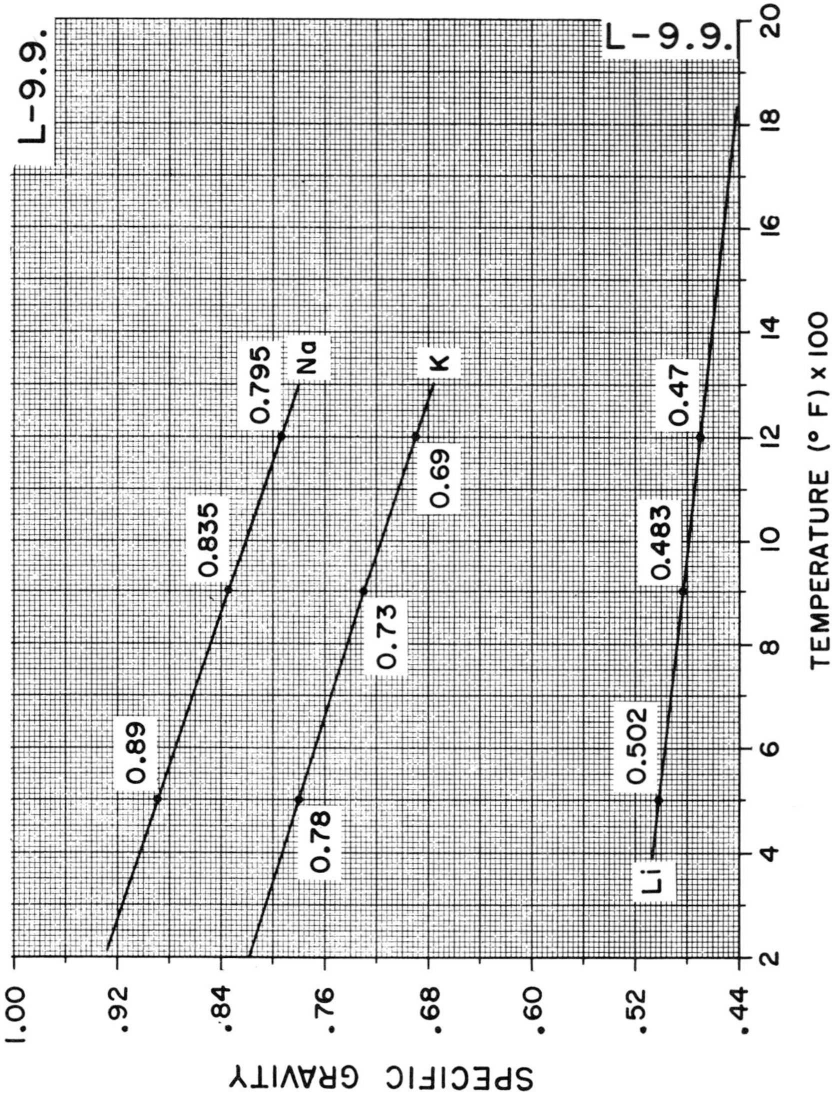

Specific Gravity vs Temperature

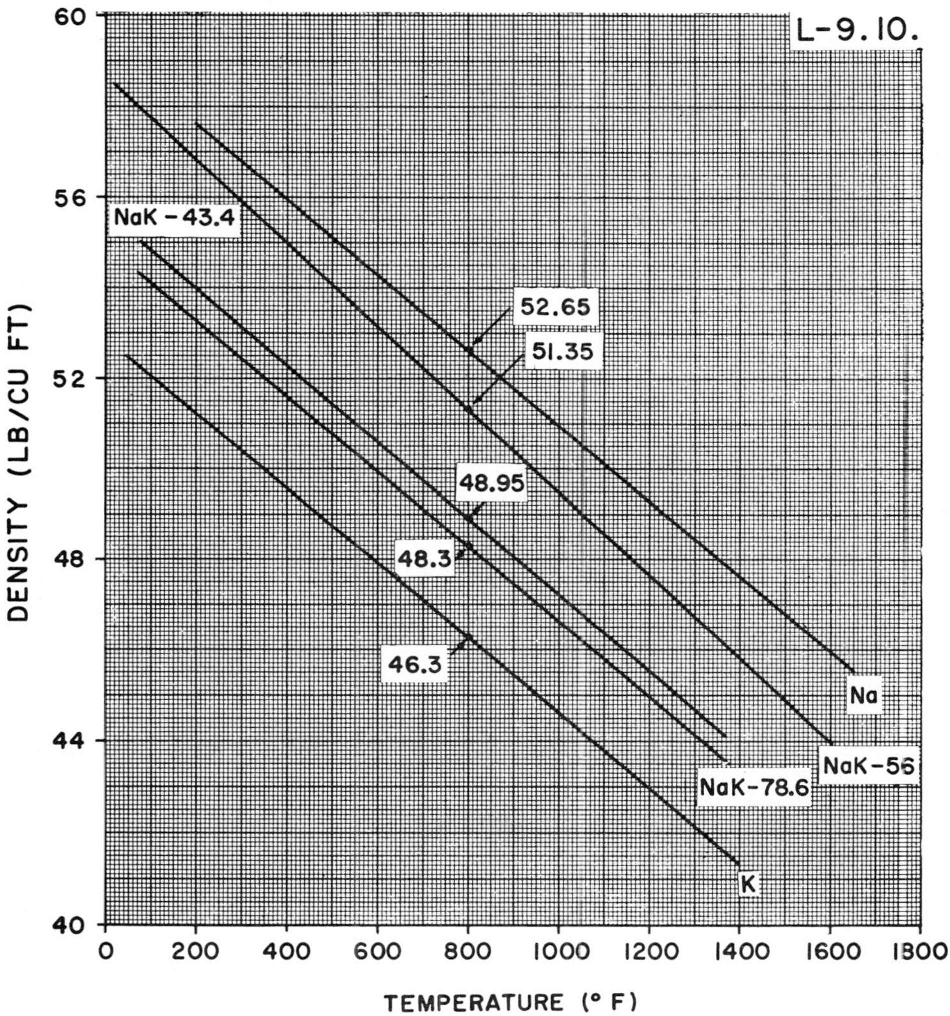

Density vs Temperature

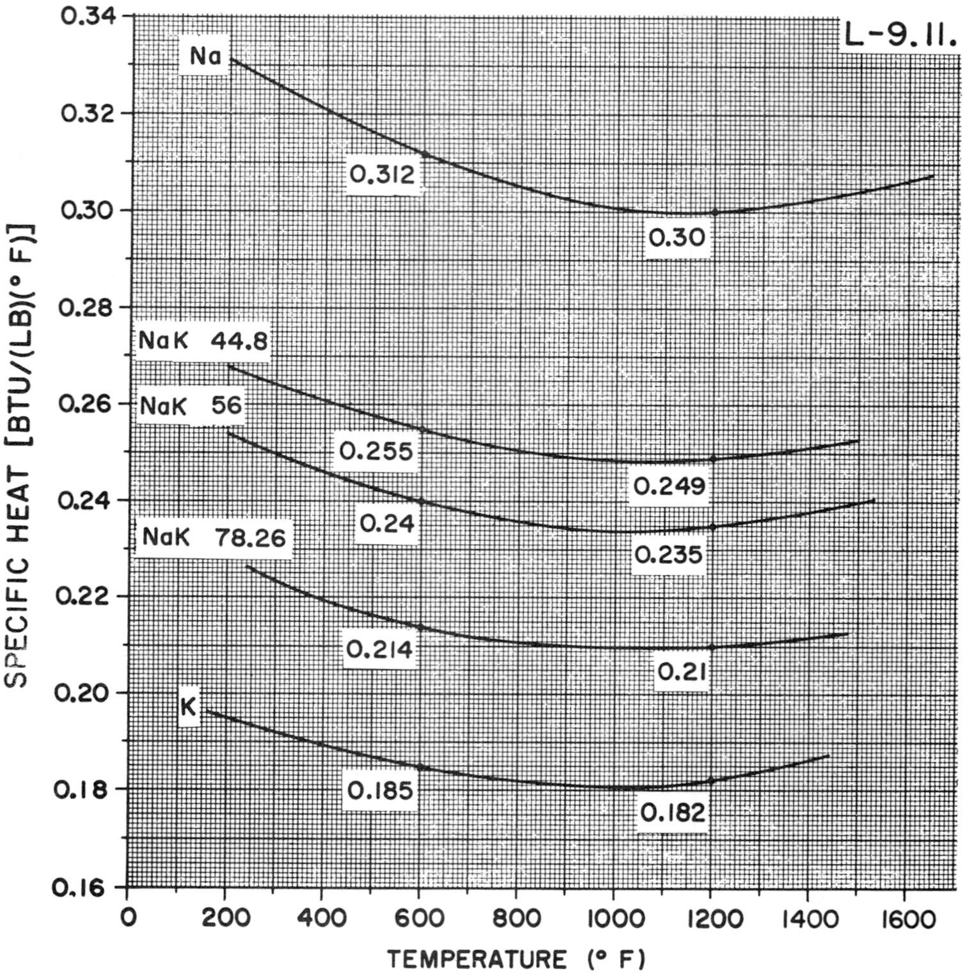

Specific Heat vs Temperature

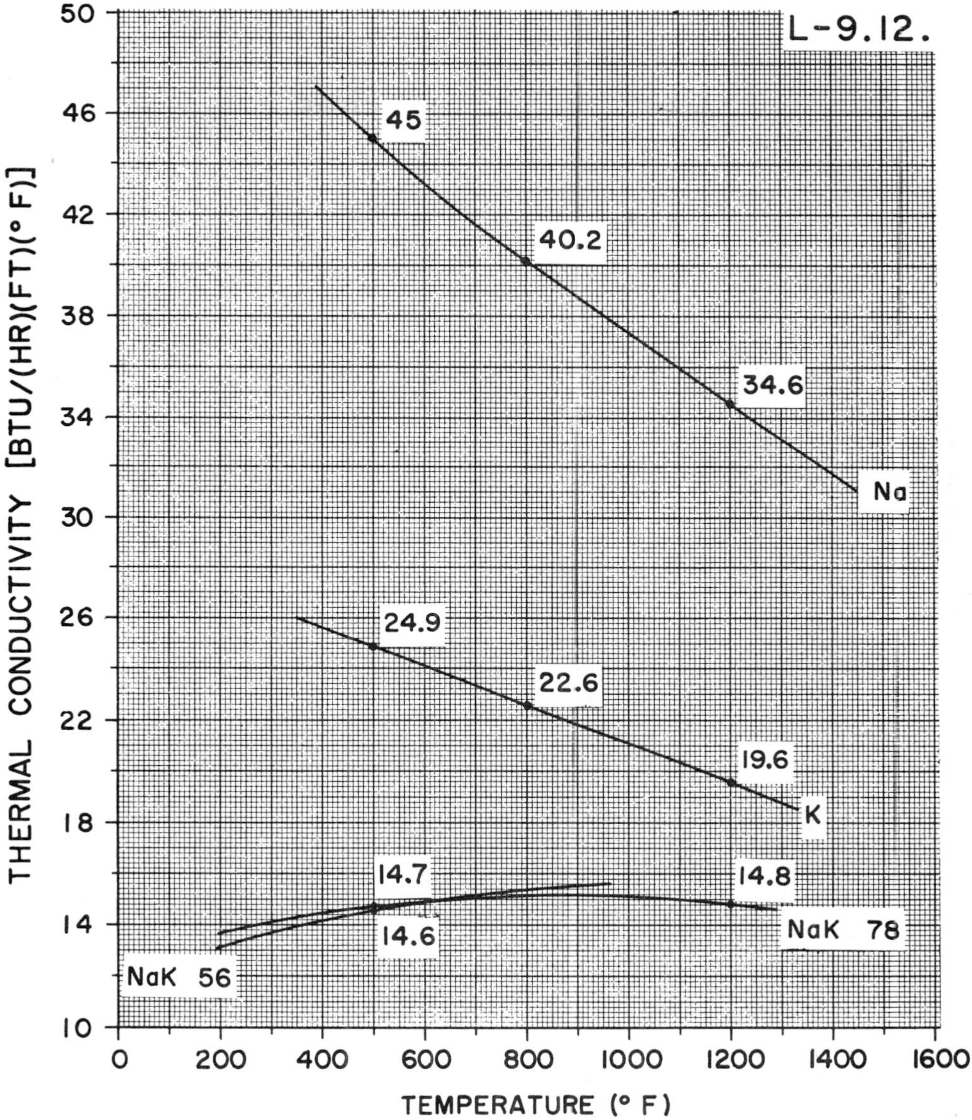

Thermal Conductivity vs Temperature

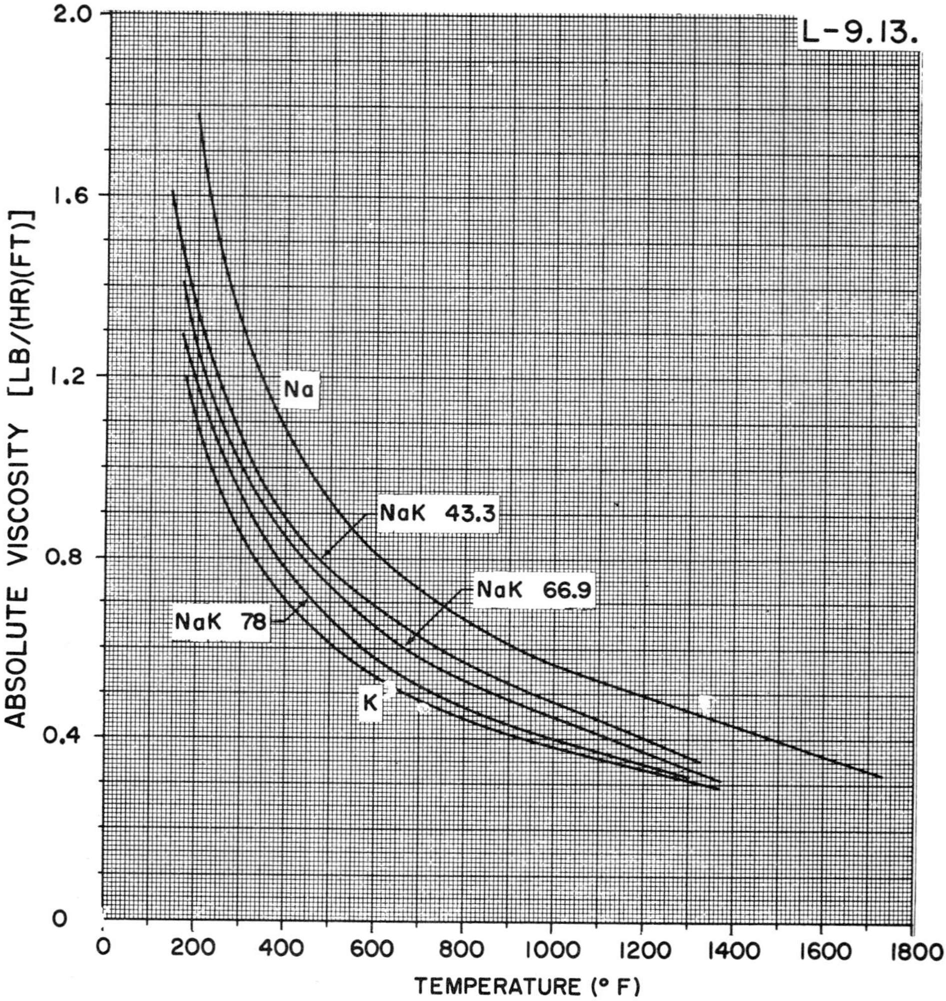

Absolute Viscosity vs Temperature

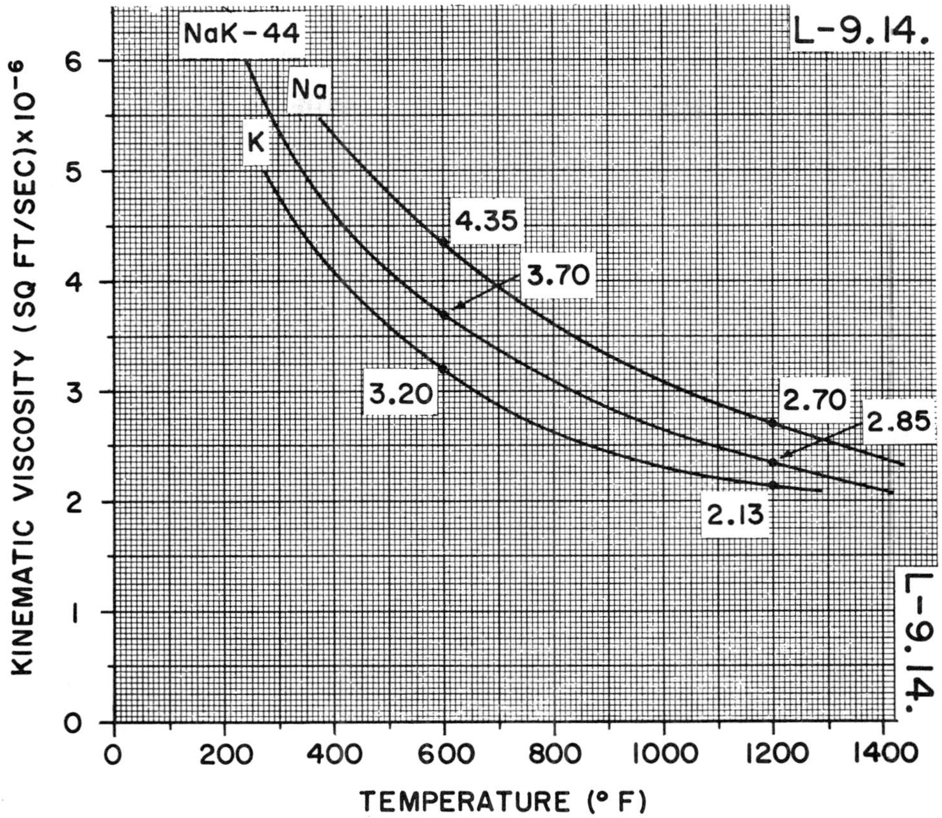

Kinematic Viscosity vs Temperature

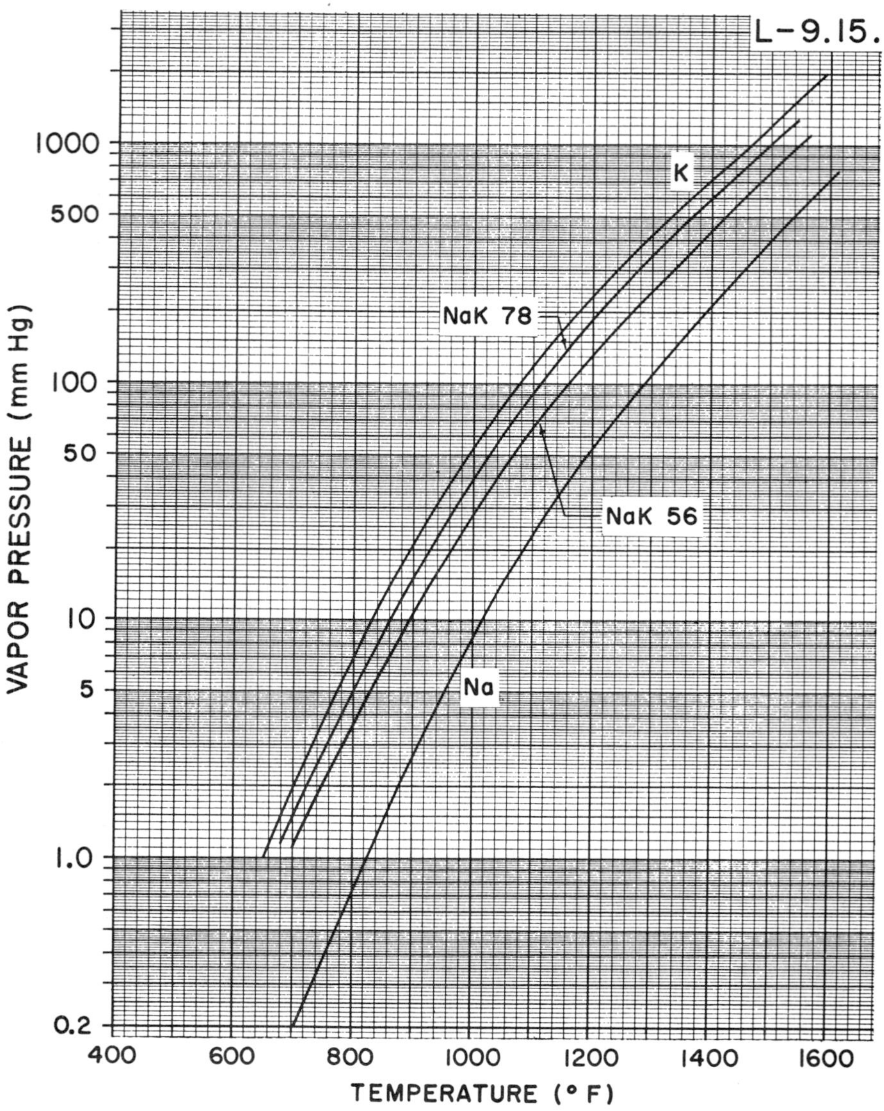

Vapor Pressure vs Temperature

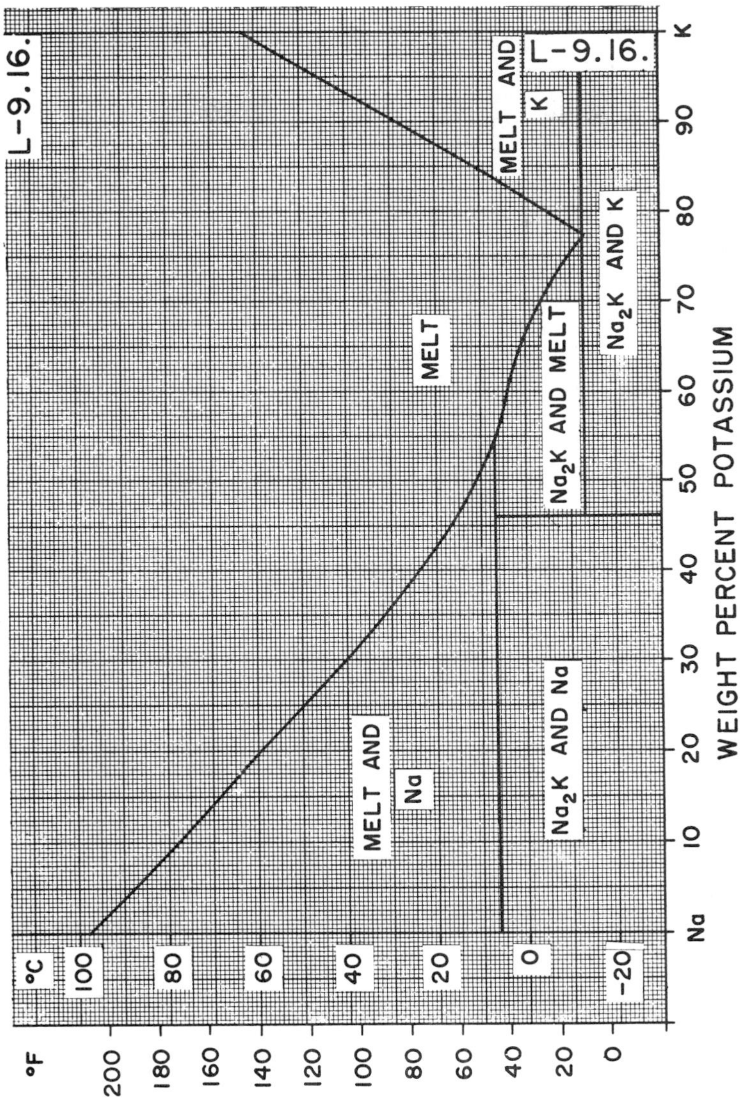

Na-K Phase Diagram

L-10. INORGANIC SALT MIXTURES

Molten salts have been employed for hundreds of years in the treatment of steel and in the manufacture of pottery. Therefore, it is not surprising that engineers have tried to find salts and salt mixtures for use in heat transfer systems.

The use of inorganic salt mixtures as thermal liquids has been the source of a number of investigations. Some rather intriguing work is taking place at Oak Ridge National Laboratory where a design study on a 30 MW., one region experimental molten salt reactor is under way. While multi-component systems such as NaF-BeF_2-UF_4 and LiF-BeF_2-ThF_2-UF_4 are being studied as fuels, molten salt mixtures such as KF-LiF-BeF_2 and LiF and $RbCl$ have aroused interest as reactor coolants. KF-LiF-BeF_2, with a eutectic mixture existing at 887°F, has a lower gamma activity than LiF-$RbCl$. The most familiar member of this family is the inorganic salt mixture known as HTS (Heat Transfer Salt). HTS is a molten, water-soluble eutectic mixture used for heat transfer at atmospheric pressure within a 300°-1000°F temperature range. The composition of HTS can vary within wide limits with only slight effects on physical properties other than freezing points. Two mixtures which have received widest attention consist of 40 per cent $NaNO_2$, 7 per cent $NaNO_3$, and 53 per cent KNO_3 (HTS_1); and 45 per cent $NaNO_2$ and 55 per cent KNO_3 (HTS_2). Both mixtures have approximately the same freezing points and may be employed interchangeably. However, HTS_2 is rarely used at the present time.

The earliest reference to HTS appears in a paper by W. E. Kirst, W. M. Nagel, and J. B. Castner *Trans. Am. Inst. Chem. Engrs.* **36**, 371 (1940), in which its major chemical and physical properties are presented. Molten salts are well suited for continuous high temperature systems in which the liquid is drained out prior to shutdown.

L-10.1. HEAT TRANSFER SALT

HITEC

Hitec, HTS_1, the eutectic mixture of sodium nitrite (40 per cent), sodium nitrate (7 per cent), potassium nitrate (53 per cent), is a registered trademark of E. I. du Pont de Nemours & Company, Inc.

Properties of Importance

Density: See Figure L-10.1.1.

Specific Heat: The specific heat of solid Hitec is 0.32 and that of molten Hitec is 0.373 Btu/(lb)(°F).

Thermal Conductivity: See Figure L-10.1.2. The data given are based on measurements by Vargaftik, Neimark and Oleschuck [*U.S.S.R. Bulletin of the V.T.I.*, No. 9 (September 1952)], for molten salt mixtures having the same composition as Hitec. However, [W. D. Powers, *ANP Quar. Prog. Rep.*, (September 30, 1957), ORNL-2387, 108] obtained a relatively constant value of 0.35 Btu/(hr)(ft)(°F) over a temperature range of 468°-489°F. Unpublished data from Battelle Memorial Institute for the range of 400°-900°F confirm that of Powers. Thus, no definite values for thermal conductivity at high temperatures may be quoted at this time.

Viscosity: See Figure L-10.1.3.

Physical Stability: The freezing point of fresh Hitec is 288°F. However, on exposure to air and other contaminants, changes in composition may occur, and the ensuing deviation from the pure eutectic causes an increase in melting points. Crystals may precipitate out at temperatures as high as 450°F.

Thermal Stability: Operating at 300°-850°F, the salt mixture is quite stable and can be expected to give many years of excellent service. Between 850°-1000°F, the salt undergoes moderate pyrolytic breakdown, with the slow emission of nitrogen gas and a gradual rise in the freezing point of the mixture. The thermal breakdown of sodium nitrite proceeds as follows:

$$[5NaNO_2 \longrightarrow 3NaNO_3 + Na_2O + N_2]$$

Above 1500°F, the reaction is so vigorous that the molten salt appears to boil. Between the temperatures 750° and 1000°F, Hitec furnishes an effective, durable and economic heat transfer medium in a region that is beyond the reach of most organics and accessible to the liquid metals only under conditions of much higher cost and greater difficulties in use. Freezing of the salt mixture can be avoided (Reference 3).

Chemical Stability: Several other chemical reactions may take place simultaneously at elevated temperatures. On contact with the atmosphere, the nitrite is subject to slow oxidation:

$$[2NaNO_2 + O_2 \longrightarrow 2NaNO_3]$$

Also, carbon dioxide is absorbed to form insoluble carbonates and water vapor is absorbed to form alkali metal hydroxides. These reactions can be eliminated by blanketing the molten salt with an inert gas such as dry nitrogen. It has been observed that molten Hitec reacts with CO or CO_2 in flue gas to form water soluble carbonates.

Corrosion: Equipment made from plain carbon steel may be used satisfactorily with Hitec to temperatures of 850°F. More resistant alloys are recommended for higher temperatures, as indicated by the corrosion data in Table L-10.

Flammability: Hitec by itself offers no fire hazard, but inflammable liquids will generally vaporize and burn at the salt surface, leaving the salt composition unchanged. Combustible solids offer considerably greater fire danger. Magnesium should come into contact with Hitec only at very low concentrations, and all contact between the molten salt and carbonaceous materials should be avoided lest violent explosions result.

Physiological Effects: Adequate protection from burns should be provided when handling the hot fluid. Hitec alone releases no toxic vapors. However, minor combustion products such as carbon monoxide and oxides of nitrogen may be quite toxic, thus adequate ventilation should be supplied.

Electrical Resistivity: The electrical resistivity of molten Hitec is 1.7 ohms/cu cm.

Color: Freshly prepared Hitec is a white granular solid. In the molten state it has a pale yellow color.

Price Class: A per lb.

References

1. E. I. Du Pont De Nemours and Co., Inc. "Hitec Heat Transfer Salt."
2. Rottenburg, P. A., *Trans. Inst. Chem. Engrs.* (London) 35, February 1957.

L-10.2. N.S. SALT

HTS is not the only inorganic salt mixture which has been utilized for heat transfer in a closed system. In 1931 a molten salt mixture, known as N. S., was patented in Germany by a firm called Heinrich and Holzhuter, of Berlin-Charlottenburg. It was based on the work of Ernest Sander and Waldemar Nocon (accounting for the letters N.S.). N.S. molten salts are composed of six parts by weight of $AlCl_3$, three parts by weight of NaCl, and one part $FeCl_3$, yet in general terms the mixture may consist of

TABLE L-10. CORROSION OF METALS BY HITEC

Metal	Corrosion Rate inches penetration per month		
	850°F	1000°F	1100°F
Steel—open hearth (ASME #S-17)	0.0003	0.001 to 0.002	0.01 to 0.05
Alloy Steels			
15-16% Chromium iron	–	0.0000	–
"Alcrosil" #3	0.0002	0.001	0.002 to 0.003
"Alcrosil" #5	0.0002	0.0005	0.001 to 0.002
Stainless steel Type 309 (annealed)	0.0000	–	–
Stainless steel Type 347	–	0.0000	–
Stainless steel Type 316	–	0.0000	–
"Hastelloy" B	0.0000	–	–
Monel	–	0.0001	–
Inconel	–	0.0000	–
Bronze	0.0001	–	–
Phosphorized Admirality	0.0001	–	–
Copper	–	0.03	–

the chlorides of any univalent and trivalent metal.

In the absence of air, N.S. softens at about 248°F and melts at about 302°F, while no signs of degradation occur below 1832°F. The recommended liquid range is from 392°-1472°F. and as high as 1832°F without decomposition. These particular proportions of Al, Na and $FeCl_3$ may well react to form a hitherto unknown thermally stable complex.

The salts have a normal boiling point of over 1472°F, a specific heat of about 0.70, and a specific gravity of 1.95 to 2.00. N.S. salts are purported to be nontoxic, incombustible, and inert towards iron and steel. Despite an extraordinary future envisioned by early investigators, it is not known whether any further study of commercial development of N.S. salts has been undertaken.

High Temperature Salt Solution: A recent patent covers a development for the use of soluble salts in water to eliminate the complication of the high freezing point of salt. These systems do not need tracing of the lines.

References

1. Kayser, *J. Chem. Met. Eng.* **49**, 353, 1933.
2. Brownlie, D., Steam Eng., 115, 1941.
3. Hygrotherm Engineering, Ltd., London, England.

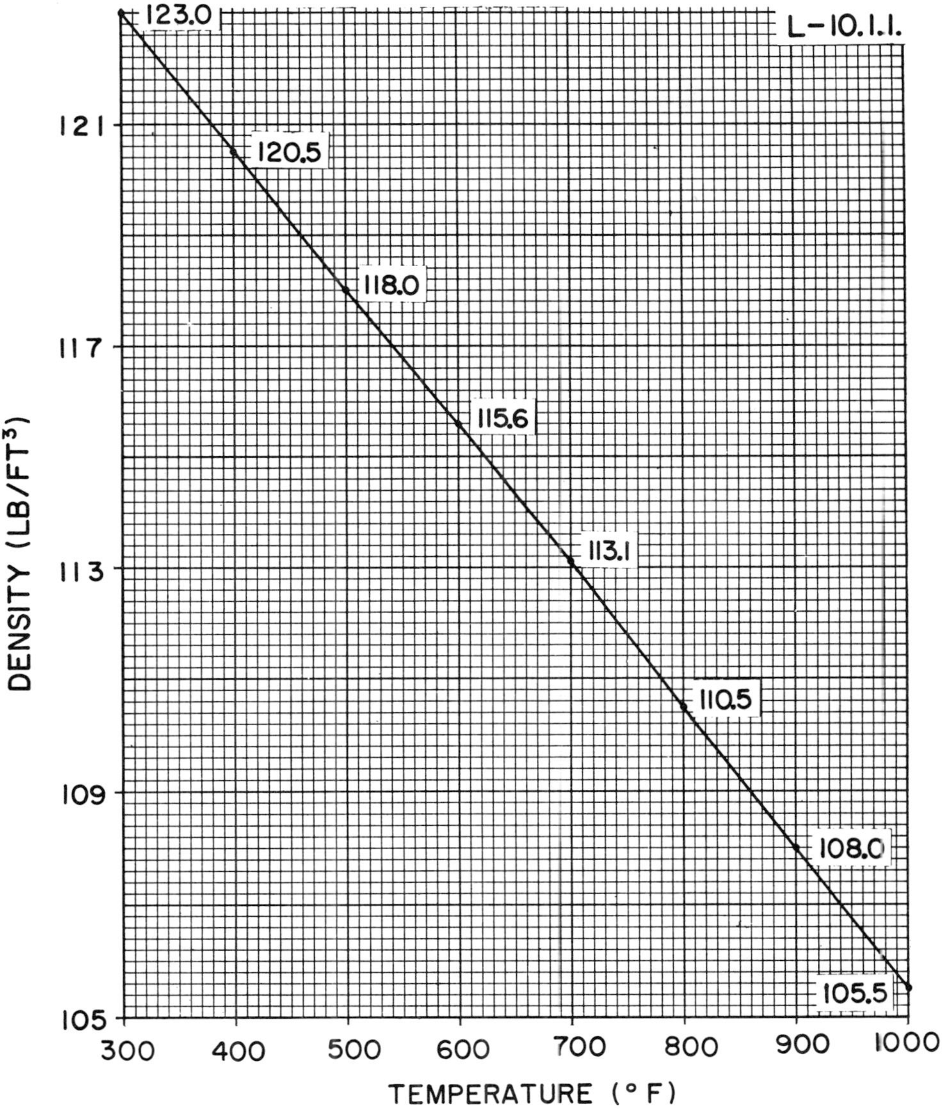
L-10.1.1.
123.0
120.5
118.0
115.6
113.1
110.5
108.0
105.5
DENSITY (LB/FT³)
121
117
113
109
105
300
400
500
600
700
800
900
1000
TEMPERATURE (° F)

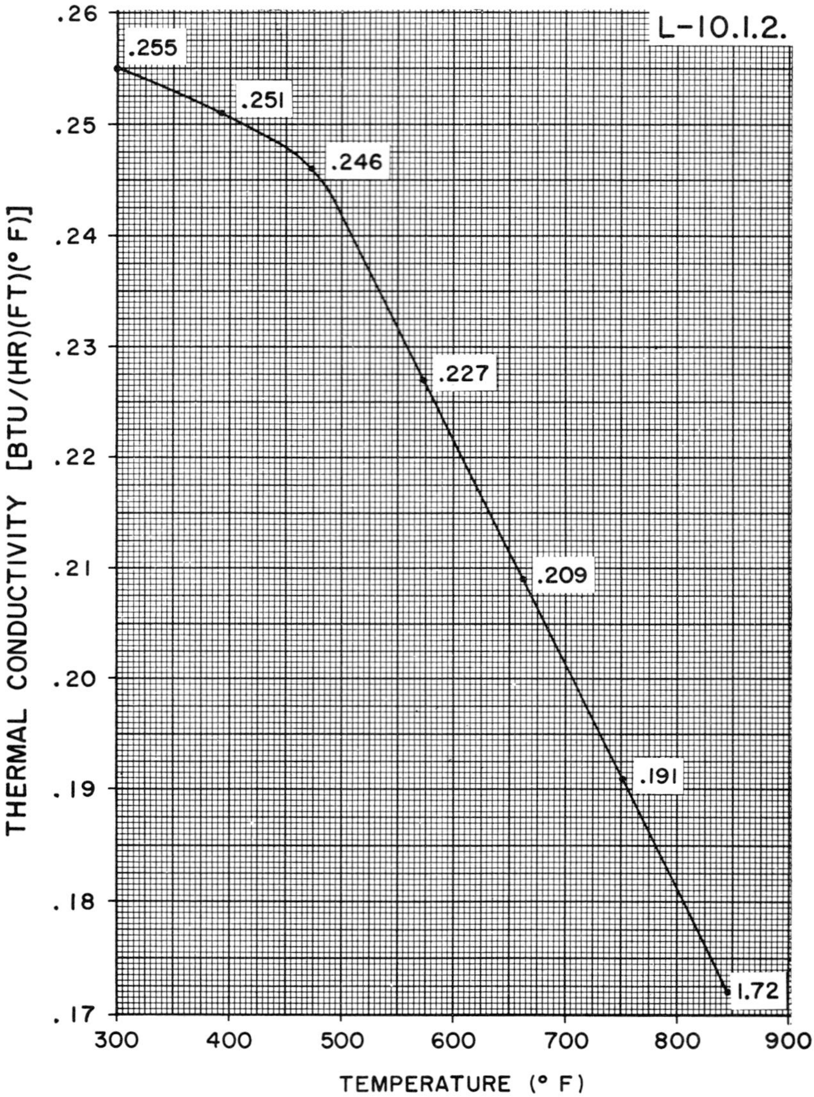

Thermal Conductivity vs Temperature

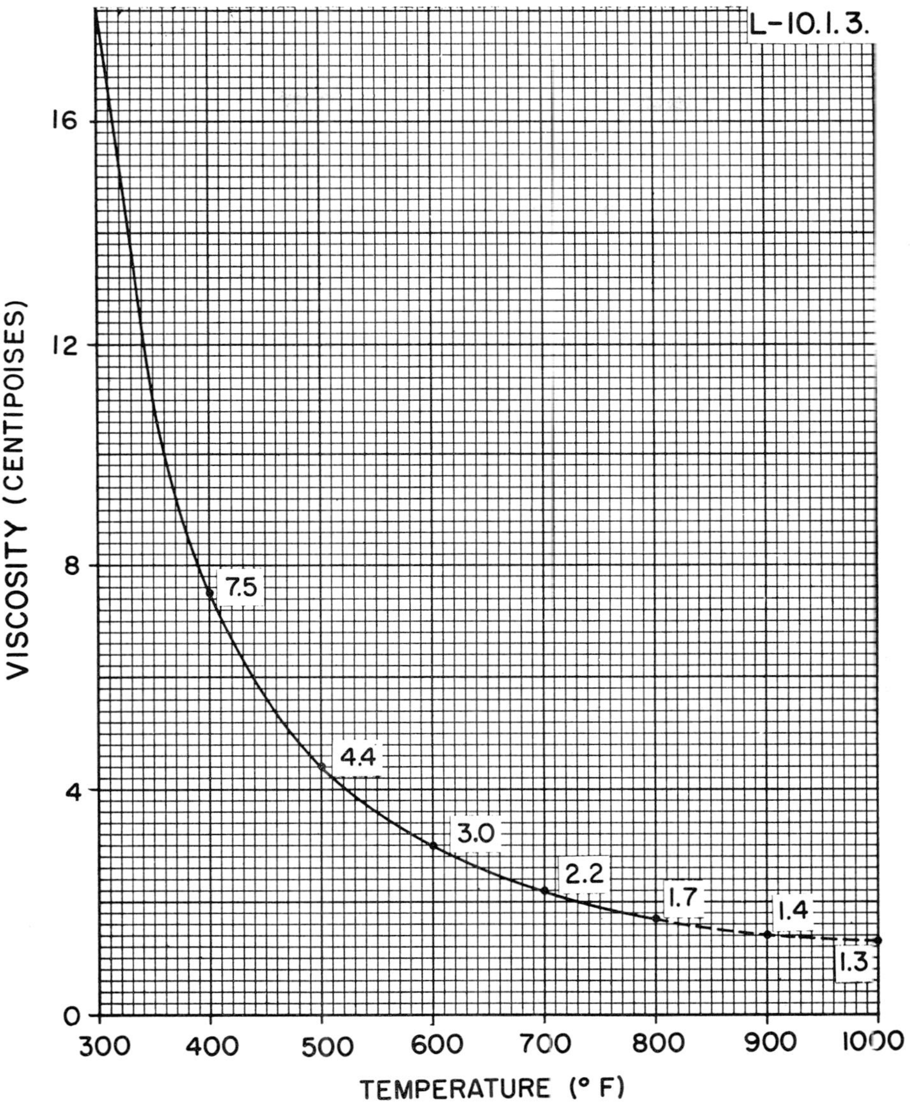

Viscosity vs Temperature

L-11. SILICON COMPOUNDS

Many different types of silicon compounds are now used as heat transfer media. However, the field of silicon chemistry is still relatively unexplored, and consequently, there is much experimentation and new developments every day. The types of compounds presently of interest are silanes, silicones, and silicates.

Silanes are substituted hydrides of silicon. Silane itself has the formula SiH_4. An example of a more complicated silane is $(C_6H_5)_3SiCl$, which would be called triphenyl chlorosilane. Some of the recent developments with silanes are discussed below.

The *silicone fluids* in use are usually polymers. The basic structure of silicones is of the form Si-O-Si. Organic radicals are attached directly to the silicon atoms.

Silicates are simply the salts or esters of the silicon acid in which the central atom is silicon. The organic heat transfer fluids are generally esters. The organic constituents customarily comprise alkyl, aryl, or alkaryl groups.

L-11.1. SILANES

(*Terphenyl-p-Biphenyl Silane*)
(*Tetra-p-Biphenyl Silane*)

A series of comprehensive studies aimed at formulating new extreme-temperature lubricants has been carried out at Wright Air Development Center, (WADC) Dayton, Ohio. Although research has been primarily directed toward fulfilling the demands of advanced gas turbine technology, the elevated temperatures withstood by these materials may inspire their future application as heat transfer liquids.

Perarylated silanes, consisting of a central silicon atom surrounded by four aryl radicals, are thermally stable for short periods of time in the 900°-1000°F range. Characterization and synthesis of phenyl biphenylyl silanes and mixed isomeric biphenylyl silanes by the Stauffer Chemical Company indicate a high degree of thermal stability. One typical member of the family, tetraphenyl silane, has an initial decomposition temperature (isoteniscope) in the neighborhood of 900°F. Although they exist as solids at room temperature, low melting points can be obtained by combining isomeric compositions of the series. Physical properties of two representative phenyl biphenylyl silanes are depicted in Table L-11.

TABLE L-11

	Triphenyl-p-Biphenylyl Silane		Tetra-p-Biphenylyl Silane	
Specific gravity	1.16 at 68°F		1.07 at 68°F	
Specific heat	0.2975		0.2248	
Surface tension (dynes/cm)	350°F – 42.26 500°F – 36.51		400°F – 40.67 475°F – 37.48	
Flash point (Cleveland Open Cup ASTM)	531°F		705°F	
Fire point (Cleveland Open Cup ASTM)	608°F		936°F	
Melting point	318°F		539°F	
Viscosity	t (°F)	μ (cp)	t (°F)	μ (cp)
	275	500	425	Solid
	300	370	450	500
	325	200	475	90
	350	100	500	15
	375	50	525	5
	400	40	550	2
	425	25	575	1
	450	12		
	475	1		

Development of commercial silanes is proceeding at a rapid rate. Metal and Thermit Company recently announced the preparation of a new high temperature lubricant purporting to have excellent resistance to stringent thermal and oxidizing conditions. This compound, diphenylbis-n-dodecylsilane, is a colorless, mobile fluid possessing a low vapor pressure and a good temperature-viscosity characteristic at operating temperatures. The compound is recommended for use as a base stock for elevated temperature gas turbine lubricating oils in the 500°-700°F range.

A novel approach to the compounding of heat carriers can be found in a paper presented before the American Society of Lubricating Engineers. Base stocks consisting of the esters of polyhydric alcohols and the silicones had fine potentials as high temperature lubricants, yet all underwent deterioration (sludge formation, viscosity rise, and increased acidity) when subjected to heavy oxidation. WADC tested numerous compounds for use as high temperature additives. Early in 1957 it was found that an amino function was an essential part of any proposed antioxidant molecule.

A heterocyclic organic compound named phenothiazine was one of the more valuable antioxidants synthesized. However, phenothiazine offered effective protection only up to oil bulk temperatures of 300°F, and it was suspected that the sulfur atom was the deleterious portion of the structure. To incorporate both desirable properties into the same chemical structure, a system containing the extreme thermal stability of tetraphenyl silane and the antioxidizing properties of phenothiazine was derived and synthesized, as described below.

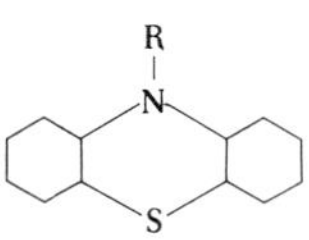

Phenothiazine

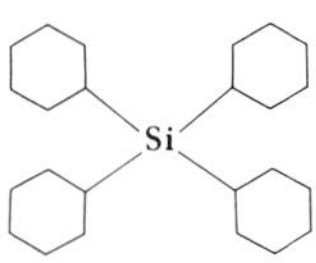

Tetraphenylsilane

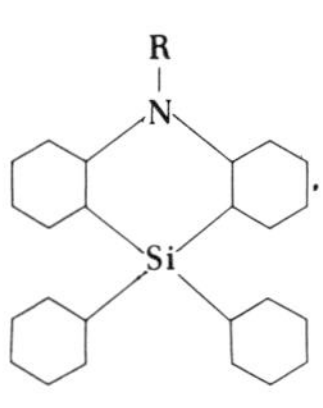

10,10-Diphenylphenazasiline

For this study *R* was an ethyl group. The final product, 5-ethyl-10, 10-diphenylphenazasiline, has been manufactured by the Merck Chemical Company, and is marketed under the name of Merck 5, 10, 10.

Further investigation revealed that a combination of Merck 5-10-10 with phenyl alphanaphthyl amine (PANA) produced the best antioxidant. Future research may be concerned with the properties of spiro compounds, such as:

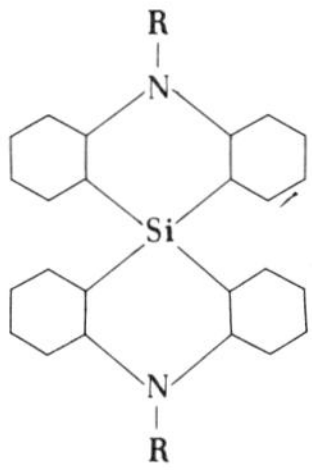

Antioxidant nitrogen centers are doubled and other metallic elements such as lead, germanium, and tin, may be substituted for silicon in the cyclic skeleton.

The John B. Pierce Foundation have synthesized disproportionated mixtures of aryl silane liquids which have low pour points (−30° to 5°F) and high, long time thermal stability up to 750°F.

References

1. Merck and Co., Inc., Chemical Div., "5, 10, 10, New Product Suggestions."
2. Adams, H. W., American Society of Lubricating Engineers, Paper, (April 1960).
3. Anderson, A. R., Lerner, R. W., Smith, W. E., Towers, R. S., "Production of Perarylated Monosilanes," Anderson Chemical Co.; Wright Air Development Center, Technical Report 57-620, September 1957.
4. Stauffer Chemical Co., "Silicate Data Sheet," Anderson Chemical Div.
5. *Chem. Eng. Progr.* **56,** 114, May 1960.
6. Correspondence with New Haven, Connecticut, John B. Pierce Foundation.

L-11.2. SILANES (DIPHENYLBIS-n-DODECYLSILANE)

$(C_6H_5)_2Si(C_{12}H_{25})_2$

Properties of Importance

Density: The specific gravity at 77°F is 0.895.

Specific Heat: 0.59 Btu/(lb)(°F)

Viscosity: See Figure L-11.2.1.

Physical Stability:
Vapor Curve: 536 °F – 0.8 mm
464 °F – 0.2 mm
Pour Point: –25 °F
Melting Point: –40 °F

Thermal Stability: Test in closed tube at 700 °F for 9 hours:

Viscosity change centistokes at 210 °F	–0.50
Change in neutralization number	+0.7
Change in fluid appearance	none

Surface Tension: 75 °F 41.88 dynes/cm

Corrosion: The fluid does not attack metal surfaces when functioning at high temperatures, as indicated by the following data:

Oxidation-Corrosion Test at 700 °F, 24 hr, 5 liters/hr of 95%O_2 plus N_2

Viscosity change, centistokes at 210 °F	+1.67
Change in neutralization number	+1.3
Change in fluid appearance	very dark
Corrosion of Cu, Ag, Ti and steel	nil

Flammability: Flash Point: 530 °F.

Refractive Index: 77 °F-1.5053.

Color: Colorless to slightly yellow.

Price Class: D.

References

1. Metal and Thermit Corp., "Diphenylbis-n-Dodecylsilane," Technical Data, December 1959, and News Release, 1960.

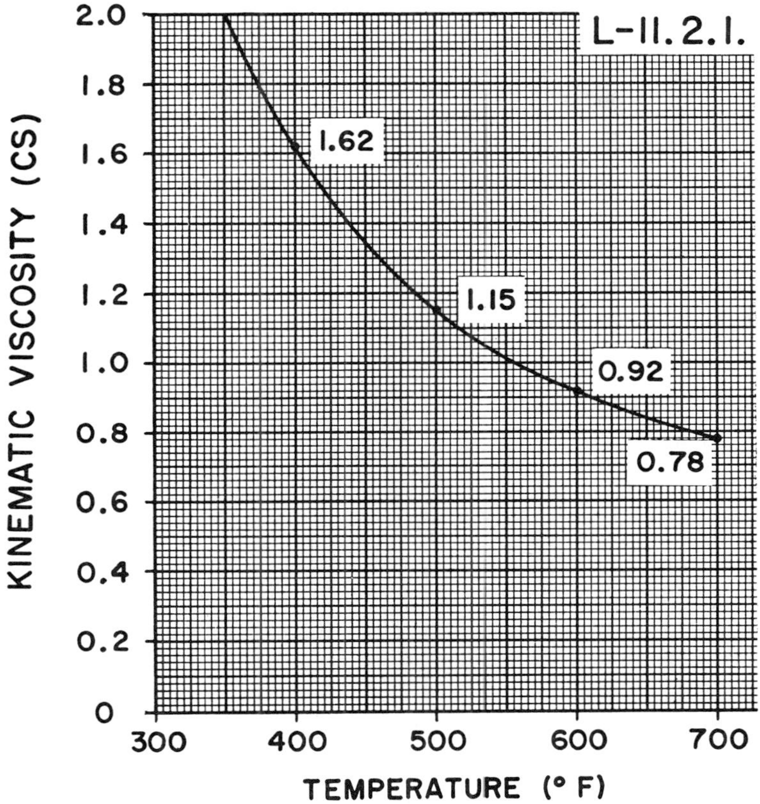

Kinematic Viscosity vs Temperature

L-11.3. SILANES

*Hydrotherm 750–200**

Hydrotherm 750-200 is an aryl aryloxy (mixed) silane used at temperatures up to 700°F. It is recommended for high temperature applications, in the temperature range of 145° to 700°F. Hydrotherm 750-200 is a synthetic organometallic liquid developed by the John B. Pierce Foundation specially for heat transfer purposes. Its outstanding thermal stability is attributed, chemically, to a predominantly aromatic structure (ratio of aromatic to total C – atoms > 0.8), and to its inorganic substituent (ca. 14% SiO_2).

Properties of Importance

Density: See Figure L-11.3.1.

Specific Heat: See Figure L-11.3.2.

Thermal Conductivity: See Fig. L-11.3.3.

Viscosity: See Figures L-11.3.4 and L-11.3.5.

Physical Stability: See Figure L-11.3.5 for vapor pressure. The pour point is 5°F.

Surface Tension: 37 dynes/cm at 77°F.

Corrosion: Corrosion is negligible in ferrous metals, copper and copper alloys; there is slight attack on aluminum and magnesium metals.

Flammability:
Flash Point: 475°F.
Fire Point: 555°F.
Spontaneous Ignition Temperature: 1190°F.

Physiological Effects: There are no chronic or other pathological effects on animals from severe exposure to vapors; skin tests show dermatitic effects similar to those of phenol. Lethal on injection or ingestion.

Electrical Properties:
Volume Resistivity: 1.3×10^5 megohm-inches, 25°C and 500 v D.C.
Dielectric Strength: 43.4 kv at 60 cps, 0.1" gap
Dielectric Constant: 3.41 at 1000 cps.

Refractive Index: 1.6015 at 25°C.

*Reg. Trademark American Hydrotherm Corporation.

Appearance: Amber transparent fluid with a slightly phenolic odor.

Price Class: D.

Reference

1. Correspondence with John B. Pierce Foundation, New Haven, Conn.

L-11.4. SILANES

*Hydrotherm 750–210**

Hydrotherm 750-210 is a mixed aryloxy, arylaryloxy silane developed by the John B. Pierce Foundation. It is recommended for very high temperature applications: 100°-750°F working temperature range. It has extraordinary heat stability and long-expected service life, compared to other organic liquids over an equivalent temperature range. It has a low pour point and low viscosity characteristics. Hydrotherm 750-210 has been evaluated for possible use in the field of nuclear reactors.

Properties of Importance

Specific Gravity: 1.160 at 100°F, 0.880 at 700°F.

Specific Heat: 0.76 at 800°F.

Thermal Conductivity:
7.52×10^{-2} Btu/(hr)(ft)(°F) at 100°F.
5.26×10^{-2} Btu/(hr)(ft)(°F) at 800°F.

Viscosity: 22.1 cs at 100°F, 0.32 cs at 800°F.

Physical Stability: The vapor pressure is 375 mm Hg at 700°F and the pour point is −30°F.

Flammability:
Flash Point: 435°F.
Fire Point: 510°F.

Surface Tension: 32.0 dynes/cm.

Corrosion: Corrosion on ferrous and copper metals is negligible; there is a slight attack on aluminum and magnesium.

Physiological Effects: The effects are similar to those of phenol.

Appearance: Hydrotherm 750-210 is a tan, transparent liquid with a slightly phenolic odor.

Price Class: D.

Reference

1. Correspondence with John B. Pierce Foundation, New Haven, Conn.

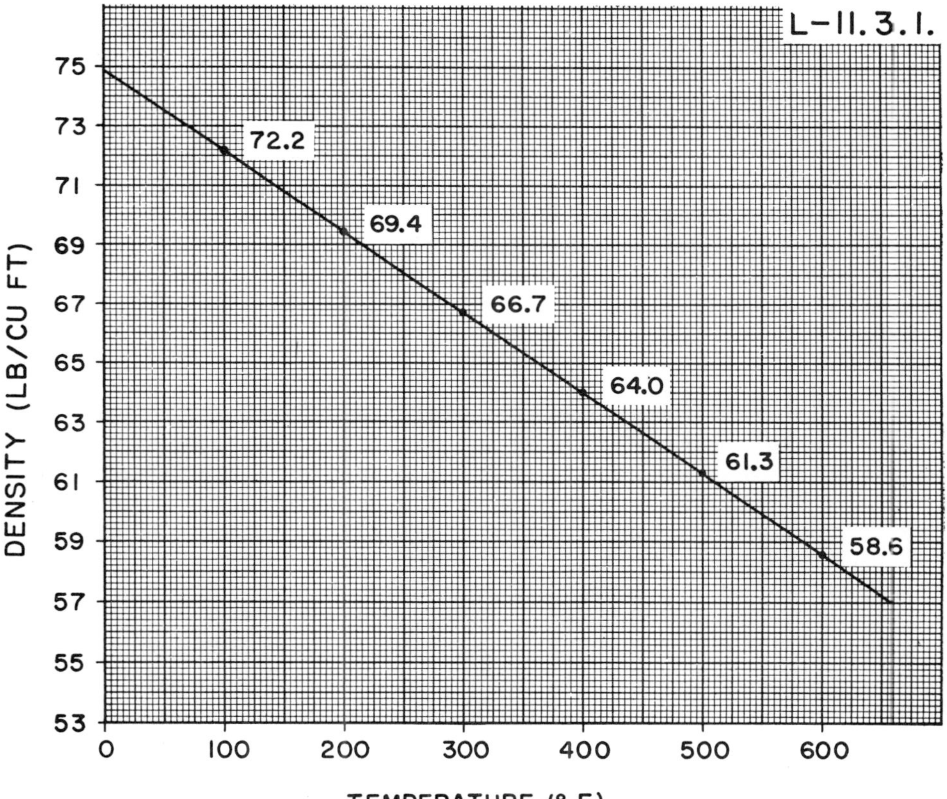

Density vs Temperature

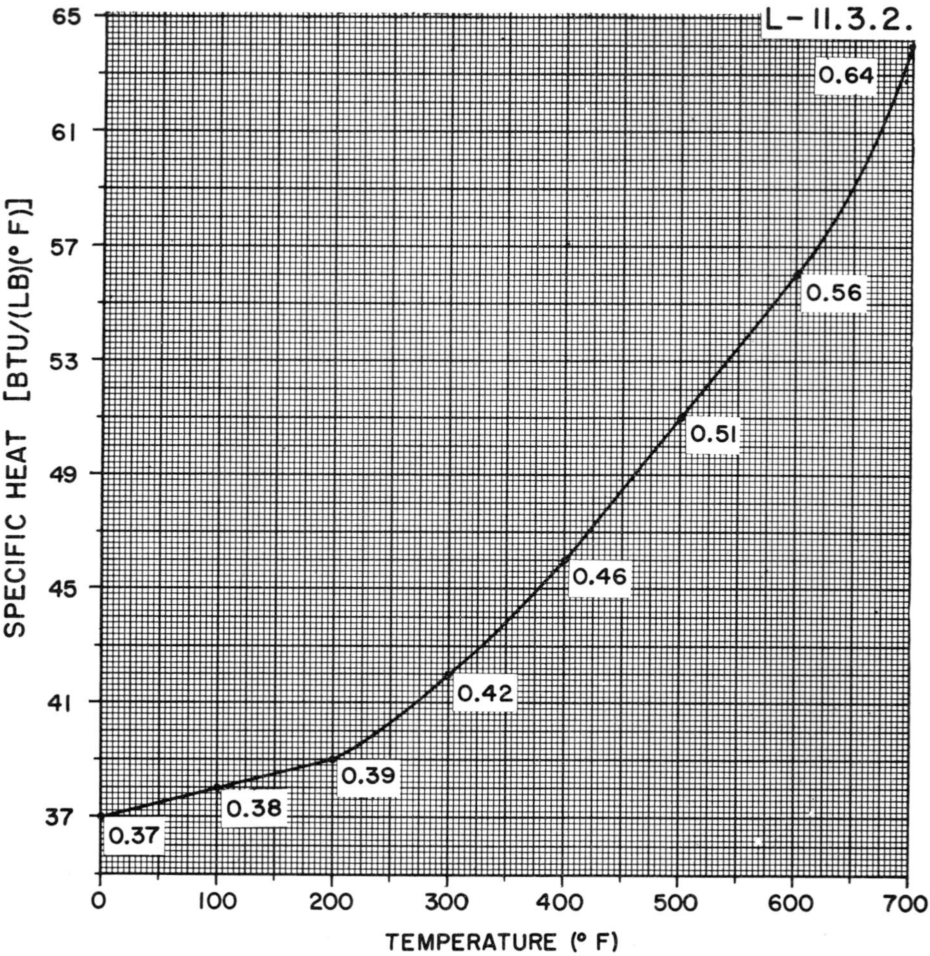

Specific Heat vs Temperature

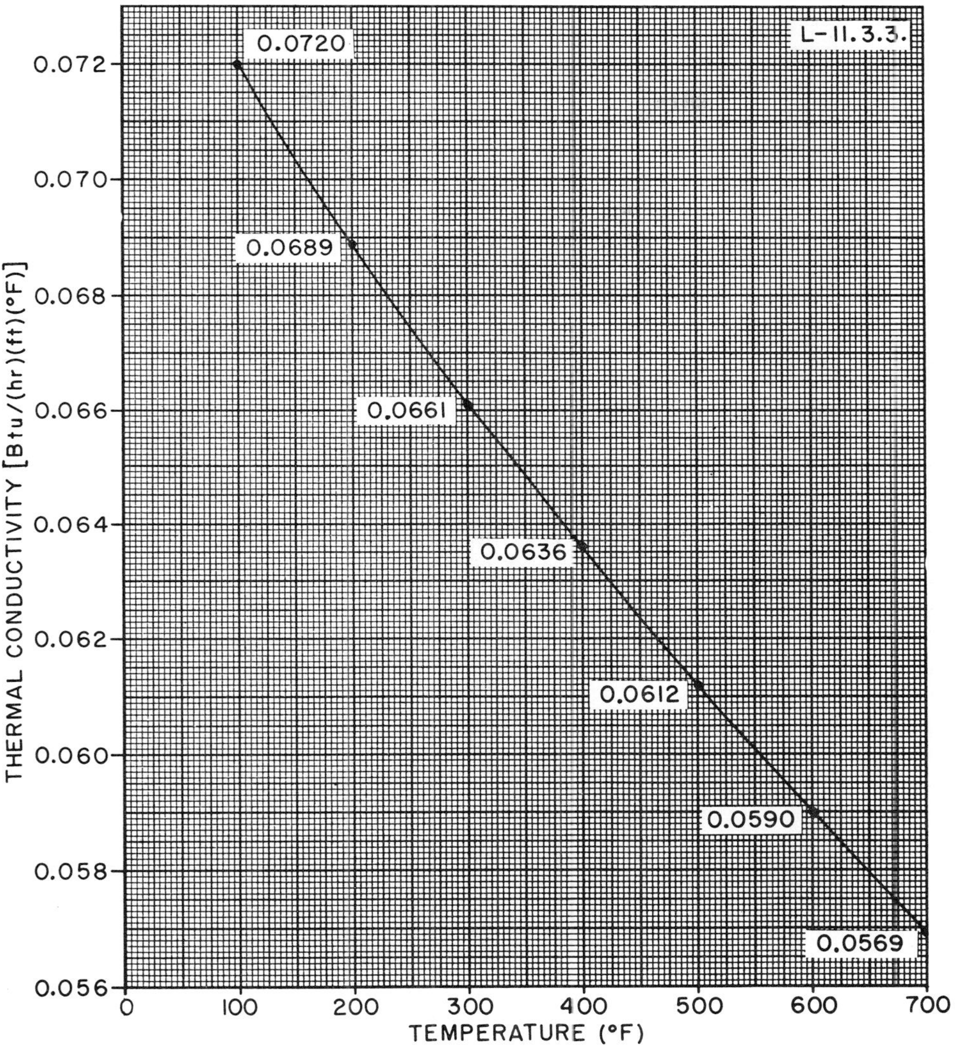

Thermal Conductivity vs Temperature

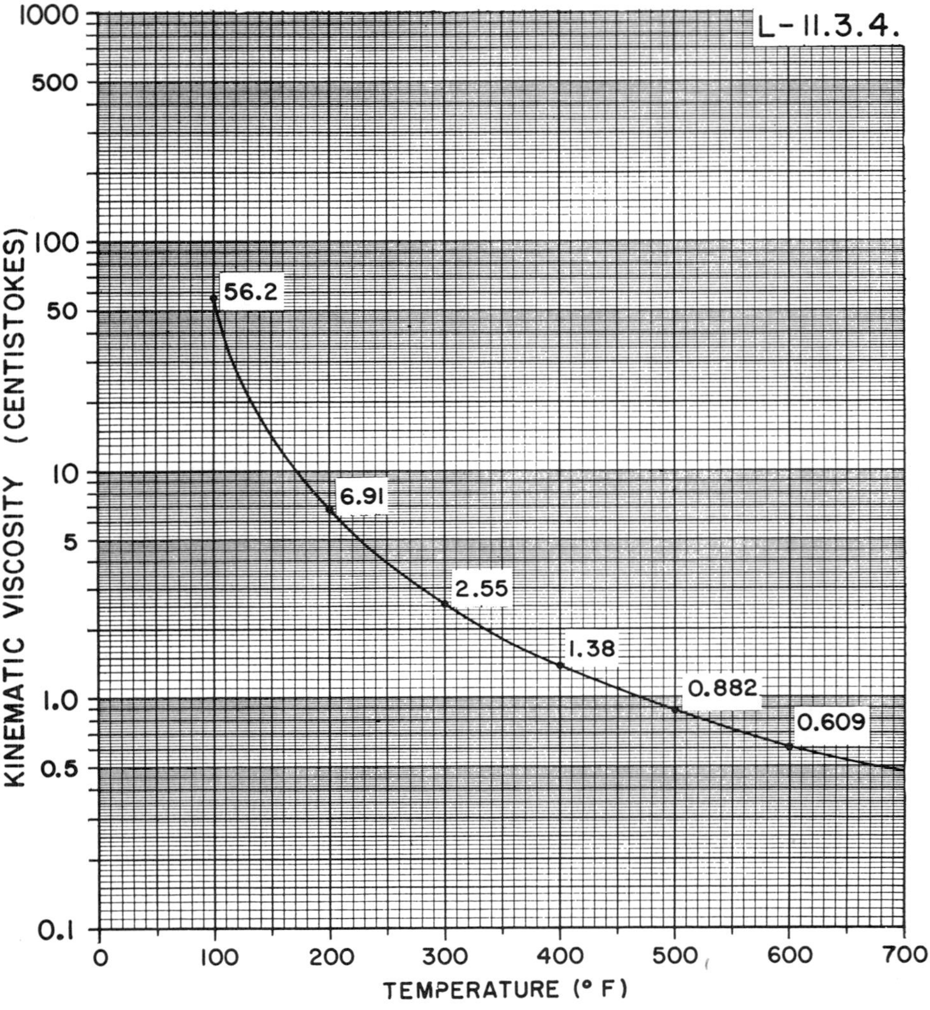

Kinematic Viscosity vs Temperature

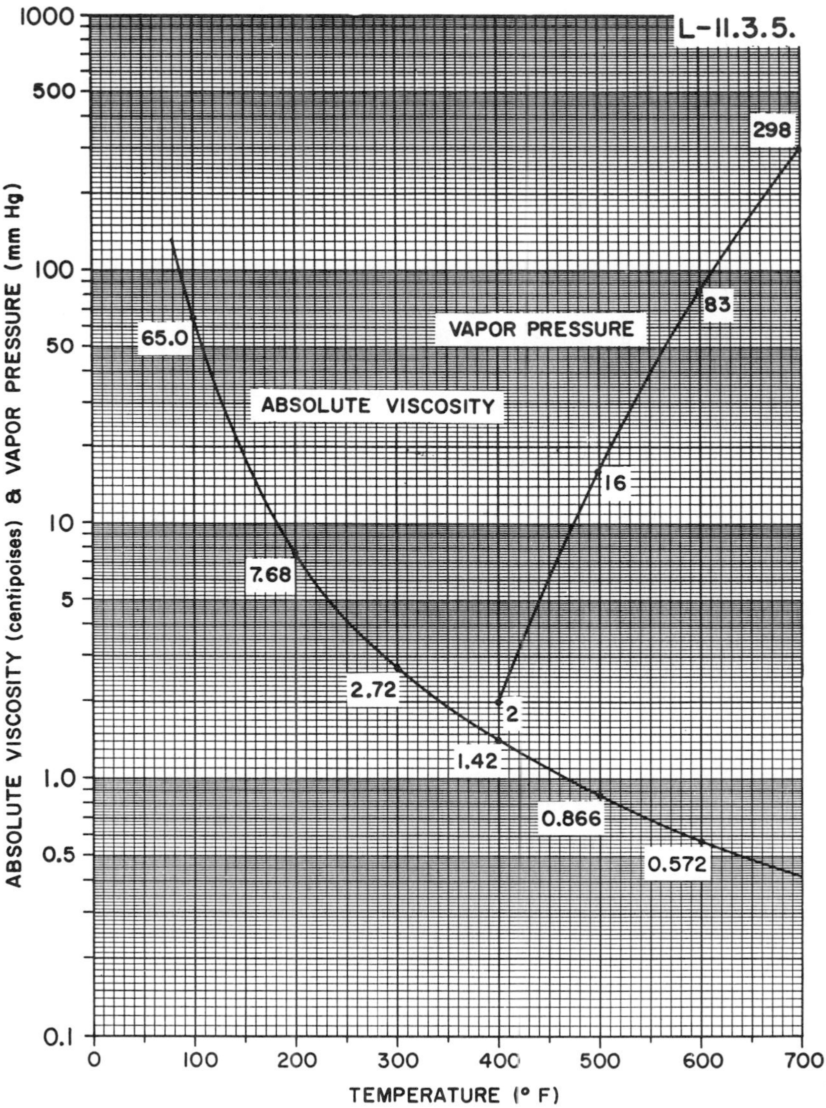

Absolute Viscosity and Vapor Pressure vs Temperature

L-11.5. SILICONE FLUIDS

(Dow Corning F-60, F-61)

Dow Corning F-60 and F-61 are similar silicone fluids. The difference between them is that F-61 has a higher viscosity. These two materials can be blended to obtain a liquid of intermediate viscosity. Both substances are very good high temperature lubricants and hydraulic fluids.

Properties of Importance

Specific Gravity: At 77°F, the specific gravity of F-60 is 1.04; and that of F-61 is 1.06.

Specific Heat: The specific heat of F-60 is 0.356 +(0.0002239)(t-32) where *t* is expressed in °F for the range from 30° to 570°F. The specific heat of F-61 has not been determined.

Thermal Conductivity: The conductivity of F-60 is 0.087 Btu/(hr)(ft)(°F). No tests have been run on F-61.

Viscosity: See Figure L-11.5.1.

Physical Stability: The vapor pressure of F-60 is below atmospheric pressure at temperatures as high as 700°F. At 400°F, the pressure is approximately 0.019 psia. No tests have been run on F-61. The freezing point of both fluids is below −94°F.

Thermal Stability: Tests show that the viscosity of F-60 decreases by 40 per cent, and the acid number increases by 10 per cent, when the material is heated for 72 hours at 700°F in a glass tube purged with nitrogen. No tests have been run on F-61.

Chemical Stability: Tests on F-61 indicate that it has good hydrolytic stability and good oxidative stability.

Corrosion: F-60 can be used with steel, copper, aluminum and magnesium.

Flash Point: (Open Cup) F-60 has a flash point of 550°F; F-61 has a flash point of 570°F.

Ignition Point: The ignition temperature of F-60 is 850°F; tests have not been run on F-61.

Refractive Index: The refractive index of both fluids is 1.43.

References

1. Dow-Corning Corp., "Dow-Corning F-60 and F-61 Fluids," Silicone Notes, (October 1958).
2. Dow-Corning Corp., "Engineering Guide to Silicone Fluids," (1958).

L-11.6. SILICONE FLUIDS

(Dow-Corning 550)

Dow-Corning 550 is a clear silicone oil used as a liquid dielectric and as a heat exchange fluid in a temperature range from −40° to over 450°F.

Properties of Importance

Specific Gravity: (77°F) 1.06 to 1.08.

Specific Heat: See Figure 11-6.1.

Thermal Conductivity: Approximately 8.5×10^{-2} Btu/(hr)(ft)(°F).

Viscosity: See Figure 11-6.2.

Physical Stability: The freezing point of DC-550 is −60°F.

Thermal Stability: Within its specified range, DC-550 has good thermal stability and low volatility, as evidenced by the following tests:

Volatility at 482°F	% By Weight
4 hrs max	3.0
48 hrs max	15.0

Volatility	
after 12 months @ 302°F	4.0
after 12 months @ 492°F	15.0
after 48 hours @ 482°F	9.0

Chemical Stability: DC-550 has a high resistance to oxidation and gumming.

Surface Tension: (77°F) 24.5 dynes/cm.

Corrosion: DC-550R, a modification of DC-550, contains a heat stable rust inhibitor.

Flammability: DC-550 has a minimum flash point of 575°F.

Physiological Effects: None.

Electrical Properties:
Volume Resistivity: 2×10^{13} ohm-cm
Dielectric Constant: 77°F @ 100 cps 2.90
@ 10^6 cps 2.90

Refractive Index: 77°F 1.50.

Price Class: D.

Appearance: Colorless to light straw.

References

1. Dow-Corning Corp., "Dow-Corning Fluid 550," Silicone Notes, November 1958 .
2. Dow-Corning Corp., "Engineering Guide to Silicone Fluids," (1958).

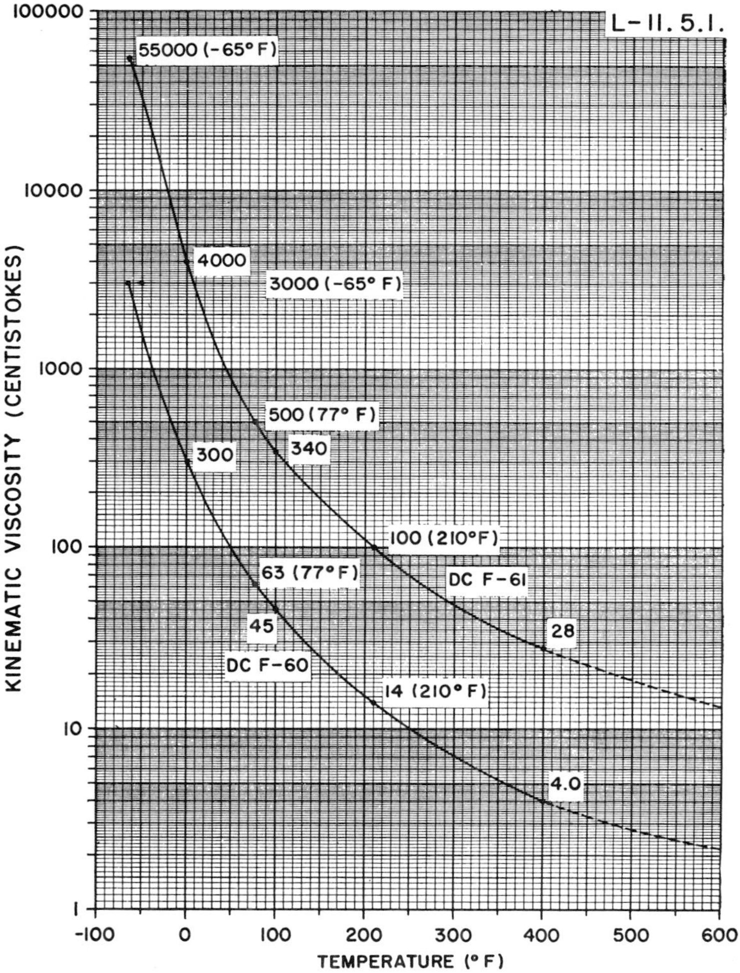

Viscosity vs Temperature

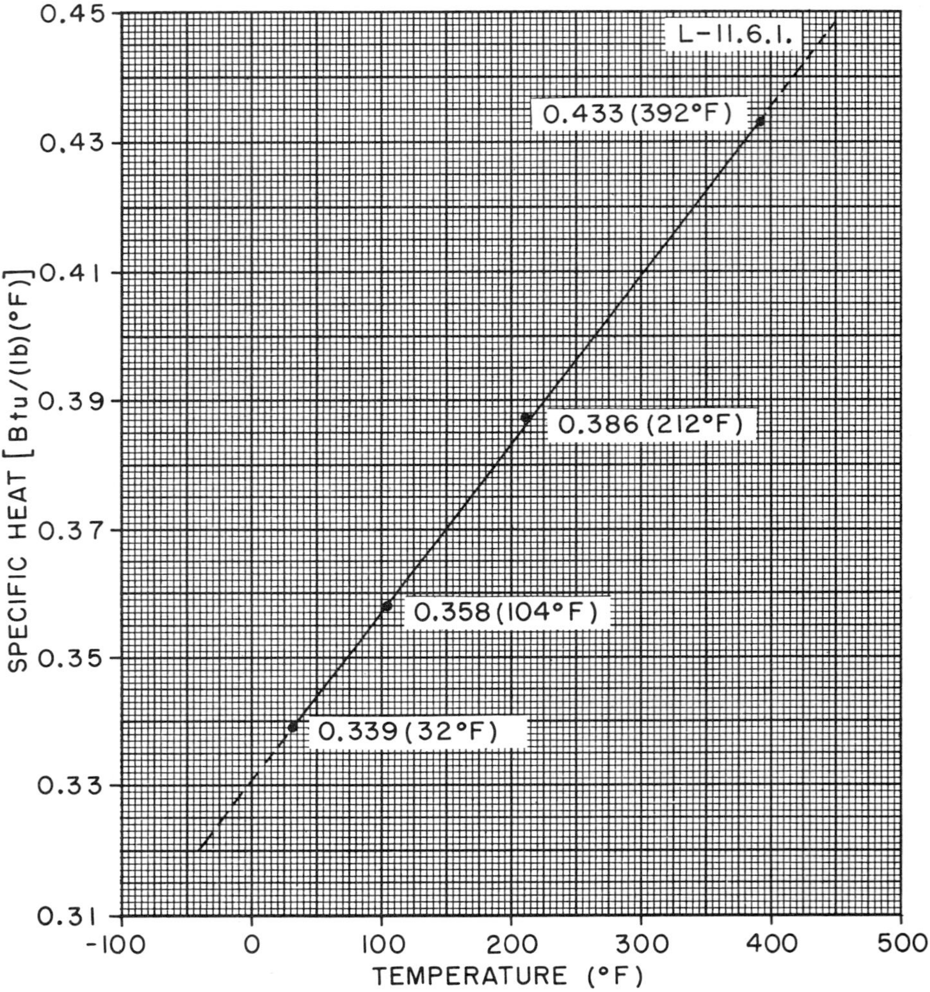

Specific Heat vs Temperature

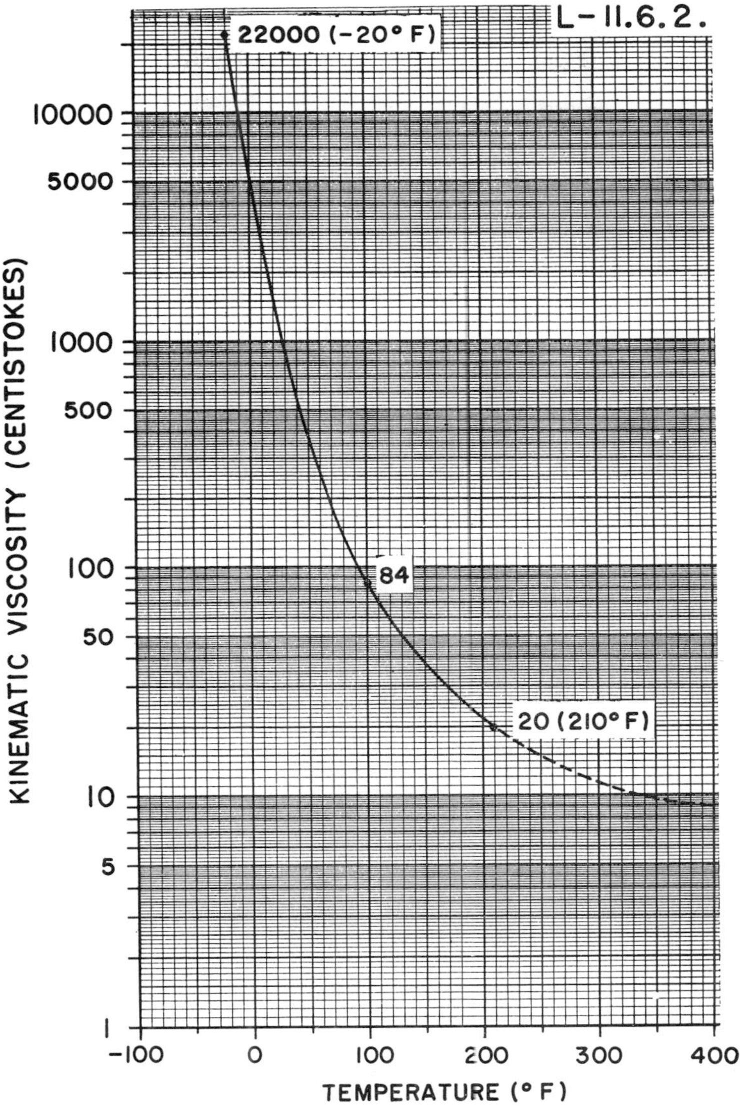

Viscosity vs Temperature

L-11.7. SILICONE FLUIDS

(*Dow-Corning* 710)

Dow-Corning 710 fluid, the most heat stable of the silicone oils, is used as a heat transfer medium at temperatures from 0° to 500°F.

Properties of Importance

Specific Gravity: (77°F) 1.10 – 1.12.

Specific Heat: See Figure L-11.7.1.

Thermal Conductivity: Approximately 8.5×10^{-2} Btu/(hr)(°F)(ft).

Viscosity: See Figure L-11.7.2.

Physical Stability: The freezing point of DC-710 is –8°F.

Thermal Stability: DC-710 has high thermal stability within its operating range, as shown by the following tests:

Volatility at 482°F	% by wt
4 hr maximum	3.0
48 hr maximum	15.0
12 months at 302°F	2.0

Chemical Stability: DC-710 has a high resistance to oxidation and gumming.

Surface Tension: 77°F 28.5 dynes/cm.

Corrosion: A modification, DC-710R, contains a heat stable rust inhibitor.

Flammability: DC-710 has a minimum flash point of 575°F (Open Cup, ASTM D92-57).

Physiological Effects: None

Price Class: D.

References

1. Dow-Corning Corp., "Dow-Corning 710 Fluid," Silicone Notes, (May 1958).
2. Dow-Corning Corp., "Engineering Guide to Silicone Fluids," 1958.

L-11.8. SILICATE ESTER

(*Coolanol 45*)

Coolanol 45 is the trade name of a fluid with a silicate ester base, manufactured by Monsanto Chemical Company. Gas and evaporative liquid cooling of electronic components are sometimes inadequate because of stringent demands for equipment reliability and the concomitant requirement of exact temperature control. Thus, liquid coolants such as Coolanol 45 were developed.

This fluid, because of its excellent dielectric strength and its capacity to dissipate heat rapidly and efficiently has found widespread application in cooling the complex electronic systems of missiles and aircraft. Coolanol 45 makes a one-fluid package feasible since, besides being a coolant, it can also function as a hydraulic fluid. Another important advantage stems from its extremely wide temperature range, –65° to 400°F, making Coolanol 45 suitable for high and low temperature applications. In summation, its dielectric, hydraulic, heat transfer, and lubricating properties are quite favorable.

For a further discussion of Coolanol 45, see page 119.

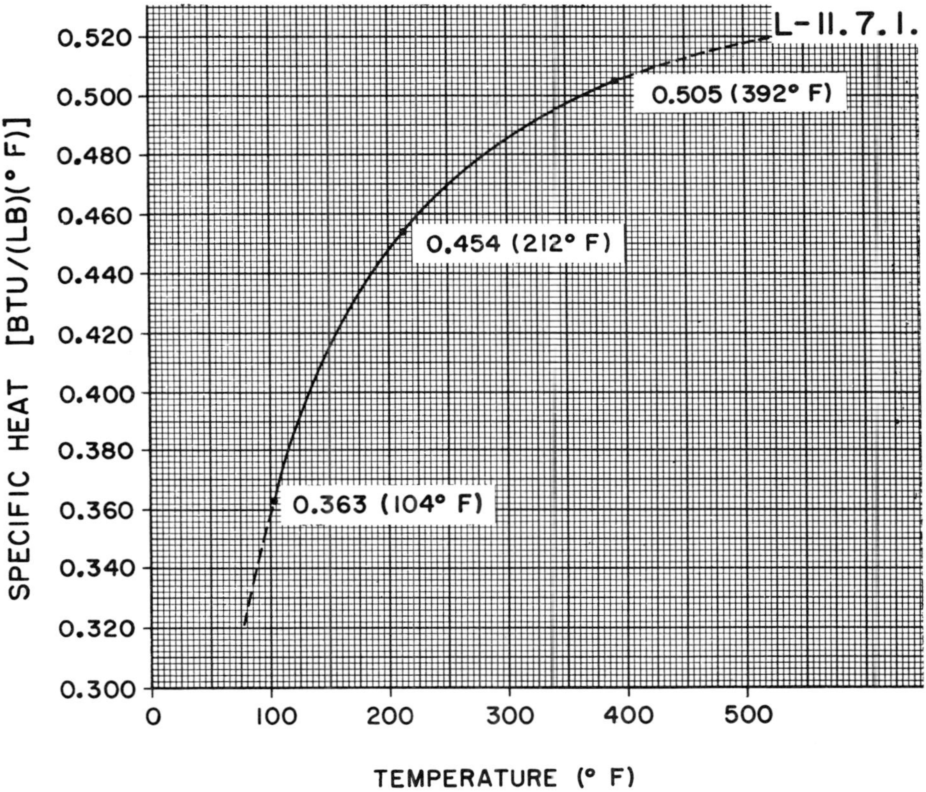

Specific Heat vs Temperature

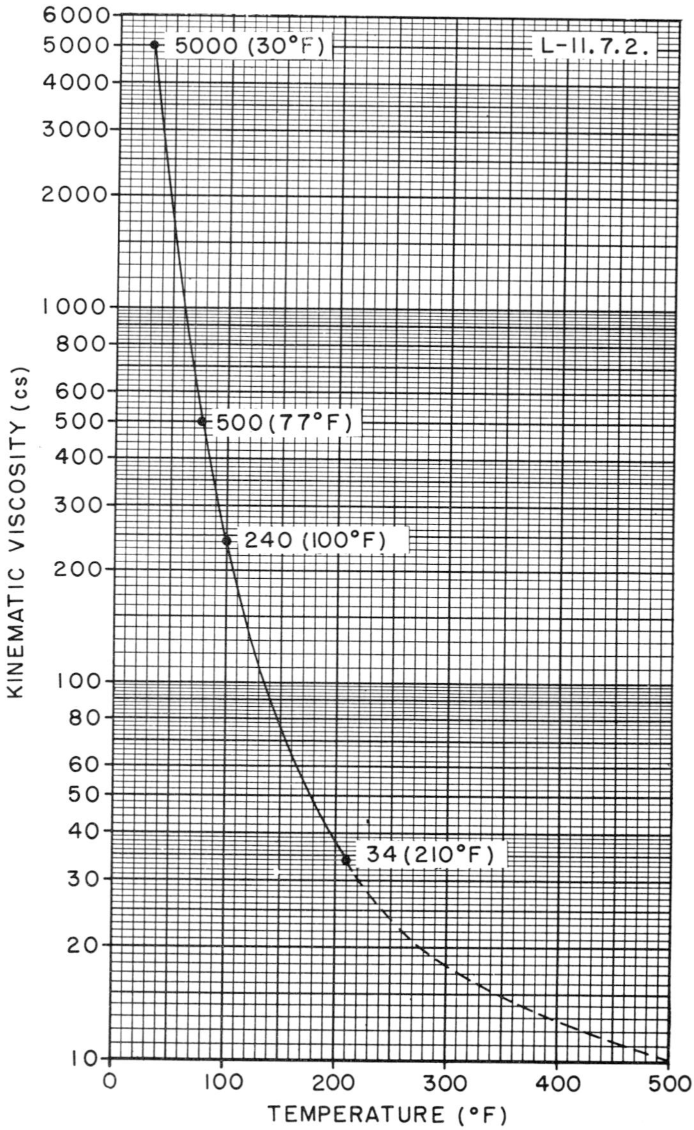

Kinematic Viscosity vs Temperature

L-11.9. SILICATES

Hydrotherm organo-silicate heat transfer fluids, sometimes referred to as TAS fluids, are applicable for heating and cooling operations from −50° to 675°F. These fluids were first introduced in 1941. Hydrotherm 700-10A, Hydrotherm 700-160 and Hydrotherm 700-190 are aryl silicates, while Hydrotherm 700-130 and Hydrotherm 700-155 are alkyl compounds.

This group of heat carriers has been utilized with notable success in pressureless systems. Organo-silicate liquids do not freeze and stay pumpable at temperatures down to 0°F and below. Therefore, neither reheating nor steam tracing is required. In fact, the recommended peak operating temperature of 675°F for Hydrotherm 700-190 is more than 100°F below the boiling point of the liquid.

Tetra aryl silicates have excellent thermal and chemical stability. On page 20, Section II, it is shown that the aryl silicates are in the highest stability range for organic compounds, as measured by the Isoteniscope method. They react with oxygen; therefore, contact with the atmosphere should be carefully controlled. In the presence of moisture hydrolysis takes place; free phenols are released and a silicate sludge is deposited. As for thermal decomposition, a reforming reaction occurs at elevated temperatures (above 675°F). Low boiling products, primarily mono or dihydric phenols (for the aryl compounds) and alcohols (for the alkyl compounds) are formed and a gradual rise in viscosity also ensues. A novel method for rejuvenating spent silicate fluids has recently been devised by Hygrotherm Engineering Ltd.

Handling properties are quite favorable.

Hydrotherm silicate heat transfer fluids are noncorrosive towards mild steel, copper and its alloys, even at high temperatures. However, magnesium, aluminum, zinc, titanium, and their alloys are attacked. Organo-silicate liquids do not present a potentially serious fire hazard. They ignite only at high temperatures when the latent heat of vaporization is furnished. Hydrotherm fluids do not possess any offensive or toxic odors.

The following pages are a more detailed survey of the properties of the silicates.

References

(For all Hydrotherm Fluids)

1. American Hydrotherm Corp., "Hydrotherm Liquid Heat-Transfer Systems," Bulletin 200, (1957).
2. American Hydrotherm Corp., "Hydrotherm TAS Liquid Heat Systems."
3. Hygrotherm Engineering Ltd., London, England, "Hygrotherm," 1958.

L-11.9. SILICATES

*Hydrotherm 700-10A in the U.S.**
*TAS-10A in the U.K.***

Hydrotherm 700-10A, or tetracresyl silicate plus an additive, is used as a liquid phase heat transfer medium without pressure between 50° and 600°F. It is also employed as a hydraulic fluid over a wide temperature range.

Properties of Importance

Specific Gravity: See Figure L-11.9.1.

Specific Heat: See Figure L-11.9.1.

Thermal Conductivity: The heat conductivity is approximately 0.072 Btu/(hr)(ft)(°F).

Viscosity: See Figure L-11.9.2.

Physical Stability: See Figure L-11.9.2 for the vapor pressure curve. Hydrotherm 700-10A, has a pour point of −30°F.

Corrosion: Aluminum, magnesium, zinc, titanium, and their alloys are the only familiar structural metals with which Hydrotherm 700-10A should not be used.

Price Class: C.

*Reg. Trademark American Hydrotherm Corporation.

**Reg. Trademark Hygrotherm Engineering, Ltd., London, England.

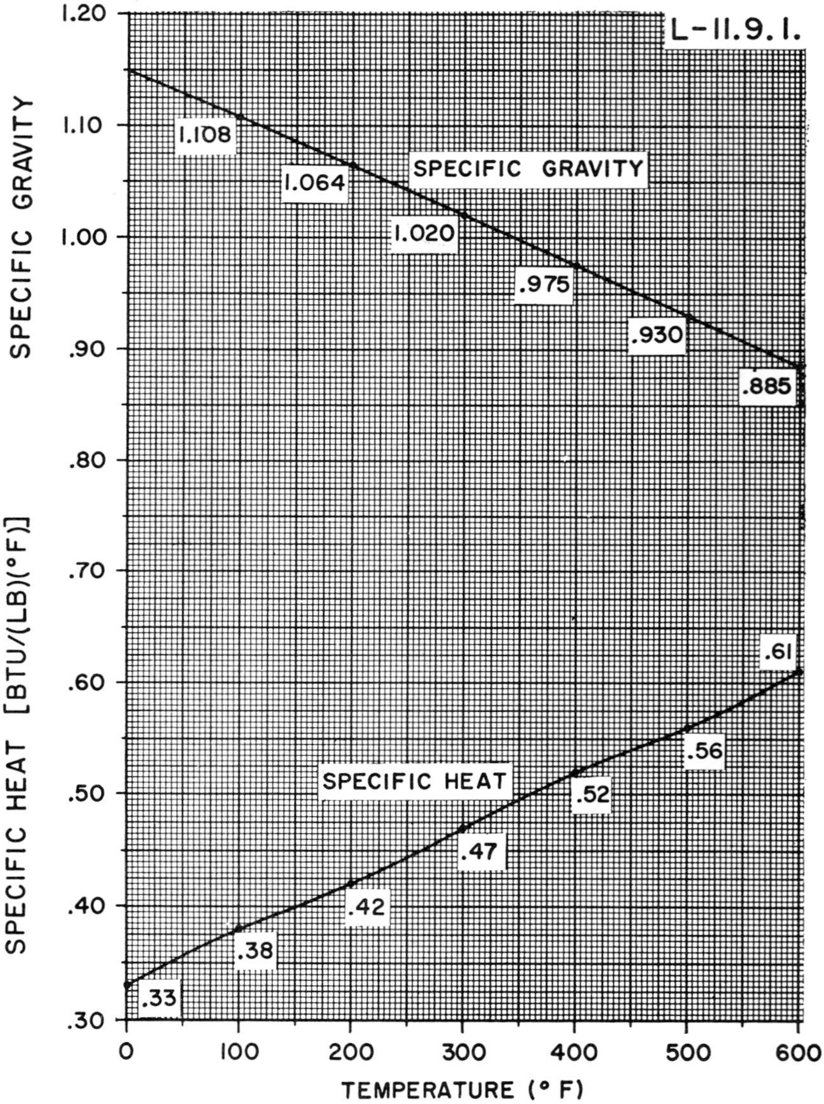

Specific Heat and Specific Gravity vs Temperature

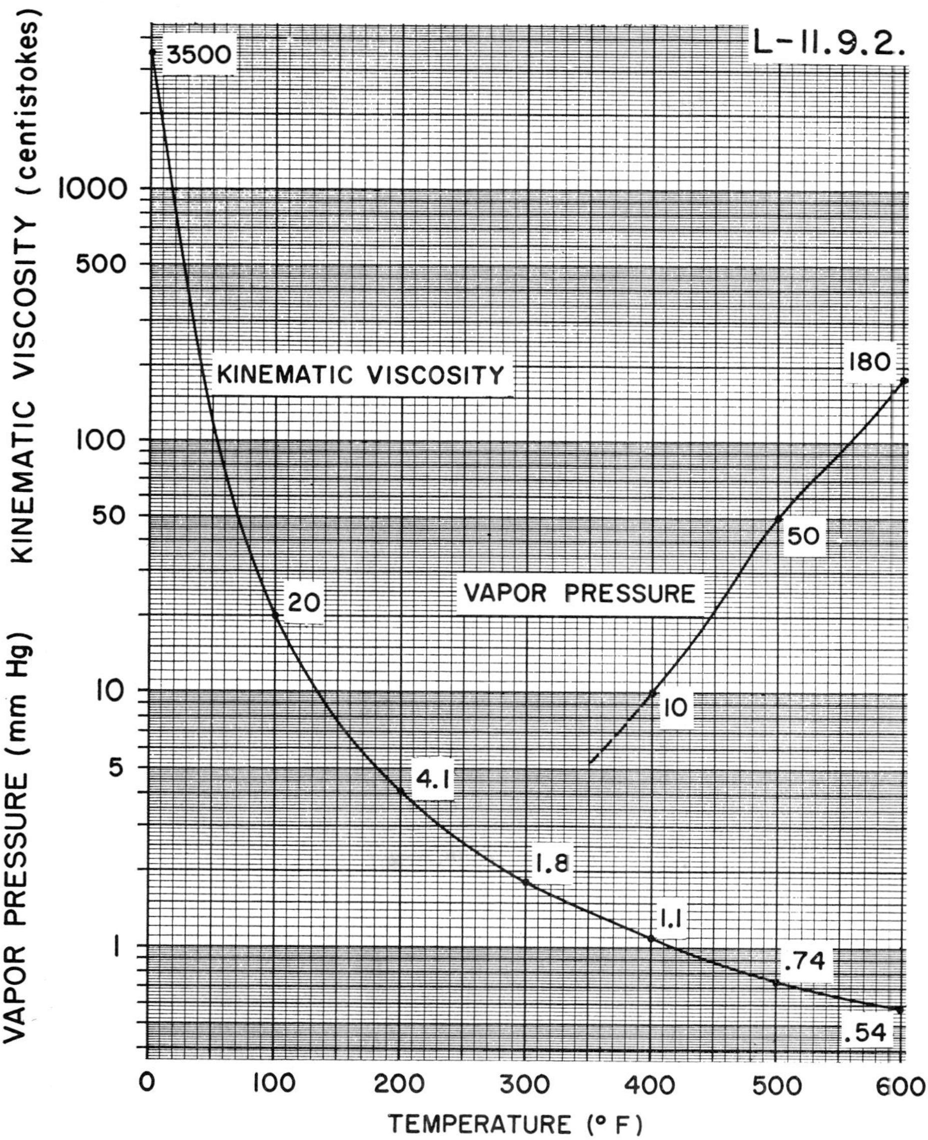

Vapor Pressure and Kinematic Viscosity vs Temperature

L-11.10. SILICATES

*Hydrotherm 700-130 in the U.S.**
*TAS-130 in the U.K.***

Hydrotherm 700-130 is an isopropyl isobutylsilicate. Its suggested operating range is –100° to 350°F. This temperature range recommends the fluid for use in systems where medium temperature process heating is followed by low temperature cooling. Hydrotherm 700-130 also finds application as a hydraulic fluid, particularly for sub-zero conditions.

Properties of Importance

Specific Gravity: See Figure L-11.10.1.

Specific Heat: See Figure L-11.10.1.

Thermal Conductivity: The conductivity is 0.0616 Btu/(hr)(ft)(°F) at 100°F and 0.0572 Btu/(hr)(ft)(°F) at 350°F.

Viscosity: See Figure L-11.10.2.

Physical Stability: See Figure L-11.10.2 for the vapor pressure curve. The fluid has no freezing point and a pour point below –100°F.

Corrosion: Aluminum, magnesium, zinc, titanium, and their alloys are the only common structural metal with which Hydrotherm 700-130 should not be used.

Appearance: Hydrotherm 700-130 is a pale straw-colored liquid with a pleasant odor.

Price Class: C.

L-11.11. SILICATES

*Hydrotherm 700-155 in the U.S.**
*TAS-155 in the U.K.***

Hydrotherm 700-155 is used as a heat transfer medium between 0° and 500°F. Chemically, this fluid is *n*-octyl silicate.

Properties of Importance

Specific Gravity: See Figure L-11.11.1.

Specific Heat: See Figure L-11.11.1.

Thermal Conductivity: The conductivity is 0.0815 Btu/(hr)(ft)(°F) at 0°F and 0.0624 Btu/(hr)(ft)(°F) at 500°F.

Viscosity: See Figure L-11.11.2.

Physical Stability: See Figure L-11.11.2 for the vapor pressure. The pour point is below –65°F.

Corrosion: Aluminum, magnesium, zinc, titanium, and their alloys; and tin, brass, bronze and copper should not be used in Hydrotherm 700-155 systems.

Appearance: Hydrotherm 700-155 is a pale yellow liquid.

Price Class: C.

*Reg. Trademark American Hydrotherm Corporation.
**Reg. Trademark Hygrotherm Engineering, Ltd., London, England.

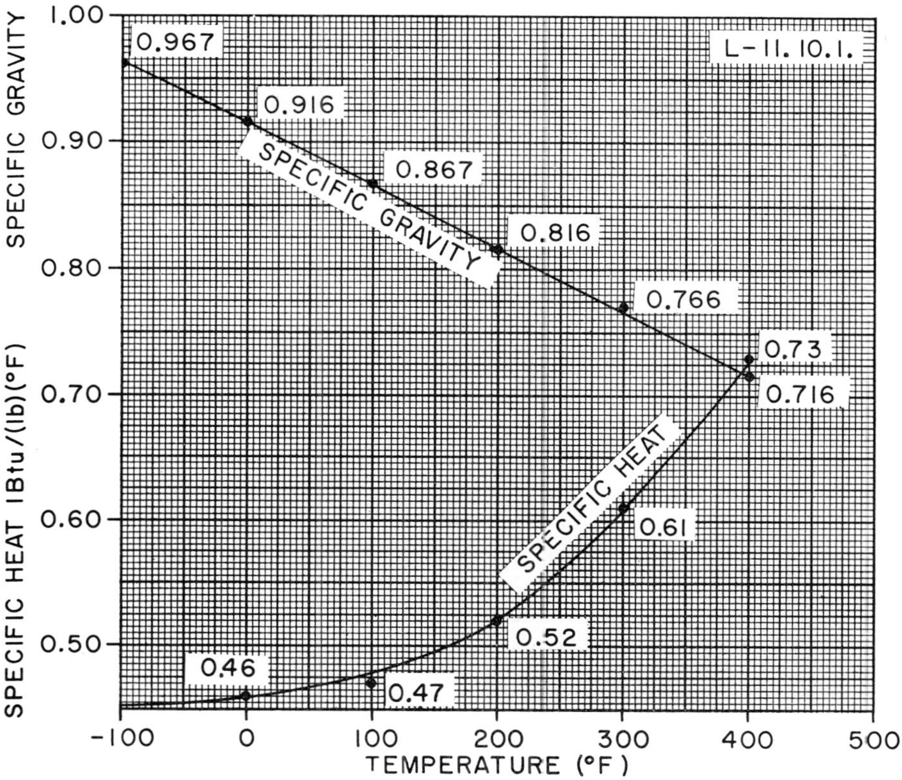

Specific Gravity and Specific Heat vs Temperature

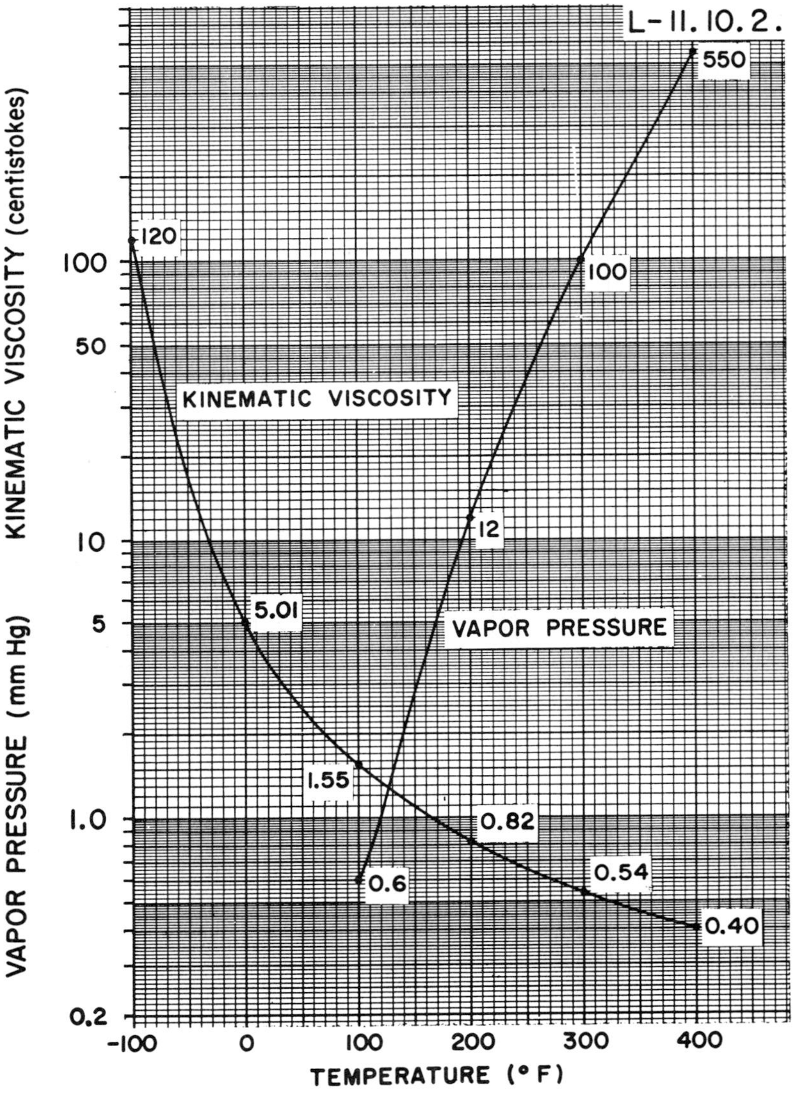

Vapor Pressure and Kinematic Viscosity vs Temperature

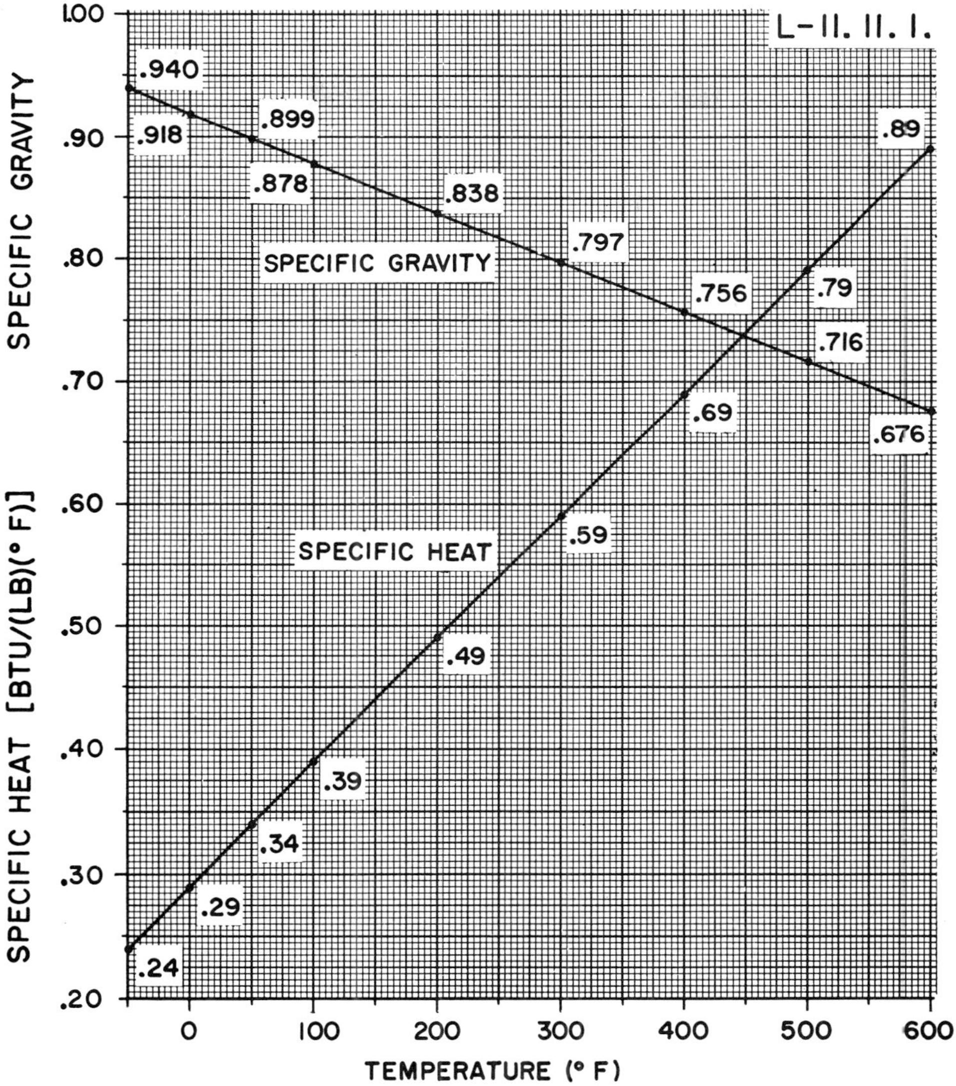

Specific Heat and Specific Gravity vs Temperature

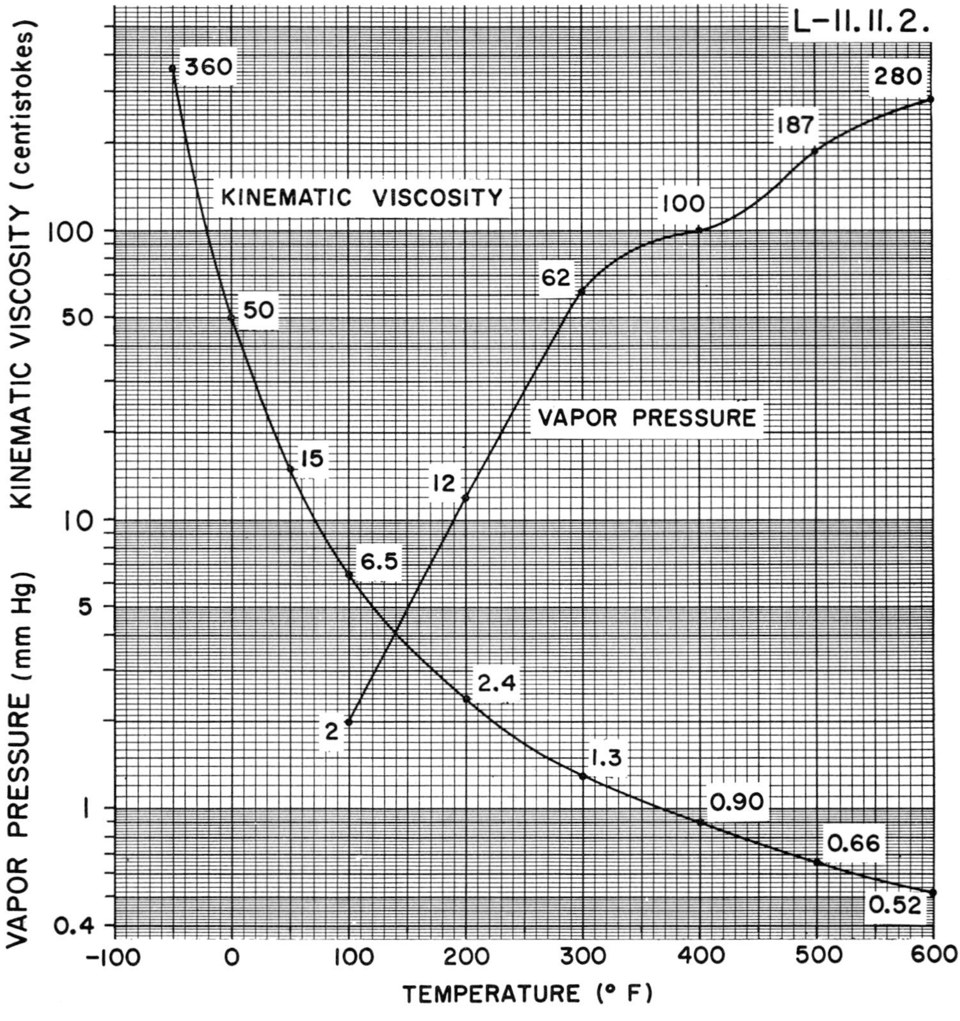

Viscosity and Vapor Pressure vs Temperature

L-11.12. SILICATES

*Hydrotherm 700-160 in the U.S.**
*TAS-160 in the U.K.***

Hydrotherm 700-160 is a tetra-aryl silicate used for pressureless heat transfer between 0° and 600°F. It is also used as a hydraulic fluid.

Properties of Importance

Specific Gravity: See Figure L-11.12.1.

Specific Heat: See Figure L-11.12.1.

Thermal Conductivity: The heat conductivity is approximately 0.072 Btu/(hr)(ft)(0°F).

Viscosity: See Figure L-11.12.2.

Physical Stability: See Figure L-11.12.2 for the vapor pressure curve. Hydrotherm 700-160 has a pour point of −40°F.

Corrosion: Aluminum, magnesium, zinc, titanium, and their alloys should not be used in conjunction with Hydrotherm 700-160.

Appearance: Hydrotherm 700-160 is an amber-colored liquid with a slight phenolic odor.

Price Class: C.

L-11.13. SILICATES

*Hydrotherm 700-190 in the U.S.**
*TAS-190 in the U.K.***

Hydrotherm 700-190 is a phenyl resorcinol silicate, which can be employed without pressure at temperatures between 50° and 675°F. Due to its excellent thermal stability, Hydrotherm 700-190 is the most frequently utilized member of the tetra-aryl silicate family.

Properties of Importance

Specific Gravity: See Figure L-11.13.1.

Specific Heat: See Figure L-11.13.1.

Thermal Conductivity: Approximately 0-072 Btu/(hr)(ft)(°F).

Viscosity: See Figure L-11.13.2.

Physical Stability: Hydrotherm 700-190 has a boiling point well above 800°F. Its pour point is −20°F.

Thermal Stability: The maximum bulk operating temperature is 675°F. The figure of 675°F is a conservative one and has been obtained through a series of rigorous thermal stability tests. One such set of experiments was conducted in glass apparatus heated by immersion in thermostatically-controlled salt baths. The samples were blanketed with dry nitrogen gas. The system was maintained at a temperature of 775°F, far above temperatures which the liquid encounters in the plant. The kinematic viscosity at 25°C was used as a yardstick of decomposition; the effective life was assumed as the time required for kinematic viscosity to reach 1000 centistokes. Data for a typical run were as follows:

Production Batch	TAS-190	Initial Viscosity	After 29 Days	After 63 Days	After 81 Days	LIFE (Time to Reach 1,000 Centistokes)
X	N.71	60.0	146.4	528	1,060	71 days
Y	Q.98	58.7	109.8	576	1,992	70 days
Z	Q.97	60.7	143.9	577	1,430	70 days

For a continuous process operating at 675°F, the life of the fluid was estimated to be two years. When the fluid is spent, solid blocks of tetraphenyl silicate (TPS) are introduced into the expansion tank, a reaction takes place, the fluid is regenerated, and its viscosity falls towards a normal level. The use of TPS make-up produces substantial savings in liquid costs.

Surface Tension: 36.6 dynes/cm at 77°F.

Corrosion: No corrosion is encountered with commonly used ferrous and nonferrous metals, with the exception of aluminum, magnesium, zinc, titanium, and their alloys.

Flammability:
Flash Point-460°F ASTM D92.
Fire Point-545°F ASTM D92.

Physiological Effects: Similar to phenol.

Appearance: Brown liquid, with a slightly phenolic odor.

Price Class: C.

*Reg. Trademark American Hydrotherm Corporation.
**Reg. Trademark Hygrotherm Engineering, Ltd., London, England.

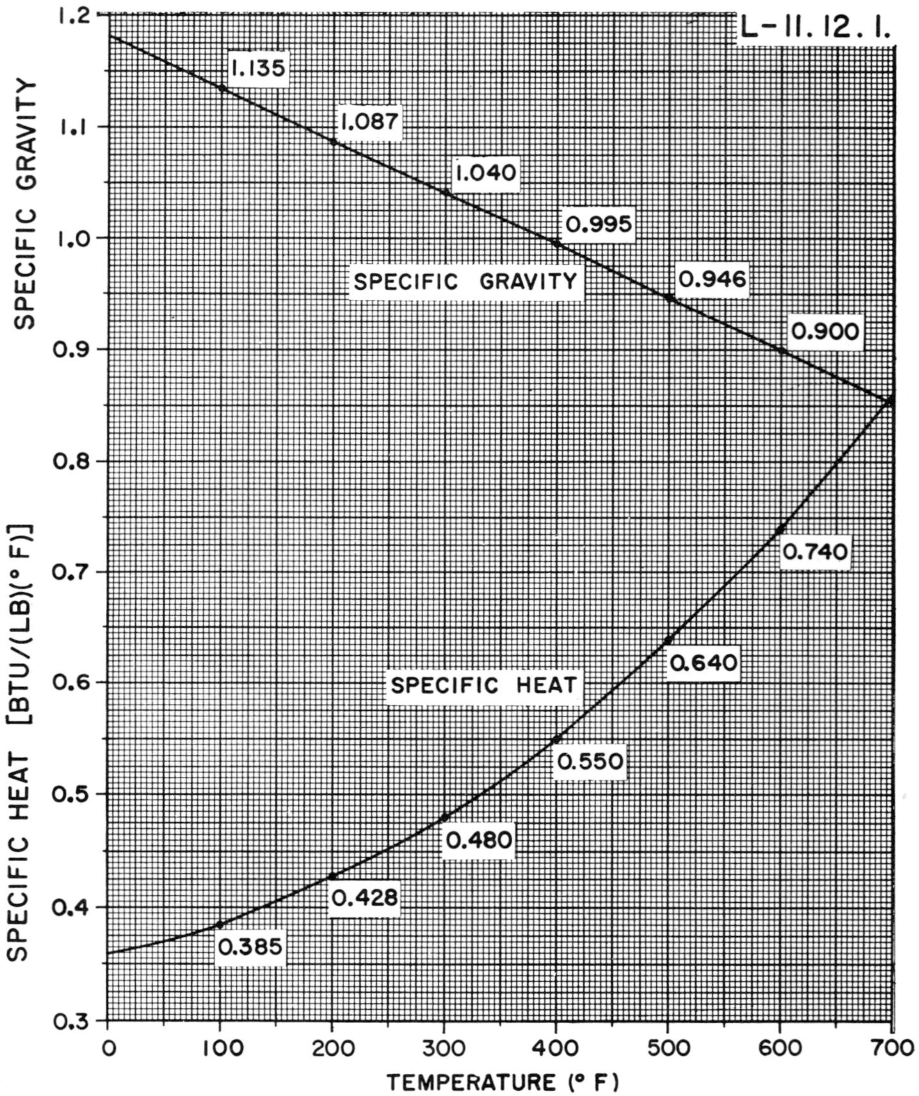

Specific Heat and Specific Gravity vs Temperature

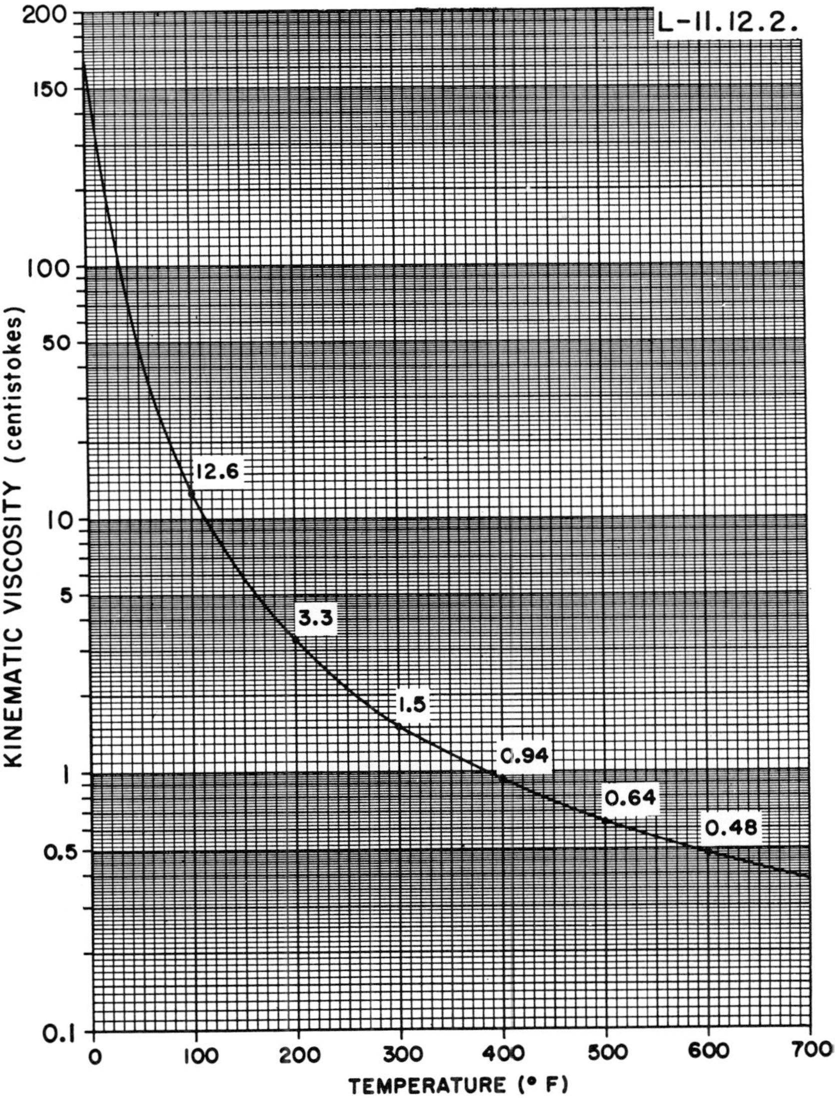

Kinematic Viscosity vs Temperature

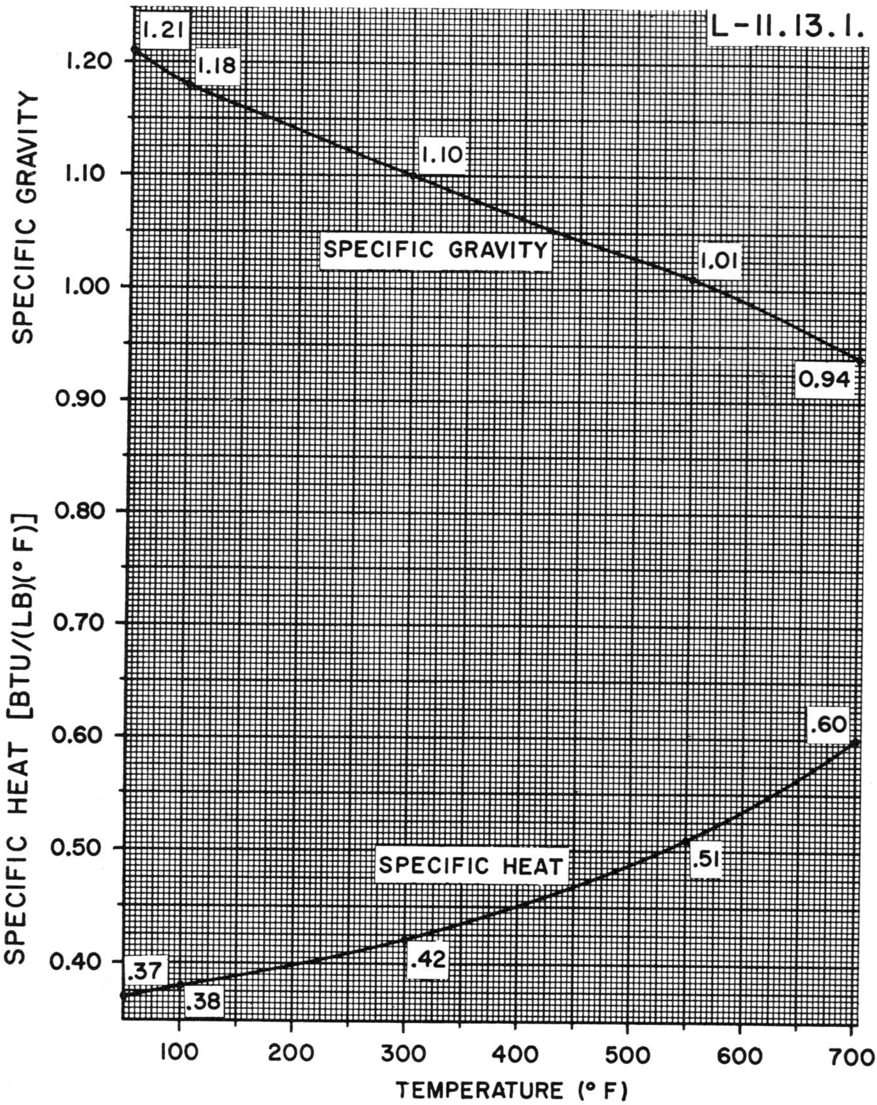

Specific Gravity and Specific Heat vs Temperature

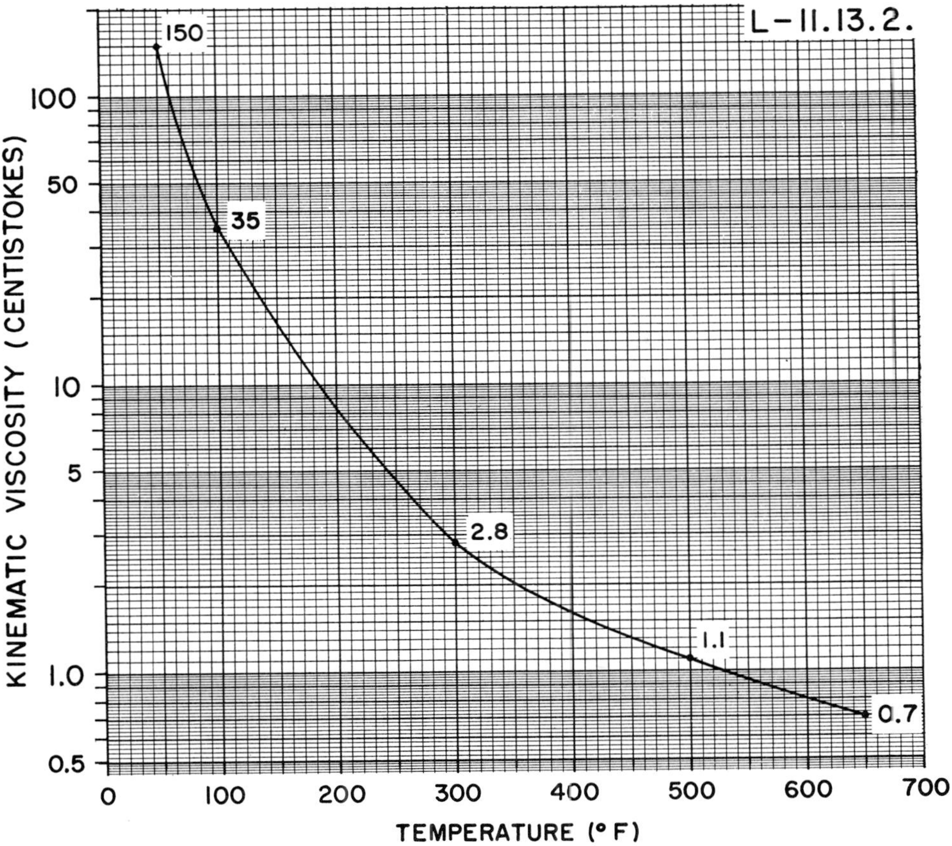

Kinematic Viscosity vs Temperature

Section IV

APPLICATION OF HEAT TRANSFER MATERIALS

A. APPLICATIONS FOR PROCESS AND SPACE HEATING AND COOLING

1. *GENERAL*

Every phase of industrial technology employs carriers in the form of gases or vapors to transfer heat from the fuel to air or solids, i.e., the ultimate consumer. A secondary heat carrier is often utilized when the primary heat transfer medium does not have the desired heat transfer properties acceptable to the ultimate heat transfer process.

All fuels, solid, liquid and gaseous, produce heat by direct combustion. The combustion gases formed transmit their thermal energy by radiation, convection, or conduction to other heat carriers, such as vapors or liquids.

A certain amount of losses in this process is unavoidable, but the progress of technology has been able to reduce these losses, which were 30 to 50 per cent sixty or seventy years ago, to about 10 to 20 per cent, so that it is possible today to transfer, by means of combustion and flue gases, 80 to 90 per cent of the heat contained in the fuel to other fluids.

If liquids are used as secondary heat carriers, heat energy can be re-transferred with only nominal losses, resulting in highly efficient processes. These losses of secondary conversion should never exceed 10 per cent and can even be reduced to below 1/2 per cent. Any waste is primarily the result of pipe conduction and radiation to the surroundings.

The application of water vapor as a heat carrier has attained a high degree of perfection, but certain difficulties in equipment design and equipment maintenance, and the problem of close control of surface temperature make its use undesirable for numerous applications. Losses up to 20 and 35 per cent have been experienced in many plants, so that for many process and space heating applications the utilization of steam is not the most economical method of heat transmission. For power production, however, gases and vapors still remain the unchallenged leaders in the field.

Liquid heating systems are more suitable for applications requiring exact temperature control, alternate heating and cooling cycles, or transmission of heat over any distance. Vapor flowing in a pipe does not have a uniform film coefficient of heat transfer around the periphery of the pipe, because condensate collects on the bottom surface inhibiting the heat transfer, whereas a uniform film coefficient will prevail when the fluid is in the liquid state. In general, liquids have higher specific heat, thermal conductivity and density than gases or vapors. Heat losses in liquid systems are minimized. For the majority of heat transfer processes, therefore, the use of fluids in their liquid state is preferable. However, there are certain exceptions. Some applications require the use of fluids in the vapor phase; in processes requiring temperatures in excess of 1000 °F gases should be used.

Heat transfer materials may be classified in accordance with their permissible peak bulk temperature and their application.

The most important criterion for the selection of a heat transfer material is the temperature level at which the transfer of heat is desired, as this heat potential strictly determines which heat carrier can and should be used.

On the following pages, the heat carriers discussed in Section III are categorized in accordance with their applicability within the various temperature ranges.

During the heating phase, the fluid film temperature along the heated walls is higher than the bulk temperature. The difference between these temperatures depends on the design of the system. In classifying the liquids, the bulk temperature is taken into account, and therefore it is necessary to place some of the fluids into lower temperature groups than those assigned to them in the literature. Of course, under favorable design conditions, the fluids can be subjected to higher temperatures for longer periods, without impairing the useful life of the fluid, which is generally set at 12,000-25,000 hours of operating time, depending on the type of service in which it is used. In classifying the heat carriers, it is assumed that no continuous filtering, treating, or replenishing of the fluids takes place.

2. CLASSIFICATION OF HEAT CARRIERS IN ACCORDANCE WITH TEMPERATURE

GROUP I. Temperatures from 1000° to 2000°F and Higher. In this range, the selection of heat carriers is rather limited and only gases and liquid metals may be used. Above 2000°F only such heat carriers as hydrogen, or a few of the noble gases, argon, etc., can be utilized. Most gases dissociate at such high temperatures. In the range of 1000°–2000°F gases at high and low pressure are generally used as heat carriers. Vapors are seldom employed in this range because they are not stable. For example, water vapor starts to dissociate above 2500°F.

Liquid metals have been used up to 1500°F, but are usually employed only up to 1000°F, because of corrosion difficulties. Therefore, compressed gases are best suited for this range. Carbon dioxide seems to be the best, but nitrogen and air can be considered as well. For higher temperature ranges, hydrogen and the noble gases are applicable.

The fluids recommended for this range are:

	Page
Air	*28*
Aluminum Chloride	*33*
Carbon Dioxide	*36*
Flue Gases	*41*
Helium	*48*
Hydrogen	*51*
Nitrogen	*57*
Gallium	*193*

GROUP II. Temperatures from 750° to 1000°F. In this range, not many heat transfer materials are available. Only heat transfer salt, biphenyl, terphenyl, and certain ethers and organometallic compounds can be mentioned. Most of these materials, excluding the heat transfer salt and the polyphenyls fall into the high-priced class, limiting their application for economic reasons. Another drawback of these liquids is their high freezing points (280°F for the heat transfer salt). It should be kept in mind that the liquid metals and gases discussed in Group I, can cover this temperature range quite well.

This high temperature range also involves large heat losses due to the high waste gas temperatures. Therefore, the heat carrier should be kept at this high temperature only if the process temperature actually requires such high heat potential.

The following fluids can be used in this temperature range:

	Page
Bismuth	*181*
Heat Transfer Salt	*208*
Lead	*183*
Lead-Bismuth Eutectic	*183*
Lithium	*196*
N. S. Salt	*209*
Sodium	*197*
Sodium-Potassium Alloys	*198*

GROUP III. Temperatures from 600° to 750°F. Several liquids are available. One can consider:

	Page
Biphenyl*	172
Diphenyl-Diphenyl Oxide* (Dowtherm A)	77
Heptacosafluorobutylamine (FC-43)	82
Aryl aryloxy silane (Hydrotherm 750-200)	218
Aryloxy arylaryloxy silane (Hydrotherm 750-210)	218
Phenyl resorcinol silicate (Hydrotherm 700-190)	239
Perfluoro-2-Butyltetrahydrofuran (FC-75)	86
Polyphenyl Ethers (OS-124)	132
o-Terphenyl (Santowax 0)	172
m-Terphenyl (Santowax M)	172
p-Terphenyl (Santowax P)	172
Steam*	96,250,251

The upper temperature limit of 750°F can be considered as being rather high for *some* of the fluids in this group, and a lower temperature level should be carefully investigated before design is started, particularly where long service life is required.

GROUP IV. Temperatures from 500° to 600°F. In this range, many liquids can be used to advantage. They are listed below. This list excludes those which are also applicable in the higher ranges, and consequently have certain drawbacks with respect to price, freezing point, flammability, safety, etc.

	Page
Certain Aryl Silicates (Hydrotherm 700-10A, 700-160)	231,239
Certain Chlorinated Biphenyls (Aroclors 1242, 1248, 1254)	154
Animal Oil (Thermofluid)	126
Certain Hydrocarbon Oils (Sun 21 Oil, Humble Therm 500, Mobiltherm 600)	172,164
Silicones (DC F-60, F-61)	224
Triaryl Borate	126
Mercury	182

*These fluids need pressurization for use in this temperature range.

GROUP V. Temperatures from 500° to 600°F. Again, certain liquids are especially suited if a pressureless system is desired. If the system is to be operated under pressure, high-temperature water and its application can also be considered.

Below 400°F, water is by far the best heat carrier for a system under pressure.

A choice can be made from a number of heat carriers which are available at these temperatures for pressureless systems. They are:

	Page
Certain Alkyl Silicates (Hydrotherm 700-155)	234
p-Cymene	91
o-Dichlorobenzene (Dowtherm E)	72
Hydrocarbon Oil (Mobiltherm Light)	164
Polyalkylene Glycol Derivatives (Ucon Fluids)	150
Certain Silicones (DC-550, DC-710)	224

The questions of safety, design and other considerations play an important role in the selection of the proper fluid. There is no general panacea and the application of the various liquids should be decided according to their merits in respect to the particular process.

3. LATEST DEVELOPMENTS OF HIGH TEMPERATURE HEAT CARRIERS FOR SPACE AND PROCESS HEATING APPLICATIONS

Thermal Heaters (Figures IV.1 and IV.1.a.) The aspect of danger and certain complications in the operation and maintenance of high pressure steam plants for industrial and space heating applications have led to the use of high-temperature heat carriers in the temperature range of 400°-600°F, to produce high pressure steam or high-temperature water *indirectly* in special heat exchangers. A pressureless high temperature primary system has great advantages for small and medium-sized plants, providing simpler maintenance and lower operating costs.

Figure IV.I Thermal Heater. (*Courtesy of American Hydrotherm Corporation*)

HIGH-TEMPERATURE LIQUIDS IN RADIATORS FOR SPACE HEATING: Another development is the direct application of high temperature water or other liquids for space heating. Special high-temperature radiators with extended surface (Figure IV.2 p. 250), overcome the disadvantage of the higher temperatures of the heat carrier and preserve the advantages of smaller lines by employing temperature drops of 150°-200°F within the radiators, permitting supply temperatures up to 400°F.

SNOW MELTING: Snow melting designs have been developed for operating temperatures between -30° and 350°F.[(2)] Figure IV.3 shows a test area of 20,000 sq ft heated by a supply temperature of 350°F.

Specially developed light oils, as well as glycol-water mixtures with inhibitors, have been tested for these temperatures.

TEST CHAMBERS: Some of the silicates are well suited for high as well as low temperature applications. They can be used for extremely low and extremely high temperatures around -100° and +500°F.

4. GENERAL CONSIDERATIONS ON THE APPLICATION OF LIQUIDS OTHER THAN WATER AS HEAT CARRIERS

In the application of these liquids, consideration must be given even at low temperatures and pressures to the various properties of these substances. When designing systems where water is circulated at low temperature or pressure, little or no care or special thought need be given to the heat carrier itself, since water can withstand considerable abuse or mistreatment in a heating or cooling process. There is no problem of decomposition, flammability or toxicity. In the design of low pressure water systems the problem of deposits from the liquid is negligible.

These problems will present themselves, however, when using heat carriers other than water, and only the experienced designer knows how to avoid them and to bring the favorable properties of the heat carrier to the forefront. These liquids, unlike water, have only limited periods of use. However, well-designed systems should operate with the same liquid at least two years, and possibly three to five years. A change of liquid every three or four years is normal. It is possible to achieve such long operation time only if the system is operated below the temperature limit of the liquid, and the fluid is maintained in accordance with the recommendations of the designer or manufacturer. No continuous purification or rejuvenation of the heat carrier, as a general rule, is acceptable, since the complexity of such a design could be disadvantageous. However, under certain conditions when the fluid is operated close to or beyond its recommended temperature limit, it becomes a necessity. In the case of many of the new and expensive specially synthesized materials, the "running of the fluid to destruction" is unnecessarily wasteful and uneconomical, particularly in view of the technical feasibility and demonstrated economy of intermittent or on-stream purification.

The periods of useful fluid life expressed here are based on many years of experience in this field. They can be achieved only when the

(2) U. S. Patent No. 2,570,226

system is properly maintained, and the liquid is used below its maximum temperature limit.

The pressureless system for the indirect production of high pressure steam or for other high temperature processes has progressed with the development of thermal heating systems permitting previously unknown simplification in design, so that there is little or no danger connected with this type of operation.

The "danger factor" inherent in high-pressure steam boilers is reduced to a minimum by the elimination of pressure, but often increased again by the use of inflammable heat carriers. The introduction of a nonflammable pressureless heat carrier having a high- or no-fire point is therefore the best and safest solution. A high fire point is nearly as valuable as *no* fire point, because the system can be easily cooled down if the heat carrier is ignited at some point in the system, so that the temperature of the circulating liquid can be reduced below the fire point and the fire can be easily extinguished.

Careful consideration of all properties makes possible the correct selection and application of the best heat transfer fluid for a particular industrial process. It is hoped that the properties of the many heat transfer fluids listed in this book will convey such knowledge.

Figure IV.I.a. Thermal Heater. (*Courtesy of American Hydrotherm Corporation*)

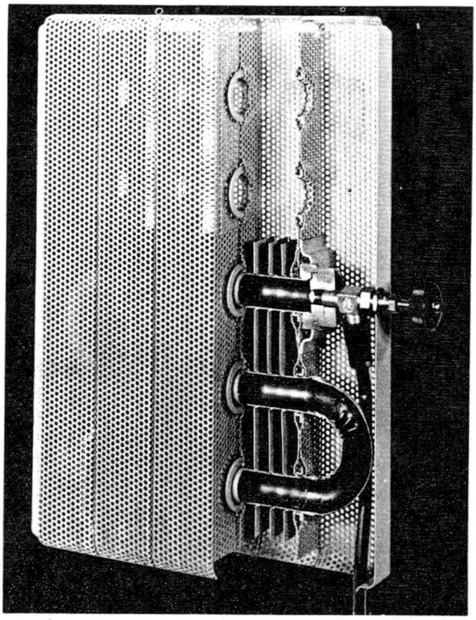

Figure IV.2. High-Temperature Radiator.
(*Courtesy of Shaw-Perkins Corporation*)

Figure IV.3. High Temperature Snow Melting System. (*Courtesy of American Hydrotherm Corporation*)

B. APPLICATIONS FOR NUCLEAR REACTORS

The development of nuclear reactors is intimately connected with the development of heat carriers able to transmit the heat produced in the core by fission to the place of application, and to employ this heat for production of steam, hot air, or other secondary heat carriers such as steam or hot water, etc.

In most nuclear applications, the fluid not only cools the reactor channels, but also makes a substantial contribution to neutron moderation, by the slowing down of the neutrons within the reactor. The following heat transfer materials are presently used in reactors:

1. **Natural Water**, which for differentiation in reactor language, is called 'light water'. It is used with or without pressurization.
2. **'Heavy Water'** (D_2O), primarily used under medium or high pressure.
3. **Water Mixed With Steam** (Boiling Water).
4. **Liquid Metals** (Sodium, Sodium Potassium).
5. **Organic Liquids** (Polyphenyls).
6. **Pressurized Gases** (mainly CO_2 and He).
7. **Atmospheric Air.**
8. **Molten Salts.**

1. NATURAL, OR LIGHT WATER is used in the so-called Swimming Pool Reactors without pressurizing except for static head, and in the Pressurized Water Reactor at pressures between 1200 and 2000 psi and temperatures up to 570° and 650°F, depending upon the size and design of the reactor.

Many medium size *pressurized water reactors* operate at temperatures of 550° to 570°F and 1500 psi pressure; smaller reactors may utilize pressures of only 1000 psi. This high pressure water serves as the heat carrier for the indirect production of steam in tubular steam producers. Most of these plants are designed for constant temperature operation, and the negative temperature coefficient of reactivity adjusts reactor power for short time load transition. This negative temperature coefficient of reactivity depends upon the coefficient of thermal expansion of the liquid water, which varies smoothly with temperature, constituting one of the advantages of this system. This design has been frequently duplicated and used in the United States for land and marine applications. The light water acts simultaneously as the moderator, since hydrogen is the most effective moderating atom, its mass being the same as that of the neutron.

2. HEAVY WATER, corresponding to the formula D_2O, is used with the same success as a moderator and coolant in similar applications. Though deuterium does not slow down neutrons as rapidly as does hydrogen, it absorbs them far less. Thus, heavy water has a moderating ratio (slowing down power divided by absorption cross-section) far greater than most fluids. The smaller value of the slowing down power of heavy water compared to that of light water requires and makes possible core lattice designs with a high ratio of D_2O to uranium, which in a sense simplifies the thermal design. The low moderating ratio of heavy water increases the neutron slowing-down time and thus the safety of the reactor. In spite of these many advantages, only a few designs have been completed employing this liquid as a moderator and coolant.

The chief disadvantage of D_2O is the high cost of the liquid, producing high carrying charges for the reactor. In 1960, the official AEC price was $28 per pound. D_2O is used both in pressurized water and boiling water designs. The few power reactor designs presently employing heavy water operate at pressures up to 1500 psi. The pressure of the indirectly produced steam is of course lower, generally 400 or 600 psi. The Savannah River production reactors, however, used D_2O below its boiling point and at only moderate pressures.

3. STEAM-WATER MIXTURE. A steam-water mixture is a good coolant and moderator for reactor channels. In the first experimental reactors, control of the reactor due to the 'voids' created by the steam bubbles was expected to produce problems, particularly if they were suddenly to collapse, as in the "cold water accident." However, the control and neutron flux regulation of the Boiling Water Reactor, as this design is called, has turned out quite successfully.

There is hope that by using water and steam as a moderator and coolant, an ultimate simplified solution for reactor design can be found.

At present, however, this design is potentially less stable and necessitates the use of a more complex control system, in part overshadowing the numerous excellent features of this relatively simple design, particularly the lower reactor pressure.

The Atomic Energy Division of American-Standard, with the cooperation of the author,[1] developed a Boiling Water Reactor design especially suited for industrial heating systems for space and process application at moderate temperatures up to 400°F.

This design offers the possibility of installing a relatively inexpensive low temperature (450°F) reactor unit for heat output of 20 to 100 megawatts (thermal), corresponding to 66 million and 330 million Btu/hr. This type of reactor can become competitive with fossil fuels in certain northern areas of the United States and Canada.

4. LIQUID METALS. The use of liquid heat carriers with more favorable characteristics than water, especially the high thermal conductivity and low vapor pressure in the higher temperature ranges, brought about the use of liquid metal (and other inorganic substances) as coolants and partly as moderators. The liquid metal, primarily sodium and possibly sodium potassium alloy (NaK) is most commonly used for the *Fast Breeder*-type reactor.

English and American designs in this field, namely, the Dounreay Fast Reactor; the Experimental Breeder Reactor of the Argonne National Laboratory; the Enrico Fermi Atomic Power Plant are all using liquid metals.

The Sodium Reactor Experiment, built by Atomics International in Van Nuys, California, as well as the Hallam Nuclear Facilities, designed by the same company, are reactor plants already completed. The Dounreay plant uses a sodium-potassium mixture, whereas most of the other plants employ only sodium as the heat carrier. A great disadvantage of sodium is its high radioactivity after neutron activation, and its slow decay time (14 hours). This is one reason why in the reactor coolant cycle, sodium is only used indirectly for producing steam. A second sodium circuit is intercalated before transferring heat to the water for steam production, avoiding radioactivity in the steam plant. The liquid metal is much less effective than hydrogen as a moderator, so that additional graphite, beryllium, etc., are employed as moderators.

Cooling temperatures up to 1000°F and higher are reached by the use of sodium or sodium-potassium alloy. Pressures employed in this application are generally low and usually do not exceed 50 psi. These alkali metals present, of course, another well-known problem—the slightest leakage and contact with water produces a violent, almost explosive, chemical reaction.

5. ORGANIC LIQUIDS. Organic liquids as coolants and moderators were introduced into reactor design in the earliest stages of reactor development. The use of organic liquids as heat transfer materials permits high temperatures with a low pressure, greatly increasing the safety of reactor operation by the elimination of the high pressure feature. Finding a liquid which can be operated at temperature levels up to 800° or even 900°F, with high heat flux is one of the great hurdles to overcome in the development of this reactor. The problem becomes especially difficult because not only heat, but gamma radiation and neutron flux, increase the rate of decomposition of these heat carriers. Very few compounds are able to withstand the severe demands of coolant and moderator applications. Only intensive study, and testing of numerous liquids can open avenues for the compounding of liquids that meet these difficult requirements.

In order to arrive at a solution, it was necessary to modify the stringent requirement of a minimum of several thousand hours of use generally set up for industrial heat carriers. No liquid has yet been discovered which can be employed for such a long time and can withstand the destructive forces of high temperature and atomic radiation, without being continually purified, or else partly drained and replaced, or partly rejuvenated by decontamination and distillation.

A list of literature is given at the end of this chapter which describes numerous liquids tested for these applications. At present, biphenyl and the higher polyphenyls developed by Monsanto Chemical Company, have been tested for this purpose. Other liquids are still under investigation to increase the field of these interesting fluids. The intention is to substitute better and more durable pressureless heat carriers for the ones available at present. It is desired to reduce the magnitude of the liquid losses, and the expenses for adding continual new material or rejuvenation of the fluid. The present cost (in 1960) for the make-up (in the neighborhood of 0.00368 lb/Kw-hr) is some $0.63/mwh of electricity produced. It is also hoped to simplify the apparatus used for purification and rejuvenation of the liquid. It would be a great step forward if a perfect liquid could be compounded as it would greatly simplify reactor design.

6. AIR AND CARBON DIOXIDE. The use of pressurized carbon dioxide as a coolant for reactors has been fostered mainly in Europe, and the first designs of this type were made in England, and later ones in France. The Calder Hall reactor in Calderbridge, Cumberland, England, was started in 1953 and is the first commercial installation using graphite as the moderator and CO_2 under pressure as the coolant. In Berkeley, Gloucestershire, England, and in other large plants such as the Berkeley Power Stations; the Bradwell Nuclear Power Station in Bradwell, Essex, England; and the Chapel Cross reactor station in Dumfrieshire, Scotland, etc.; the combination of graphite as moderator and CO_2 as coolant is used.

In Avoine, France, near Chinon, and at other places, designs using the same method were started. Canada, as well as some of the Iron Curtain countries, i.e., Czechoslovakia, has begun construction on similar designs.

In the United States, the use of carbon dioxide under pressure as a coolant has not received much attention, because studies prepared by American-Standard and others have proved that, based on American material prices and wages, this design would not provide low investment costs. The advantages claimed for this design do not seem sufficiently important to warrant introduction into the United States where enriched uranium is readily available and small reactor cores are thus feasible.

The maximum temperature of CO_2 reaches 700°F, pressures varying between 150 and 400 psi. At these pressures and temperatures, the heat transfer properties of carbon dioxide, as seen in Section III can be acceptable, so that cost of a heat exchanger, as well as reactor design, is not too high.

Employing Other Gases. Carbon dioxide is relatively low-priced and has good characteristics as coolant and moderator. Helium is costly, but has better characteristics and is used in several current designs. Nitrogen is not as attractive since it possesses a poor combination of heat transfer and flow properties, thus absorbing in the pumps an unduly high proportion of the energy generated. Its neutron cross-section is also somewhat unfavorable. Hydrogen offers most interesting properties and would be a fine moderator and reduce the graphite cost, provided a large enough mass could be gathered at a given location. Blower power would be reduced to 10 per cent in a typical design. However, its possible chemical reaction with uranium may cause serious trouble. It can even diffuse through steel, which would be still more dangerous.

Summary

In judging the different heat carriers which have been used as coolants and moderators in reactor design, it is not possible to assert the superiority of one material over another. At present, the main criterion seems to be one of technical experience of one group of designers as compared to another. Only after more years have passed and more experience has been gained with the different heat carriers and their optimized designs, will it be possible to determine whether any of the various heat carriers possesses such outstanding properties that it can be used for most applications. The technological selection of the heat carrier must be adapted to the ultimate use of the reactor and the function of the reactor determines the type of coolant and moderator required.

Even if methods for *direct* conversion of nuclear energy to electrical energy become practical, the use of nuclear heat energy for space and process heating applications will ensure the continued significance of heat transfer fluids.

References

1. Nuclear Reactor Plant Data, A.S.M.E. Volume I, Power Reactors, New York, McGraw-Hill Book Co., 1959.
2. Potential Applications of Nuclear Energy for Process and Space Heat in the United States, Geiringer, Paul L., and Goodfriend, Morton J., prepared for Civilian Reactors Branch, Division of Reactor Development, U.S. Atomic Energy Commission, (TID-450 p., 14th Ed.).
3. Low Temperature Nuclear Process Heat, Proceedings of the 1959 Symposium, AEC Headquarters, Germantown, Md.
4. Peaceful Uses of Atomic Energy, Proceedings of the International Conference in Geneva, August 1955, (United Nations).
5. Argonne National Laboratory, Study of 40 MW Pressurized Water, Boiling Water, and Organic Moderated Reactors for Production of Process Steam, ANL-6009, June 1959.

INDEX